# Translational Neurorehabilitation

Rocco Salvatore Calabrò
Editor

# Translational Neurorehabilitation

## Brain, Behavior and Technology

Springer

*Editor*
Rocco Salvatore Calabrò
IRCCS Centro Neurolesi Bonino-Pulejo
Messina, Italy

ISBN 978-3-031-63603-5 ISBN 978-3-031-63604-2 (eBook)
https://doi.org/10.1007/978-3-031-63604-2



This Springer imprint is published by the registered company Springer Nature Switzerland AG
The registered company address is: Gewerbestrasse 11, 6330 Cham, Switzerland

# Contents

# Introduction to Translational Neurorehabilitation

# 1

Rocco Salvatore Calabrò

Population is aging worldwide, especially in the western countries, with a great burden on the social and economic system. Neurological diseases increase with age, especially neurodegenerative ones, including Alzheimer's Disease and Parkinson's Disease.

The Global Burden of Disease Study estimated that in 2016, there were more than 80 million stroke survivors in the world, 43.8 million people with dementia, 45.9 million patients with an active epilepsy, and 6.1 million individuals with Parkinson's disease. Globally, in 2016, neurological disorders were the leading cause of disability (276 million disability-adjusted life-years) and the second leading cause of deaths (9 million) in the world [1].

Moreover, people who survive a brain injury are rising, thanks to the improvement of intensive acute care; then, the need for neurorehabilitation will double in the next few years. Motor, cognitive, and behavior approaches have changed over years and novel tools to treat brain and spinal cord injury should be validated before translating into the clinical practice.

Translational Neuroscience is aimed to integrate basic research of brain morphology and functional activity in vivo, with the needs of patients suffering from disorders of the Central Nervous System. The study of these disorders is a subject of Neurology, Psychiatry, and Neurosurgery, as well as Neurorehabilitation. In fact, Translational Neurorehabilitation is an interesting new field that seeks to produce more meaningful, applicable rehabilitation results that directly benefit human health, performance, and quality of life.

Robotics and virtual reality (VR) are the most promising tools of the last decades in the rehabilitation area [2, 3]. Robotic devices have been developed to reduce the

R. S. Calabrò (✉)
IRCCS Centro Neurolesi Bonino-Pulejo, Messina, Italy
e-mail: roccos.calabro@irccsme.it



R. S. Calabrò (ed.), *Translational Neurorehabilitation*,
https://doi.org/10.1007/978-3-031-63604-2_1

required labor and time, improve the reproducibility of the kinematics of different movements and gesture, and increase the volume of the motor exercises. They can also accurately and objectively measure patient's output in terms of joint kinematics and kinetics. The current rehabilitation robotic devices can be grouped into two main groups: exoskeleton (that fits to the patients' joints and move them through predetermined patterns) and end-effectors (moving the limb from the more distal part).

VR is the multisensory and interactive simulation of real scenarios, mainly presented in a three-dimensional way, with which the patient can interact [4, 5]. VR devices use specific software with input-output peripherals that make the experience complex and engaging, promoting the improvement of patients with motor/cognitive disabilities, as well as their well-being and participation. VR allows being at the center of rehabilitation training, through two perceptive conceptions, i.e., immersion and presence "Immersion" is the objective perception of a sense of "sensory absorption" in the three-dimensional environment, whereas "presence" is a subjective psychological state whereby the user is consciously involved in the virtual context [6, 7].

The application and implementation of robotics and VR in clinical practice is the best example of translational neurorehabilitation of the last decades. Indeed, many devices (for either rehabilitative or assistive purpose) have been developed and the way neurological patients are trained is really changing with positive results on their outcomes.

Moreover, understanding the neurophysiological underpinnings of functional recovery is also fundamental, and to this aim advanced electrophysiology and neuroimaging could be of help. The former is used within translational neuroscience/neurorehabilitation as a means of studying the electric properties of neurons in animal models as well as to investigate the properties of human neurological dysfunction and the basis of functional recovery. Neuroimaging involves a variety of techniques, including fMRI, DTI, PET/SPECT, used to observe the activity or the structures of, or within, the nervous system [8, 9].

Finally, the use of neuromodulation (TMS and tDCS), alone or combined to other innovative tools, may boost neural plasticity and therefore, improve patients' recovery and quality of life [10].

The stages of translational neurorehabilitation, as well as all the other fields of translational medicine, neuroscience research are as follows:

- S0: Basic science research
- In the neurorehabilitation field, this could be aimed at finding/experiment compounds able to improve neuroplasticity, reduce/slow neurodegeneration, and then improve functional recovery
- S1: Preclinical research
- Preclinical applications of task-specific rehabilitation include skilled reaching tasks. Skilled reaching tasks can be applied to experimental models to investigate motor behavior and sensorimotor integrations in post-injury recovery. Preclinical studies may also involve proof-of-concept works on robots/VR tools.

- S2: Clinical research
- Once a device has been made by bioengineering, this should be validated in clinical practice. First, feasibility and safety can be investigated in healthy subjects; then, pilot studies could address this issue in neurological patients, and finally RCT and real-life clinical studies must evaluate the efficacy, as compared to traditional methods.
- S3: Clinical implementation.
- This involves studies aimed at applying diagnostic or treatment devices to other patients than those for whom the tool was initially designed, approved, and commercialized.
- S4: Public health
- The device could become fundamental within a service or pathway in neurorehabilitation, especially for patients suffering from chronic disability. Politicians, CEO, and all stakeholders are fundamental at this stage.

This summary of the transitional stages allows to better understand how the approach to this new field is complex and integrated, involving several disciplines (such as biomechanics, engineering, neurology, physiology, physiotherapy, neuropsychology, biochemistry, physical and rehabilitation medicine) basic researchers, clinicians, and other non-healthcare professionals.

The book provides useful information concerning brain–behavior interactions to basic neuroscientists, neural engineers, clinical neurologists, and physiatrists.

It is aimed to expand current understanding of brain function and disease by evaluating preclinical and clinical trials on neural plasticity and functional recovery after nervous system disorders, and disseminate the knowledge coming from novel therapies, including advanced robotic and ICT/AI-based applications.

## References

1. GBD 2016 Neurology Collaborators. Global, regional, and national burden of neurological disorders, 1990–2016: a systematic analysis for the Global Burden of Disease Study 2016. Lancet Neurol. 2019;18:459–80.
2. Bertani R, Melegari C, De Cola MC, Bramanti A, Bramanti P, Calabrò RS. Effects of robot-assisted upper limb rehabilitation in stroke patients: a systematic review with meta-analysis. Neurol Sci. 2017;38(9):1561–9.
3. Maggio MG, Maresca G, De Luca R, Stagnitti MC, Porcari B, Ferrera MC, Galletti F, Casella C, Manuli A, Calabrò RS. The growing use of virtual reality in cognitive rehabilitation: fact, fake or vision? A scoping review. J Natl Med Assoc. 2019;111(4):457–63.
4. Charette C, Déry J, Blanchette AK, Faure C, Routhier F, Bouyer LJ, Lamontagne ME. A systematic review of the determinants of implementation of a locomotor training program using a powered exoskeleton for individuals with a spinal cord injury. Clin Rehabil. 2023;37:1119–38. https://doi.org/10.1177/02692155231164092.
5. Calabrò RS, Cacciola A, Bertè F, Manuli A, Leo A, Bramanti A, Naro A, Milardi D, Bramanti P. Robotic gait rehabilitation and substitution devices in neurological disorders: where are we now? Neurol Sci. 2016;37(4):503–14.
6. Tieri G, Morone G, Paolucci S, Iosa M. Virtual reality in cognitive and motor rehabilitation: facts, fiction and fallacies. Expert Rev Med Devices. 2018;15(2):107–17.

7. Zhang B, Wong KP, Qin J. Effects of virtual reality on the limb motor function, balance, gait, and daily function of patients with stroke: systematic review. Medicina (Kaunas). 2023;59(4):813.
8. Filippi M, Preziosa P, Arnold DL, Barkhof F, Harrison DM, Maggi P, Mainero C, Montalban X, Sechi E, Weinshenker BG, Rocca MA. Present and future of the diagnostic work-up of multiple sclerosis: the imaging perspective. J Neurol. 2023;270(3):1286–99.
9. Zhang JJ, Sánchez Vidaña DI, Chan JN, Hui ESK, Lau KK, Wang X, Lau BWM, Fong KNK. Biomarkers for prognostic functional recovery poststroke: a narrative review. Front Cell Dev Biol. 2023;10:1062807.
10. Li KP, Wu JJ, Zhou ZL, Xu DS, Zheng MX, Hua XY, Xu JG. Noninvasive brain stimulation for neurorehabilitation in post-stroke patients. Brain Sci. 2023;13(3):451.

# Brain Injury, Neural Plasticity, and Neuromodulation

# 2

Antonino Naro and Rocco Salvatore Calabrò

## 2.1 Introduction

Any brain structural damage, with regard to stroke and traumatic brain injury, can impair behavioral skills and the correlated motor skill learning, i.e., the process of optimizing sequences of action for accomplishing specific tasks [1]. Such an impairment mainly depends on the deterioration in connectivity between sets of corticospinal neurons following changes in synaptic efficacy [2], whose spatial and temporal organization is known as the "connectivity map."

Both connectivity maps and behavioral skills can at least be partially restored through intense motor practice and rehabilitation [3], ultimately aimed at restoring functions essential to independence in daily activities. Actually, motor training triggers the principles-pillars governing the organization of connectivity maps, including fractured somatotopy (i.e., the representation of any individual skill is highly distributed across different cortical regions), interconnectivity (i.e., adjacent cortical areas are densely interconnected via white matter bundles), and area equals dexterity (i.e., the more demanding the skill, the larger the proportion of the map is involved in the skill's representation) [4, 5], as a strong correlation between connectivity maps and its synaptic plasticity and performance of skills exists through a learning-dependent motor cortical map organization [6]. Therefore, motor training affects plastic changes in synaptic efficacy within the motor cortex, with consequent changes in map topography and, eventually, behavioral skills. Particularly, motor recovery through rehabilitation strategies (including neuromodulatory pharmaceutical agents and stimulation techniques such as exercises, pharmacological interventions, and brain stimulation) is achieved by activating a variety of neuroplastic

A. Naro
Stroke Unit, Policlinico Universitario, Messina, Italy

R. S. Calabrò (✉)
IRCCS Centro Neurolesi Bonino-Pulejo, Messina, Italy



R. S. Calabrò (ed.), *Translational Neurorehabilitation*,
https://doi.org/10.1007/978-3-031-63604-2_2

processes (including synapse strengthening, neural circuit rewiring, axonal sprouting, spinogenesis, and neurogenesis) that comprehensively allow the brain to change and adapt its functions to brain damage [7–11].

To achieve such neurobiological changes, motor practice must be intensive to induce effective brain reorganization, according to an experience-dependent neuroplasticity principle [12]. This consists in the entrainment of long-term potentiation (LTP) and inhibition of long-term depression (LTD) mechanisms, which both promote behavioral motor learning [12, 13]. These are hallmarks of recovery processes as compared to simple compensation processes, which do not entrain plasticity mechanisms, as well as the activation of some biochemical cascades, including myokines, neurotrophic factors, neuropeptides, growth factors (GF) and GF-like molecules, and cytokines, which have all been recognized as crucial for recovery. Interestingly, all such mechanisms can be further potentiated by adding external stimuli through brain stimulation devices, including repetitive transcranial magnetic stimulation (rTMS) and transcranial current stimulation (TCS). Finally, the use of technological devices, as robots, has been of significant help to let the patient to perform an intensive, repetitive, assisted-as-needed, and task-oriented motor training, which are all essential factors to modify synaptic efficacy within the motor cortex, with consequent changes in map topography and, eventually, behavioral skills.

This chapter provides an overview of the neuroplasticity mechanisms related to brain injury and neuromodulation, focusing on how motor practice and brain stimulation can drive neural plasticity processes to facilitate functional recovery, also paving the way for next-generation strategies for brain injury rehabilitation.

## 2.2 Brain Injury

The acute disruption of brain tissue by a cortical contusion, ischemia, hemorrhage, or axonal injury [14] often causes irreversible damage to the central nervous system (CNS) [15]. Focal brain damage determines a degeneration process initiated by unrestrained neuronal depolarization (excitotoxicity) [16], increasing the likelihood of neuronal dissolution [17], axon disintegration [18], cell lysis and neuronal necrosis (19), apoptosis and postsynaptic receptor modification [14]. These neuronal damages trigger an inflammatory process and an accumulation of reactive oxygen species [19], with consequential DNA fragmentation and lipid peroxidation, all causing further neuronal disconnection [20] and a future harmful effect on neurogenesis [21].

Secondarily to this damage, a cascade of events can be triggered (including metabolic processes [22], decrease in energy transduction and lack of adenosine triphosphate [23], excitotoxicity and inflammation processes, vasogenic and cytotoxic edema (the former resulting from BBB damage, the latter from cell metabolic derangements) [21, 24], disruption of the blood–brain barrier (BBB) [21], damage to the vasculature (which favor pro-inflammatory processes and release excitatory amino acids, creating a more vulnerable environment) [25–27], and white matter destruction, ultimately leading to a collapse of brain tissue [14, 28], implying impairment of cell functions, cell death [29], and consequent dissemination of damage [24]. In addition, owing to neuroinflammation and BBB loss of autoregulation, cerebral blood flow (CBF) results impaired, placing the brain at an increased risk of ischemic injury during the first hours after the injury.

All these changes can affect the connectivity of thousands of neurons, resulting in an impairment of their functional interactions. In this way, a relatively localized injury can result in widespread damage to the brain [30]. All these aspects determine the degree and magnitude of long-term deficits [22].

### 2.2.1 The Role of Plasticity in Brain Damage Recovery

Plasticity is defined as the intrinsic property of the nervous system to reorganize itself in response to an injury [31, 32], beyond its role in neural development and homeostasis [33]. Brain plasticity may be neuronal (synaptic or non-synaptic) or non-neuronal [34], as well as activity- or time-dependent [35], including modulation of synaptic transmission [36], integrative properties of individual neurons [37] and neuronal networks [38], neurotransmitters and ions [39, 40], gap junctions [41], and glial cells [42, 43]. All these mechanisms ultimately result in anatomical and functional modifications [44].

Brain damage recovery develops through three main stages: activation of cell repair [45], including subsidence of inflammation and edema, functional cell plasticity changes (including changes in the amount of excitation or inhibition induced and in the strength of specific synapses, also known as short-term plasticity), and anatomical plasticity changes [46–49]. Rehabilitation strategies therefore aims at targeting activity-dependent synaptic plasticity [50] (mainly based on long-term potentiation (LTP) and long-term depression (LTD) phenomena) [51, 52] that has a significant role in brain injury recovery [53–58] through the induction of anatomical neuronal changes [59–62], including using redundant connections or forming new connections among residual neurons, which all support functional recovery [63–65]. However, such plastic changes can also have negative effects, potentially leading to maladaptive outcomes (including spasticity [34], pathological pain [66, 67], schizophrenia [68, 69], dystonia [70], cognitive impairment [71, 72], and seizure foci [73]. It is therefore essential in the rehabilitation processes to suppress functionally maladaptive changes while enhancing favorable processes (and their related outcomes), leading to a better recovery of motor and cognitive function thus decreasing disability burden. To this end, artificially coupling neuronal pool discharges in a spiking time-dependent plasticity manner through brain stimulation, so to eventually modify functional and/or structural network connectivity (i.e., through activity-dependent plasticity), represents a promising therapeutic intervention in the rehabilitation setting [5].

## 2.3 Brain Stimulation

Transcranial magnetic stimulation (TMS) and transcranial direct current stimulation (tDCS) use, respectively, magnetic field to induce electric fields (consistently with the principle of electromagnetic induction) [74] and direct low-intensity currents in cortical tissue to affect cortical excitability with potentially relevant neuroplastic effects consistently with LTP and LTD phenomena [75, 76] and, eventually, functional and structural changes of neuronal networks [8, 77, 78]. Magnetic stimuli can either directly or indirectly (by interneurons) depolarize cortical neurons [79], depending on stimulus configuration (including intensity, frequency, pattern, and morphology) and the different thresholds to electrical stimulation of neuronal populations [80].

The electric current yields, first, ionic changes around targeted neurons by flowing ions through the cell membrane, mostly at axonal-soma and axonal-bouton boundaries [81], thus inducing neuronal depolarization or hyperpolarization [82], and then a stimulation-induced storage of charge, which ultimately lead to modifications in neural excitability [83–85]. The repeated application of magnetic stimuli corrupts the ongoing neuronal activity, whose summatory effects result in an increase or a decrease in cortical excitability, which is when the after stimulation neuromodulatory effects occur. Generally, low frequencies (below 1 Hz) tend to decrease cortical excitability probably due to preferential stimulation of GABAergic neurons [86, 87], while high frequencies (above 5 Hz) have the opposite effect [88]. Both approaches can induce both local and distant aftereffects, harnessing large scale cortical networks, as also suggested by specific changes in CBF, which are not however necessarily coherent with excitability changes direction (i.e., cortical excitability increase, CBF increase) [79].

Neuromodulation may affect cortical plasticity by providing peripheral stimuli when timilgly collimated each other (namely, Paired Associative Stimulation—PAS), consistently with a spike-timing dependent plasticity principle [89–91]. Additionally, stimuli can be timingly sequenced in specific patterns, for example, low-frequency pulses can be preceded by a short train of high-frequency pulses (i.e., priming), leading to a stronger inhibitory effect [92]. Alternatively, brief simple or patterned trains of stimuli (three 50-Hz stimuli at 5 Hz) can have relevant consequences on cortical excitability whether administered continuously (cTBS, GABAergic-dependent excitability decrease) [93, 94] or intermittently (iTBS, excitability increase) [95]. The duration of TMS aftereffect is variable depending on the individual's physiology and the stimulation setup. It can range from minutes to weeks [96, 97]. It has been proposed that the repeated modulation of cell polarization (namely, short-term effects) can affect the modulation of NMDA glutamatergic receptors [98, 99], which could account for LTP/LTD induction. These effects, in turn, pave the way for long-term effects, including early genes expression associated with neuronal activation (including c-Fos) [100–102] and neurotrophic factors (such as brain-derived neurotrophic factor), which altogether account for structural plasticity changes (which is proposed to the most relevant for the neurorehabilitation aftereffects) [100].

Less information is available for tDCS aftereffect. The weak direct currents employed in tDCS paradigms can modify the resting membrane potential, which impacts the level of spontaneous neuronal excitability and activity [103, 104]. Noteworthy, tDCS does not interrupt neuronal activity like TMS does; in other words, it does not induce action potentials through a rapid depolarization of neurons, like TMS does, but it rather modulates neuronal membrane excitability [103]. The direction of tDCS effects depends on the stimulation setup adopted, particularly including electrode polarity (i.e., anodal stimulation increases cortical excitability, whereas cathodal stimulation decreases it) [105–111], besides current intensity, stimulation duration, and current density, i.e., the quotient of current strength and electrode size. Polarity changes (without determining action potential) is achieved by changing the activity of transmembrane proteins and hydrogen ions [107, 110, 111] at the interneuron level [112, 113], as well as sodium and calcium channels [107, 110, 111], whereas NMDA, $GABA_A$, and glutamate-dependent mechanisms seem less relevant [107, 110–112, 114]. Interestingly, a significant polarization of neuronal membrane can result in the inactivation of voltage-gated channels, thus reverting tDCS aftereffects [107, 110, 111].

Prolonging current exposure can then induce ionic changes and modifications in transmembrane proteins, which in turn cause long-lasting changes of neural membrane function [115], calling into account synaptic mechanisms (contrarily to what observed for the immediate effects of tDCS), which likely involve GABAAergic and glutamatergic synapses [93, 94, 112, 113] related to LTP- and LTD-like mechanisms [116]. Particularly, the activation of NMDA receptors by glutamate entrains LTP phenomena [117, 118], whereas cathodal stimulation may be able to contribute to the development of LTD-like phenomena.

## 2.4 NBS as a Therapeutic Tool

A brain injury can yield a variety of neurobehavioral consequences, including seizures [119, 120], headache [121], movement disorders, motor impairment, language and visual deficits, sleep [122], memory [123, 124], and attention disorders as well as concentration impairment [125]. These can occur soon after a brain injury as well as up to months. Similarly, the course of recovery may last months to years. This means that specific interventions at different time periods can be used in the attempt to foster brain function recovery, mainly depending on the phase (acute, subacute, and chronic) of brain damage. Basically, we can try (i) to limit the extent of the initial injury to minimize further neurological deficits, and (ii) to promote reorganization of neural networks, allowing for the relearning/vicariation of lost functions [12, 126–128].

Modifying cortical excitability by using NIBS could help to counteract the acute inflammatory phenomena of a brain lesion and to favor an adaptive rewiring of damaged neural connections, thus ultimately enhancing behavioral recovery [129–133]. Specifically, NIBS aims at reducing the excessive glutamatergic activity resulting from neuronal damage and the loss of surround inhibition mechanisms through enhancing residual inhibition mechanisms, ultimately resulting in a suppression of cortical excitability. In this regard, daily cathodal tDCS, low-frequency rTMS, and cTBS may prove to be useful, as suggested by the remodulation of the glutamatergic and GABAergic systems in murine models [134]. Moreover, NIBS can limit the oxidative stress and apoptosis process following brain injury [135]. Finally, NIBS can modulate plastic changes even in the acute phase, to avoid maladaptive consequences, including effects on axonal sprouting and synaptogenesis [136]. Following glutamatergic-mediated neurotoxicity, a prevailing inhibitory tone takes step, which both silence neural networks [137] and affect LTP/LTD plasticity processes [137, 138]. Therefore, facilitatory interventions such as high-frequency rTMS or anodal tDCS could increase cortical excitability and counteract GABAergic inhibition, so to facilitate, for example, motor function recovery as shown in stroke models [139].

Consistently with the issue that function recovery occurs through a series of distributed cortical activation in different brain areas of both hemispheres, the strategy of contemporary modulating both hemispheres seem reasonable. Indeed, the modulation of interhemispheric balance using a combination of NIBS parameters has been shown promising [107, 110, 111, 140–149], with particular regard to motor learning [147] by using high-frequency rTMS, iTBS, and anodal tDCS [88, 107, 110, 111, 140, 144] over the affected hemisphere or low-frequency rTMS, cTBS, and cathodal tDCS on the contralesional hemisphere [150, 151] by modulating transcallosal inhibition [152]. To enhance NIBS aftereffects, coupling with Physical Therapy Motor training involving skill learning (as opposed to simple exercises) may be critical to induce plastic changes in the CNS via increased synaptogenesis, LTP/LTD-like mechanisms, and reorganization in the thalamo-cortical motor maps [53, 153, 154]. NBS and motor learning seem to share similar mechanisms for inducing neuroplasticity; thus, their individual therapeutic effects may be enhanced by their combination. For instance, physical exercise can improve motor and cognitive outcomes by improving motor cortical representations [55, 155]. NBS delivered prior to a motor task may prime neuronal networks in the cortex, whereas simultaneous application may recruit specific sets of synapses involved with motor performance. In this regard, physical therapy coupled with high-frequency rTMS on the injured hemisphere [156] or low-frequency stimulation on the contralesional hemisphere [157], as well as CIMT coupled with active tDCS or cathodal tDCS of the intact M1 and anodal stimulation of the affected M1 [158], induce structural neuroplasticity changes and a modulation of transcallosal inhibition from the undamaged to the affected hemisphere sustaining functional outcomes after ischemic stroke compared to sham stimulation. However, the specific anatomic and neurophysiologic derangements of each patient, the time elapsed after brain injury, the severity of paresis, the type of intervention, the technique of NIBS used and its parameters, the target area, the type of physical training performed, and its timing in relation to stimulation [116] may all influence the success of treatment [159], as different tDCS setups coupled to robotic therapy showed non-significant outcome difference. All such above-mentioned factors must be thus individually tailored and the need for further research in the field is obviously emphasized.

## 2.5 Conclusions

The knowledge of the complex mechanisms and timing of neuroplasticity processes developing after a brain injury are the fundamental prerequisite to build a patient-tailored rehabilitation program. This would help enhance recovery and decrease the burden of disabling sequelae after the injury. Although plasticity processes sustain functional recovery, they might also lead to additional injury and negative outcomes if developing as maladaptive. Targeting plasticity processes in an adequate manner becomes therefore critical to enhance the former while suppressing the latter. This has been demonstrated as possible through rehabilitation strategies which are furthermore fostered by NIBS, so to precisely target specific neural networks and potentiate plasticity processes aftereffects. However, further research is needed to understand the mechanistic interaction between the aforementioned techniques, as well as to establish their safety and define optimal stimulation parameters.

## References

1. Monfils MH, Plautz EJ, Kleim JA. In search of the motor engram: motor map plasticity as a mechanism for encoding motor experience. Neuroscientist. 2005;11(5):471–83. https://doi.org/10.1177/1073858405278015.
2. Hess G, Donoghue JP. Long-term potentiation of horizontal connections provides a mechanism to reorganize cortical motor maps. J Neurophysiol. 1994;71(6):2543–7. https://doi.org/10.1152/jn.1994.71.6.2543.
3. Kleim JA, Bruneau R, VandenBerg P, MacDonald E, Mulrooney R, Pocock D. Motor cortex stimulation enhances motor recovery and reduces peri-infarct dysfunction following ischemic insult. Neurol Res. 2003;25(8):789–93. https://doi.org/10.1179/016164103771953862.
4. Nudo RJ, Wise BM, SiFuentes F, Milliken GW. Neural substrates for the effects of rehabilitative training on motor recovery after ischemic infarct. Science. 1996;272(5269):1791–4. https://doi.org/10.1126/science.272.5269.1791.
5. Nudo RJ. Neural bases of recovery after brain injury. J Commun Disord. 2011;44(5):515–20.
6. Conner JM, Culberson A, Packowski C, Chiba AA, Tuszynski MH. Lesions of the basal forebrain cholinergic system impair task acquisition and abolish cortical plasticity associated with motor skill learning. Neuron. 2003;38(5):819–29. https://doi.org/10.1016/s0896-6273(03)00288-5.
7. Boddington LJ, Reynolds JNJ. Targeting interhemispheric inhibition with neuromodulation to enhance stroke rehabilitation. Brain Stimul. 2017;10:214–22.
8. Caglayan AB, Beker MC, Caglayan B, Yalcin E, Caglayan A, Yulug B, Hanoglu L, Kutlu S, Doeppner TR, Hermann DM, Kilic E. Acute and post-acute neuromodulation induces stroke recovery by promoting survival signaling, neurogenesis, and pyramidal tract plasticity. Front Cell Neurosci. 2019;13:144.
9. Murphy TH, Corbett D. Plasticity during stroke recovery: from synapse to behaviour. Nat Rev Neurosci. 2009;10:861–72.
10. Silasi G, Murphy TH. Stroke and the connectome: how connectivity guides therapeutic intervention. Neuron. 2014;83:1354–68.
11. Szelenberger R, Kostka J, Saluk-Bijak J, Miller E. Pharmacological interventions and rehabilitation approach for enhancing brain self-repair and stroke recovery. Curr Neuropharmacol. 2020;18:51–64.
12. Kleim JA, Jones TA. Principles of experience-dependent neural plasticity: implications for rehabilitation after brain damage. J Speech Lang Hear Res. 2008;51:S225–39.

13. Luft A, Macko R, Forrester L, Goldberg A, Hanley DF. Post-stroke exercise rehabilitation: what we know about retraining the motor system and how it may apply to retraining the heart. Cleve Clin J Med. 2008;75(Suppl 2):S83–6. https://doi.org/10.3949/ccjm.75.suppl_2.s83.
14. Park E, Bell JD, Baker AJ. Traumatic brain injury: can the consequences be stopped? CMAJ. 2008;178:1163–70.
15. Laurer HL, Meaney DF, Margulies SS, McIntosh TK. Modeling brain injury/trauma. In: Ramachandran VS, editor. Encyclopedia of the human brain, vol. 3. San Diego, CA: Elsevier Science and Technology Books; 2002. p. 93–5.
16. Choi DW. Ionic dependence of glutamate neurotoxicity. J Neurosci. 1987;7:369–79.
17. Goforth PB, Ellis EF, Satin LS. Enhancement of AMPA-mediated current after traumatic injury in cortical neurons. J Neurosci. 1999;19:7367–74.
18. Kampfl A, Posmantur R, Nixon R, et al. mu-calpain activation and calpain-mediated cytoskeletal proteolysis following traumatic brain injury. J Neurochem. 1996;67:1575–83.
19. Maciel EN, Vercesi AE, Castilho RF. Oxidative stress in Ca(2+)-induced membrane permeability transition in brain mitochondria. J Neurochem. 2001;79:1237–45.
20. Maxwell WL, Watt C, Graham DI, Gennarelli TA. Ultrastructural evidence of axonal shearing as a result of lateral acceleration of the head in non-human primates. Acta Neuropathol. 1993;86:136–44.
21. Zink BJ, Szmydynger-Chodobska J, Chodobski A. Emerging concepts in the pathophysiology of traumatic brain injury. Psychiatr Clin North Am. 2010;33:741–56.
22. Garcia AN, Shah MA, Dixon CE, Wagner AK, Kline AE. Biologic and plastic effects of experimental traumatic brain injury treatment paradigms and their relevance to clinical rehabilitation. PM R. 2011;3(6 Suppl):S18–27.
23. Xiong Y, Gu Q, Peterson PL, Muizelaar JP, Lee CP. Mitochondrial dysfunction and calcium perturbation induced by traumatic brain injury. J Neurotrauma. 1997;14:23–34.
24. Greve MW, Zink BJ. Pathophysiology of traumatic brain injury. Mt Sinai J Med. 2009;76:97–104.
25. Hua Y, Keep RF, Hoff JT, Xi G. Brain injury after intracerebral hemorrhage: the role of thrombin and iron. Stroke. 2007;38(2 Suppl):759–62.
26. Perez de la Ossa N, Sobrino T, Silva Y, et al. Iron-related brain damage in patients with intracerebral hemorrhage. Stroke. 2010;41:810–3.
27. Xi G, Keep RF, Hoff JT. Mechanisms of brain injury after intracerebral haemorrhage. Lancet Neurol. 2006;5:53–63.
28. Gentry LR. Imaging of closed head injury. Radiology. 1994;191:1–17.
29. Stoica BA, Faden AI. Cell death mechanisms and modulation in traumatic brain injury. Neurotherapeutics. 2010;7:3–12.
30. Povlishock JT, Katz DI. Update of neuropathology and neurological recovery after traumatic brain injury. J Head Trauma Rehabil. 2005;20:76–94.
31. Cramer SC, Sur M, Dobkin BH, O'Brien C, Sanger TD, Trojanowski JQ, Rumsey JM, Hicks R, Cameron J, Chen D, Chen WG, Cohen LG, deCharms C, Duffy CJ, Eden GF, Fetz EE, Filart R, Freund M, Grant SJ, Haber S, et al. Harnessing neuroplasticity for clinical applications. Brain. 2011;134:1591–609.
32. Pascual-Leone A, Amedi A, Fregni F, Merabet LB. The plastic human brain cortex. Annu Rev Neurosci. 2005;28:377–401.
33. Turrigiano GG, Nelson SB. Homeostatic plasticity in the developing nervous system. Nat Rev Neurosci. 2004;5:97–107.
34. Thickbroom GW, Mastaglia FL. Plasticity in neurological disorders and challenges for noninvasive brain stimulation (NBS). J Neuroengineering Rehabil. 2009;6:1–5.
35. Linden DJ. The return of the spike: postsynaptic action potentials and the induction of LTP and LTD. Neuron. 1999;22:661–6.
36. Malenka RC, Bear MF. LTP and LTD: an embarrassment of riches. Neuron. 2004;44:5–21.
37. Gulledge AT, Kampa BM, Stuart GJ. Synaptic integration in dendritic trees. J Neurobiol. 2005;64:75–90.

38. Destexhe A, Marder E. Plasticity in single neuron and circuit computations. Nature. 2004;431:789–95.
39. Bach-y-Rita P. Theoretical basis for brain plasticity after a TBI. Brain Inj. 2003;17:643–51.
40. Desai NS, Rutherford LC, Turrigiano GG. Plasticity in the intrinsic excitability of cortical pyramidal neurons. Nat Neurosci. 1999;2:515–20.
41. Connors BW, Long MA. Electrical synapses in the mammalian brain. Annu Rev Neurosci. 2004;27:393–418.
42. Ben Achour S, Pascual O. Glia: the many ways to modulate synaptic plasticity. Neurochem Int. 2010;57:440–5.
43. Pannasch U, Vargova L, Reingruber J, et al. Astroglial networks scale synaptic activity and plasticity. Proc Natl Acad Sci USA. 2011;108:8467–72.
44. Matsuzaki M, Honkura N, Ellis-Davies GC, Kasai H. Structural basis of long-term potentiation in single dendritic spines. Nature. 2004;429:761–6.
45. Fawcett J. Molecular control of brain plasticity and repair. Prog Brain Res. 2009;175:501–9.
46. Cao Y, Vikingstad EM, George KP, Johnson AF, Welch KM. Cortical language activation in stroke patients recovering from aphasia with functional MRI. Stroke. 1999;30:2331–40.
47. Hallett M. Neuroplasticity and rehabilitation. J Rehabil Res Dev. 2005;42:xvii–xii.
48. Heiss WD, Kessler J, Thiel A, Ghaemi M, Karbe H. Differential capacity of left and right hemispheric areas for compensation of poststroke aphasia. Ann Neurol. 1999;45:430–8.
49. Wieloch T, Nikolich K. Mechanisms of neural plasticity following brain injury. Curr Opin Neurobiol. 2006;16:258–64.
50. Hebb DO. The organization of behavior: a neuropsychological theory. New York, NY: John Wiley and Sons, Inc; 1949.
51. Bi GQ, Poo MM. Synaptic modifications in cultured hippocampal neurons: dependence on spike timing, synaptic strength, and post-synaptic cell type. J Neurosci. 1998;18:10464–72.
52. Bliss TV, Lomo T. Long-lasting potentiation of synaptic transmission in the dentate area of the anaesthetized rabbit following stimulation of the perforant path. J Physiol. 1973;232:331–56.
53. Adkins DL, Boychuk J, Remple MS, Kleim JA. Motor training induces experience-specific patterns of plasticity across motor cortex and spinal cord. J Appl Physiol. 2006;101:1776–82.
54. Griesbach GS, Hovda DA, Molteni R, Wu A, Gomez-Pinilla F. Voluntary exercise following traumatic brain injury: brain-derived neurotrophic factor upregulation and recovery of function. Neuroscience. 2004;125:129–39.
55. Liepert J, Bauder H, Wolfgang HR, Miltner WH, Taub E, Weiller C. Treatment-induced cortical reorganization after stroke in humans. Stroke. 2000;31:1210–6.
56. Maldonado MA, Allred RP, Felthauser EL, Jones TA. Motor skill training, but not voluntary exercise, improves skilled reaching after unilateral ischemic lesions of the sensorimotor cortex in rats. Neurorehabil Neural Repair. 2008;22:250–61.
57. Nudo RJ. Postinfarct cortical plasticity and behavioral recovery. Stroke. 2007;38(2 Suppl):840–5.
58. O'Callaghan RM, Ohle R, Kelly AM. The effects of forced exercise on hippocampal plasticity in the rat: a comparison of LTP, spatial- and non-spatial learning. Behav Brain Res. 2007;176:362–6.
59. Chidambaram SB, Rathipriya AG, Bolla SR, Bhat A, Ray B, Mahalakshmi AM, Manivasagam T, Thenmozhi AJ, Essa MM, Guillemin GJ, Chandra R, Sakharkar MK. Dendritic spines: revisiting the physiological role. Prog Neuro-Psychopharmacol Biol Psychiatry. 2019;92:161–93.
60. Hatakeyama M, Ninomiya I, Kanazawa M. Angiogenesis and neuronal remodeling after ischemic stroke. Neural Regen Res. 2020;15:16–9.
61. Joy MT, Carmichael ST. Encouraging an excitable brain state: mechanisms of brain repair in stroke. Nat Rev Neurosci. 2021;22:38–53.
62. Lee KF, Soares C, Béïque JC. Examining form and function of dendritic spines. Neural Plast. 2012;2012:704103.
63. Reinecke S, Dinse HR, Reinke H, Witte OW. Induction of bilateral plasticity in sensory cortical maps by small unilateral cortical infarcts in rats. Eur J Neurosci. 2003;17:623–7.
64. Rothwell JC. Plasticity in the human motor system. Folia Phoniatr Logop. 2010;62:153–7.

65. Ward NS. Neural plasticity and recovery of function. Prog Brain Res. 2005;150:527–35.
66. Flor H, Elbert T, Knecht S, et al. Phantom-limb pain as a perceptual correlate of cortical reorganization following arm amputation. Nature. 1995;375:482–4.
67. Kuner R. Central mechanisms of pathological pain. Nat Med. 2010;16:1258–66.
68. Black JE, Kodish IM, Grossman AW, et al. Pathology of layer V pyramidal neurons in the prefrontal cortex of patients with schizophrenia. Am J Psychiatry. 2004;161:742–4.
69. Hasan A, Nitsche MA, Rein B, et al. Dysfunctional long-term potentiation-like plasticity in schizophrenia revealed by transcranial direct current stimulation. Behav Brain Res. 2011;224:15–22.
70. Quartarone A, Rizzo V, Morgante F. Clinical features of dystonia: a pathophysiological revisitation. Curr Opin Neurol. 2008;21:484–90.
71. Mesulam MM. A plasticity-based theory of the pathogenesis of Alzheimer's disease. Ann N Y Acad Sci. 2000;924:42–52.
72. Van Den Heuvel C, Thornton E, Vink R. Traumatic brain injury and Alzheimer's disease: a review. Prog Brain Res. 2007;161:303–16.
73. Prince DA, Connors BW. Mechanisms of epileptogenesis in cortical structures. Ann Neurol. 1984;16(Suppl):S59–64.
74. Barker AT, Jalinous R, Freeston IL. Non-invasive magnetic stimulation of human motor cortex. Lancet. 1985;1:1106–7.
75. Fisicaro F, Lanza G, Grasso AA, Pennisi G, Bella R, Paulus W, Pennisi M. Repetitive transcranial magnetic stimulation in stroke rehabilitation: review of the current evidence and pitfalls. Ther Adv Neurol Disord. 2019;12:1756286419878317.
76. Johnson MD, Lim HH, Netoff TI, Connolly AT, Johnson N, Roy A, Holt A, Lim KO, Carey JR, Vitek JL, He B. Neuromodulation for brain disorders: challenges and opportunities. IEEE Trans Biomed Eng. 2013;60:610–24.
77. Ahn SM, Jung DH, Lee HJ, Pak ME, Jung YJ, Shin YI, Shin HK, Choi BT. Contralesional application of transcranial direct current stimulation on functional improvement in ischemic stroke mice. Stroke. 2020;51:2208–18.
78. Cambiaghi M, Cherchi L, Masin L, Infortuna C, Briski N, Caviasco C, Hazaveh S, Han Z, Buffelli M, Battaglia F. High-frequency repetitive transcranial magnetic stimulation enhances layer II/III morphological dendritic plasticity in mouse primary motor cortex. Behav Brain Res. 2021;410:113352.
79. Kobayashi M, Pascual-Leone A. Transcranial magnetic stimulation in neurology. Lancet Neurol. 2003;2:145–56.
80. Siebner HR, Funke K, Aberra AS, Antal A, Bestmann S, Chen R, Classen J, Davare M, Di Lazzaro V, Fox PT, Hallett M, Karabanov AN, Kesselheim J, Beck MM, Koch G, Liebetanz D, Meunier S, Miniussi C, Paulus W, Peterchev AV, et al. Transcranial magnetic stimulation of the brain: what is stimulated? A consensus and critical position paper. Clin Neurophysiol. 2022;140:59–97.
81. Wagner T, Rushmore J, Eden U, Valero-Cabre A. Biophysical foundations underlying TMS: setting the stage for an effective use of neurostimulation in the cognitive neurosciences. Cortex. 2009;45:1025–34.
82. Rossi S, Hallett M, Rossini PM, Pascual-Leone A. Safety, ethical considerations, and application guidelines for the use of transcranial magnetic stimulation in clinical practice and research. Clin Neurophysiol. 2009;120:2008–39.
83. Badoiu A, Mitran SI, Catalin B, Balseanu TA, Popa-Wagner A, Gherghina FL, Albu CV, Sandu RE. From molecule to patient rehabilitation: the impact of transcranial direct current stimulation and magnetic stimulation on stroke-a narrative review. Neural Plast. 2023;2023:5044065. https://doi.org/10.1155/2023/5044065.
84. Surendrakumar S, Rabelo TK, Campos ACP, Mollica A, Abrahao A, Lipsman N, Burke MJ, Hamani C. Neuromodulation therapies in pre-clinical models of traumatic brain injury: systematic review and translational applications. J Neurotrauma. 2023;40(5–6):435–48.

85. Xing Y, Zhang Y, Li C, Luo L, Hua Y, Hu J, Bai Y. Repetitive transcranial magnetic stimulation of the brain after ischemic stroke: mechanisms from animal models. Cell Mol Neurobiol. 2023;43(4):1487–97. https://doi.org/10.1007/s10571-022-01264-x.
86. Chen R, Classen J, Gerloff C, et al. Depression of motor cortex excitability by low-frequency transcranial magnetic stimulation. Neurology. 1997;48:1398–403.
87. George MS, Aston-Jones G. Noninvasive techniques for probing neurocircuitry and treating illness: vagus nerve stimulation (VNS), transcranial magnetic stimulation (TMS) and transcranial direct current stimulation (tDCS). Neuropsychopharmacology. 2010;35:301–16.
88. Pascual-Leone A, Valls-Sole J, Wassermann EM, Hallett M. Responses to rapid-rate transcranial magnetic stimulation of the human motor cortex. Brain. 1994;117:847–58.
89. Mishra AM, Pal A, Gupta D, Carmel JB. Paired motor cortex and cervical epidural electrical stimulation timed to converge in the spinal cord promotes lasting increases in motor responses. J Physiol. 2017;595:6953–68.
90. Seeman SC, Mogen BJ, Fetz EE, Perlmutter SI. Paired stimulation for spike-timing-dependent plasticity in primate sensorimotor cortex. J Neurosci. 2017;37:1935–49.
91. Ting WK, Fadul FA, Fecteau S, Ethier C. Neurostimulation for stroke rehabilitation. Front Neurosci. 2021;15:649459.
92. Iyer MB, Schleper N, Wassermann EM. Priming stimulation enhances the depressant effect of low-frequency repetitive transcranial magnetic stimulation. J Neurosci. 2003;23:10867–72.
93. Stagg CJ, Best JG, Stephenson MC, et al. Polarity-sensitive modulation of cortical neurotransmitters by transcranial stimulation. J Neurosci. 2009a;29:5202–6.
94. Stagg CJ, Wylezinska M, Matthews PM, et al. Neurochemical effects of theta burst stimulation as assessed by magnetic resonance spectroscopy. J Neurophysiol. 2009b;101:2872–7.
95. Di Lazzaro V, Pilato F, Dileone M, et al. The physiological basis of the effects of intermittent theta burst stimulation of the human motor cortex. J Physiol. 2008;586:3871–9.
96. Miniussi C, Cappa SF, Cohen LG, et al. Efficacy of repetitive transcranial magnetic stimulation/transcranial direct current stimulation in cognitive neurorehabilitation. Brain Stimul. 2008;1:326–36.
97. Romero JR, Anschel D, Sparing R, Gangitano M, Pascual-Leone A. Subthreshold low frequency repetitive transcranial magnetic stimulation selectively decreases facilitation in the motor cortex. Clin Neurophysiol. 2002;113:101–7.
98. Huang YZ, Chen RS, Rothwell JC, Wen HY. The after-effect of human theta burst stimulation is NMDA receptor dependent. Clin Neurophysiol. 2007;118:1028–32.
99. Stefan K, Kunesch E, Benecke R, Cohen LG, Classen J. Mechanisms of enhancement of human motor cortex excitability induced by interventional paired associative stimulation. J Physiol. 2002;543:699–708.
100. Hausmann A, Weis C, Marksteiner J, Hinterhuber H, Humpel C. Chronic repetitive transcranial magnetic stimulation enhances c-fos in the parietal cortex and hippocampus. Brain Res Mol Brain Res. 2000;76:355–62.
101. Ji RR, Schlaepfer TE, Aizenman CD, et al. Repetitive transcranial magnetic stimulation activates specific regions in rat brain. Proc Natl Acad Sci USA. 1998;95:15635–40.
102. Zhang X, Mei Y, Liu C, Yu S. Effect of transcranial magnetic stimulation on the expression of c-Fos and brain-derived neurotrophic factor of the cerebral cortex in rats with cerebral infarct. J Huazhong Univ Sci Technolog Med Sci. 2007;27:415–8.
103. Nitsche MA, Cohen LG, Wassermann EM, et al. Transcranial direct current stimulation: state of the art 2008. Brain Stimul. 2008;1:206–23.
104. Wagner T, Fregni F, Fecteau S, Grodzinsky A, Zahn M, Pascual-Leone A. Transcranial direct current stimulation: a computer-based human model study. NeuroImage. 2007;35:1113–24.
105. Antal A, Kincses TZ, Nitsche MA, Bartfai O, Paulus W. Excitability changes induced in the human primary visual cortex by transcranial direct current stimulation: direct electrophysiological evidence. Invest Ophthalmol Vis Sci. 2004;45:702–7.
106. Antal A, Nitsche MA, Paulus W. External modulation of visual perception in humans. Neuroreport. 2001;12:3553–5.

107. Nitsche MA, Fricke K, Henschke U, et al. Pharmacological modulation of cortical excitability shifts induced by transcranial direct current stimulation in humans. J Physiol. 2003a;553:293–301.
108. Nitsche MA, Paulus W. Excitability changes induced in the human motor cortex by weak transcranial direct current stimulation. J Physiol. 2000;527:633–9.
109. Nitsche MA, Paulus W. Sustained excitability elevations induced by transcranial DC motor cortex stimulation in humans. Neurology. 2001;57:1899–901.
110. Nitsche MA, Schauenburg A, Lang N, et al. Facilitation of implicit motor learning by weak transcranial direct current stimulation of the primary motor cortex in the human. J Cogn Neurosci. 2003b;15:619–26.
111. Nitsche MA, Liebetanz D, Antal A, Lang N, Tergau F, Paulus W. Modulation of cortical excitability by weak direct current stimulation—technical, safety and functional aspects. Suppl Clin Neurophysiol. 2003c;56:255–76.
112. Nitsche MA, Seeber A, Frommann K, et al. Modulating parameters of excitability during and after transcranial direct current stimulation of the human motor cortex. J Physiol. 2005;568:291–303.
113. Stagg CJ, Nitsche MA. Physiological basis of transcranial direct current stimulation. Neuroscientist. 2011;17:37–53.
114. Nitsche MA, Grundey J, Liebetanz D, Lang N, Tergau F, Paulus W. Catecholaminergic consolidation of motor cortical neuroplasticity in humans. Cereb Cor. 2004;14:1240–5.
115. Ardolino G, Bossi B, Barbieri S, Priori A. Non-synaptic mechanisms underlie the after-effects of cathodal transcutaneous direct current stimulation of the human brain. J Physiol. 2005;568:653–63.
116. Bolognini N, Pascual-Leone A, Fregni F. Using non-invasive brain stimulation to augment motor training-induced plasticity. J Neuroengineering Rehabil. 2009;6:1–13.
117. Liebetanz D, Nitsche MA, Tergau F, Paulus W. Pharmacological approach to the mechanisms of transcranial DC-stimulation-induced after-effects of human motor cortex excitability. Brain. 2002;125:2238–47.
118. Paulus W. Transcranial direct current stimulation (tDCS). In: Paulus W, Tergau F, Nitsche MA, Rothwell JC, Ziemann U, Hallett M, editors. Transcranial magnetic stimulation and transcranial direct current stimulation: proceedings of the 2nd international magnetic stimulation (TMS) and transcranial direct current stimulation (tDCS) symposium, Gottingen, Germany. Amsterdam: Elsevier; 2003. p. 249–54.
119. Garga N, Lowenstein DH. Posttraumatic epilepsy: a major problem in desperate need of major advances. Epilepsy Curr. 2006;6:1–5.
120. Herman ST. Epilepsy after brain insult: targeting epileptogenesis. Neurology. 2002;59(9 Suppl. 5):S21–6.
121. Hoffman JM, Lucas S, Dikmen S, et al. Natural history of headache following traumatic brain injury. J Neurotrauma. 2011;28(9):1719–25.
122. Castriotta RJ, Murthy JN. Sleep disorders in patients with traumatic brain injury: a review. CNS Drugs. 2011;25:175–85.
123. Brown AW, Moessner AM, Mandrekar J, Diehl NN, Leibson CL, Malec JF. A survey of very-long-term outcomes after traumatic brain injury among members of a population-based incident cohort. J Neurotrauma. 2011;28:167–76.
124. Skandsen T, Finnanger TG, Andersson S, Lydersen S, Brunner JF, Vik A. Cognitive impairment 3 months after moderate and severe traumatic brain injury:a prospective follow-up study. Arch Phys Med Rehabil. 2010;91:1904–13.
125. Miotto EC, Cinalli FZ, Serrao VT, Benute GG, Lucia MC, Scaff M. Cognitive deficits in patients with mild to moderate traumatic brain injury. Arq Neuropsiquiatr. 2010;68:862–8.
126. Kesikburun S. Non-invasive brain stimulation in rehabilitation. Turk J Phys Med Rehabil. 2022;68(1):1–8.
127. Li KP, Wu JJ, Zhou ZL, et al. Noninvasive brain stimulation for neurorehabilitation in post-stroke patients. Brain Sci. 2023;13(3):451. Published 2023 Mar 6. https://doi.org/10.3390/brainsci13030451.

128. Shen QR, Hu MT, Feng W, Li KP, Wang W. Narrative review of noninvasive brain stimulation in stroke rehabilitation. Med Sci Monit. 2022;28:e938298.
129. Cha TH, Hwang HS. Rehabilitation interventions combined with noninvasive brain stimulation on upper limb motor function in stroke patients. Brain Sci. 2022;12(8):994.
130. Chino T, Kinoshita S, Abo M. Repetitive transcranial magnetic stimulation and rehabilitation therapy for upper limb hemiparesis in stroke patients: a narrative review. Prog Rehabil Med. 2023;8:20230005.
131. Gao Y, Qiu Y, Yang Q, et al. Repetitive transcranial magnetic stimulation combined with cognitive training for cognitive function and activities of daily living in patients with post-stroke cognitive impairment: a systematic review and meta-analysis [published online ahead of print, 2023 Mar 31]. Ageing Res Rev. 2023;87:101919.
132. Qi S, Tian M, Rao Y, Sun C, Li X, Qiao J, Huang ZG. Applying transcranial magnetic stimulation to rehabilitation of poststroke lower extremity function and an improvement: individual-target TMS. Wiley Interdiscip Rev Cogn Sci. 2023;14(2):e1636.
133. Sánchez-Cuesta FJ, González-Zamorano Y, Arroyo-Ferrer A, Moreno-Verdú M, Romero-Muñoz JP. Repetitive transcranial magnetic stimulation of primary motor cortex for stroke upper limb motor sequelae rehabilitation: a systematic review [published online ahead of print, 2023 Mar 25]. NeuroRehabilitation. 2023;52(3):329–48. https://doi.org/10.3233/NRE-220306.
134. Yue L, Xiao-Lin H, Tao S. The effects of chronic repetitive transcranial magnetic stimulation on glutamate and gamma-aminobutyric acid in rat brain. Brain Res. 2009;1260:94–9.
135. Tasset I, Drucker-Colin R, Pena J, et al. Antioxidant-like effects and protective action of transcranial magnetic stimulation in depression caused by olfactory bulbectomy. Neurochem Res. 2010;35:1182–7.
136. Stroemer RP, Kent TA, Hulsebosch CE. Neocortical neural sprouting, synaptogenesis, and behavioral recovery after neocortical infarction in rats. Stroke. 1995;26:2135–44.
137. Jacobs KM, Donoghue JP. Reshaping the cortical motor map by unmasking latent intracortical connections. Science. 1991;251:944–7.
138. De Beaumont L, Tremblay S, Poirier J, Lassonde M, Theoret H. Altered bidirectional plasticity and reduced implicit motor learning in concussed athletes. Cereb Cor. 2012;22(1):112–21.
139. Clarkson AN, Huang BS, Macisaac SE, Mody I, Carmichael ST. Reducing excessive GABA-mediated tonic inhibition promotes functional recovery after stroke. Nature. 2010;468:305–9.
140. Boggio PS, Castro LO, Savagim EA, et al. Enhancement of non-dominant hand motor function by anodal transcranial direct current stimulation. Neurosci Lett. 2006;404:232–6.
141. Bueno VF, Brunoni AR, Boggio PS, Bensenor IM, Fregni F. Mood and cognitive effects of transcranial direct current stimulation in post-stroke depression. Neurocase. 2011;17:318–22.
142. Fecteau S, Knoch D, Fregni F, Sultani N, Boggio P, Pascual-Leone A. Diminishing risk-taking behavior by modulating activity in the prefrontal cortex: a direct current stimulation study. J Neurosci. 2007;27:12500–5.
143. George MS. Transcranial magnetic stimulation for the treatment of depression. Expert Rev Neurother. 2010;10:1761–72.
144. Kim YH, Park JW, Ko MH, Jang SH, Lee PK. Facilitative effect of high frequency subthreshold repetitive transcranial magnetic stimulation on complex sequential motor learning in humans. Neurosci Lett. 2004;367:181–5.
145. Nitsche MA, Boggio PS, Fregni F, Pascual-Leone A. Treatment of depression with transcranial direct current stimulation (tDCS): a review. Exp Neurol. 2009;219:14–9.
146. Oliveri M, Bisiach E, Brighina F, et al. rTMS of the unaffected hemisphere transiently reduces contralesional visuospatial hemineglect. Neurology. 2001;57:1338–40.
147. Reis J, Robertson E, Krakauer JW, et al. Consensus: "can tDCS and TMS enhance motor learning and memory formation?". Brain Stimul. 2008;1:363–9.
148. Shindo K, Sugiyama K, Huabao L, Nishijima K, Kondo T, Izumi S. Long-term effect of low-frequency repetitive transcranial magnetic stimulation over the unaffected posterior parietal cortex in patients with unilateral spatial neglect. J Rehabil Med. 2006;38:65–7.

149. Vines BW, Nair DG, Schlaug G. Contralateral and ipsilateral motor effects after transcranial direct current stimulation. Neuroreport. 2006;17:671–4.
150. Boggio PS, Nunes A, Rigonatti SP, Nitsche MA, Pascual-Leone A, Fregni F. Repeated sessions of noninvasive brain DC stimulation is associated with motor function improvement in stroke patients. Restor Neurol Neurosci. 2007;25:123–9.
151. Fregni F, Boggio PS, Valle AC, et al. A sham-controlled trial of a 5-day course of repetitive transcranial magnetic stimulation of the unaffected hemisphere in stroke patients. Stroke. 2006;37:2115–22.
152. Takeuchi N, Chuma T, Matsuo Y, Watanabe I, Ikoma K. Repetitive transcranial magnetic stimulation of contralesional primary motor cortex improves hand function after stroke. Stroke. 2005;36:2681–6.
153. Ding Y, Li J, Lai Q, Azam S, Rafols JA, Diaz FG. Functional improvement after motor training is correlated with synaptic plasticity in rat thalamus. Neurol Res. 2002;24:829–36.
154. Rioult-Pedotti MS, Friedman D, Donoghue JP. Learning-induced LTP in neocortex. Science. 2000;290:533–6.
155. Devine JM, Zafonte RD. Physical exercise and cognitive recovery in acquired brain injury: a review of the literature. PM R. 2009;1:560–75.
156. Kim YH, You SH, Ko MH, et al. Repetitive transcranial magnetic stimulation-induced corticomotor excitability and associated motor skill acquisition in chronic stroke. Stroke. 2006;37:1471–6.
157. Takeuchi N, Tada T, Toshima M, Chuma T, Matsuo Y, Ikoma K. Inhibition of the unaffected motor cortex by 1 Hz repetitive transcranical magnetic stimulation enhances motor performance and training effect of the paretic hand in patients with chronic stroke. J Rehabil Med. 2008;40:298–303.
158. Bolognini N, Vallar G, Casati C, et al. Neurophysiological and behavioral effects of tDCS combined with constraint-induced movement therapy in poststroke patients. Neurorehabil Neural Repair. 2011;25(9):819–29.
159. Hesse S, Waldner A, Mehrholz J, Tomelleri C, Pohl M, Werner C. Combined transcranial direct current stimulation and robot-assisted arm training in subacute stroke patients: an exploratory, randomized multicenter trial. Neurorehabil Neural Repair. 2011;25(9):838–46.

# Brain Behaviour and Cognition: From Lesion to Rehabilitation

# 3

Rosaria De Luca, Mirjam Bonanno,
Rocco Salvatore Calabrò, and Antonio Cerasa

## 3.1 Brain, Behaviour, and Cognition

The brain is the organ responsible for task-evoked responses, senses, movement, emotions, language, communication, thinking, and memory [1] and the cognition is computation over representations to yield behaviour [2]. The correlation models between brain structures and functions are studied by neuropsychology, an interdisciplinary field, that investigates the connections among cognition, behaviour, emotions, and brain [3]. In this context, anatomo-clinical approaches were crucial to allow the expansion of this field of knowledge. In fact, this method is founded on the possible correlation between clinical cognitive manifestation and the brain injury site [4]. Nowadays, scientific developments suggest more complex and interrelated connections between anatomic brain areas and its functions, and not a mere system with separate compartments, containing cognitive and behavioural functions [5]. At first, scientific evidence correlating cognitive functions and cerebral areas dates to 150 years ago. At that time, Paul Broca (1824–1880) published his findings showing the relationship between the left posterior lower frontal lobe and speech production based on clinical assessments and post-mortem anatomical studies of patients for whom this region was compromised [6, 7]. In this venue, the anatomo-clinical approach individuated that sensory and motor functions are in specific brain spots [8] (i.e. parietal and frontal lobes), but superior cognitive functions are distributed

R. De Luca · M. Bonanno · R. S. Calabrò
IRCCS Centro Neurolesi Bonino-Pulejo, Messina, Italy
e-mail: rosaria.deluca@irccsme.it; mirjam.bonanno@irccsme.it; roccos.calabro@irccsme.it

A. Cerasa (✉)
S'Anna Institute, Crotone, Italy

Institute of BioImaging and Complex Biological Systems (IBSBC), National Research Council of Italy (CNR), Milan, Italy
e-mail: antonio.cerasa@cnr.it



R. S. Calabrò (ed.), *Translational Neurorehabilitation*,
https://doi.org/10.1007/978-3-031-63604-2_3

in larger neural networks [9], such as the executive functions that are associated with the prefrontal cortex (PFC) [10].

Despite the brain's complex structure and functions, the left hemisphere is commonly associated with controlling speech and abstract thinking, while the right is responsible for spatial thinking. However, cognition is a complex concept, and it isn't a unitary notion. In fact, it includes multiple domains, including attention (focusing, shifting, divided, or sustained attention on specific stimulus/task) executive functions (planning, organising thoughts, inhibition, control), reasoning, problem-solving, visuospatial ability (visual search, drawing, and construction), memory (recall and recognition of visual and verbal information), language (expressive and receptive), social cognition, and emotions [11–13].

## 3.2 Definition, Aetiology, and Clinical Evaluation

An acquired brain injury (ABI) is an injury to the brain caused by traumatic (falls or car incidents) and vascular (i.e. ischaemic or haemorrhagic stroke) events or by tumours. In particular, the most common ABI's aetiologies are the non-traumatic causes (55.69%), including a cerebrovascular injury (40.09%), anoxic brain damage (12.04%), and other causes (3.6%), with a mean age of 56.6, while traumatic events are estimated in the 44.31% of cases, with a mean age of 43.6 [14–17]. Both aetiology, traumatic and non-traumatic brain injury, can be correlated with three main pattern of damage: (i) widespread destruction of the cortical ribbon; (ii) widespread damage of the white matter tracts; and (iii) damage of deep brain structures such as the thalamus and basal ganglia [18]. The neurological consequences of ABI, of traumatic or non-traumatic origin, are caused by an alteration of neuronal activity that compromises physical integrity or function of one or more areas of the brain [19] related to motor, sensory, cognitive, and behavioural alterations [20]. Generally, the severity of brain injury is distinguished in mild, moderate, and severe relating to the neurological severity of the impairments, including cognition, speech-language communication, memory, attention and concentration, reasoning, abstract thinking, physical functions, psychosocial behaviour, and information processing. These impairments may be either temporary or permanent and cause partial or total functional disability or psychosocial maladjustment [21–23]. Medical community traditionally classified ABI patients using Glasgow Coma Scale (GCS) to assess clinical severity and an imaging evaluation for structural damage and aetiology, using CT and/or MRI. The GCS has evolved into a universal classification system for the severity of ABI and consists of the sum score (range 3–15) of three components: eye, motor, and verbal responses. For assessment of severity in individual patients, the three components should be reported separately [24].

In this venue, the British Society of Rehabilitation Medicine has defined the severity of head injuries for three levels: (i) Mild (GCS 14–15), characterised by any period of loss of consciousness, any loss of memory for events immediately before (retrograde amnesia) or after (anterograde amnesia) the accident (collectively referred to as the period of post-traumatic amnesia, or PTA), any alteration in mental state at the time of the accident (e.g. feeling dazed, disoriented, or confused), or focal neurological deficit(s) that may or may not be transient; [25, 26]; (ii) Moderate (GCS 9–13): when a loss of consciousness lasts from a few minutes to a few hours, confusion lasts from days to weeks, and physical, cognitive, and/or behavioural impairments last for months or are permanent [27]; (iii) Severe (GCS 3–8): disorders of low level of consciousness may include coma, vegetative state (VS) also known as unresponsive wakefulness syndrome (UWS), or minimally conscious state (MCS) [28]. Another clinical tool to assess stroke severity includes the standardised National Institute of Health Stroke Scale (NIHSS) that contains 13 items, evaluating consciousness, orientation, gaze, motor function, sensation, language, speech, and inattention [29]. While clinicians often administer the modified Rankin scale to measure handicap on a scale from 1 to 6.

## 3.3 Cognitive and Behavioural Sequelae Post ABI

As mentioned above, ABI is responsible for a variety of deficits, including functional, motor, cognitive, emotional, and behavioural, that interfere with work, relationships, activities of daily living, and also economic burden. Given that, cognitive deficits caused by both TBI and/or stroke can determine short- and long-term sequelae in specific domains. The correlation between acute brain injury and severity of cognitive alterations is linked to longer duration of impaired consciousness [30, 31]. From a clinical and neuropsychological point of view, impaired consciousness, and post-traumatic amnesia (PTA) are neurobehavioural symptoms of acute TBI. According to consensus definitions, moderate and severe TBI are characterised by loss of consciousness for greater than 30 min and/or PTA persisting for at least 24 h. The term of PTA is intended as the transitory period of disorientation, confusion, and amnestic memory impairment following TBI. People in PTA also frequently experience distressed sleep-wake cycles, reduced daytime arousal, improving agitation and emotional lability [32]. In particular, mild and moderate TBI often present memory, attention, processing speed, and executive affections that can be resolved within 3–6 months after injury [33]. Notably, the executive functions are mostly impaired due to TBI, this is why frontal lobes, and their related circuitry (subcortical white matter, basal ganglia, and thalamus) are particularly exposed in traumatic events, i.e. frontal accidents and/or falls [34]. Cognitive domains additionally involved in TBI are communication, visuospatial processing, intellectual ability, and awareness alteration [35–39]. On the other hand, stroke survivors commonly present behavioural dysexecutive syndrome (BDES) [40], which itself comprises hypoactivity-apathy, anosognosia, stereotypies, hypoactivity-distractibility, and impaired social behaviour [41], associated to predicting functional disability [42]. In this context, the anatomical-function brain correlation could play a key role in the development of symptoms, based on brain injury areas. In fact, working memory and planning dysfunctions may be linked to focal injury in the dorsolateral prefrontal cortex (DLPFC) or diffuse axonal injury (DAI) affecting the projections between the lateral frontal and posterior regions [43]. In the same way, apathy has been associated with subcortical lesions and right hemisphere dysfunction [44]; impaired awareness is characteristic of patients with focal frontal injury [45] and the post injury level of self-referential insight has been associated with the integrity of the right dorsal prefrontal cortex. Indeed, emotional alterations are quite frequent in ABI patients [46], especially in post-stroke survivors, who suffer from mood symptoms that increase the cognitive complaints, in turn fuelling depressive and mood disturbances [47, 48]. Importantly, mood disorder is associated with poor long-term outcomes post ABI. Mood indicators also predict reduced independence in ADLs, with a negative influence in life activities [49, 50], and a poorer quality of life [51–53]. Each therapist should remember that cognitive impairment and mood disturbance act as significant blocks to recovery from ABI, highlighting the need for evidence-based interventions that report these limitations.

## 3.4 Psychometric Tools and Functional Neuroimaging Methods to Assess Cognitive and Behavioural Alterations

Psychometric and clinical assessment in ABI patients is fundamental to establish the degree of cognitive impairment, based on the evaluation of specific cognitive domains. The results of the cognitive evaluation can help inform the development of an individualised treatment plan for the patient. This may include cognitive rehabilitation strategies such as cognitive training, compensatory strategies, and environmental modifications. The evaluation may also be used to monitor the patient's progress over time and adjust the treatment plan as needed.

Several validated scales that are simple to use characterise the state-of-the-art in cognitive/behavioural assessment in ABI patients (Table 3.1).

1. **Global Cognitive Status**: Mini-mental state examination (MMSE) [53] and Montreal cognitive assessment (MoCA) [54]. In clinical practice, these screening tools are useful to detect rapidly cognitive impairments maximising patient cooperation and reducing the fatigue effects. Notably, the MoCA has higher sensitivity in revealing post-stroke cognitive alterations rather than MMSE, which is more administered in severe conditions and in TBI. In this way, clinicians should consider the multitude of psychometric tools [55–77] and adapt them to the specific cognitive and behavioural condition of the patient.
2. **Attention/Working Memory**: Cognitive evaluation of attentional deficits in patients with ABI typically involves a comprehensive assessment of various aspects of attention, including sustained attention, selective attention, divided attention, and attentional control. The cognitive evaluation typically involves a battery of tests that measure different aspects of attention, such as the Continuous Performance Test (CPT), the Stroop Test, Trail Making Test Part (TMT), the Digit Span and the Digit Vigilance Test (DVT). Moreover for stroke patients with spatial unilateral spatial neglect the Albert's test is one of the most employed.

**Table 3.1** Main standardised measures used to assess cognition and behaviour in people with ABI in current clinical practice

| Test/scale | Domain | Short description |
|---|---|---|
| Mini-mental state examination (MMSE) [53] | Cognition | MMSE or Folstein test is a 30-point questionnaire that is used extensively in clinical and research settings to measure cognitive impairment. Administration of the test takes between 5 and 10 min and examines functions including registration (repeating named prompts), attention and calculation, recall, language, ability to follow simple commands and orientation |
| Montreal cognitive assessment (MoCA) [54] | Cognitive functioning | The Montreal cognitive assessment (MoCA) is a widely used screening assessment for detecting cognitive impairment. It was validated as a highly sensitive tool for early detection of mild cognitive impairment (MCI) in 2000. MoCA accurately and quickly assesses:<br>– Short term memory<br>– Visuospatial abilities<br>– Executive functions<br>– Attention, concentration, and working memory<br>– Language<br>– Orientation to time and place |
| Trail making test part A and part B (TMT A and B) [55] | Visuo-executive functions<br>Attentional shift (or shift of attention) | TMT consists of 25 circles distributed over a sheet of paper. In part A, the circles are numbered 1–25, and the patient should draw lines to connect the numbers in ascending order. In part B, the circles include both numbers (1–13) and letters (A–L); as in part A, the patient draws lines to connect the circles in an ascending pattern, but with the added task of alternating between the numbers and letters (i.e. 1-A-2-B-3-C, etc.). The patient should be instructed to connect the circles as quickly as possible, without lifting the pen or pencil from the paper. Results for both TMT A and B are reported as the number of seconds required to complete the task; therefore, higher scores reveal greater impairment |
| Cancellation test (Albert's test) (CT-AT) [56, 57] | Unilateral spatial neglect (USN) visuospatial research | Albert's test is commonly a visual neglect screen that requires patients to cross out lines on a single piece of paper |

(continued)

**Table 3.1** (continued)

| Test/scale | Domain | Short description |
|---|---|---|
| Digit vigilance test (DVT) [58] | Sustained attention | The DVT is a simple task designed to measure vigilance during rapid visual tracking and accurate selection of target stimuli. It focuses on alertness and vigilance while placing minimal demands on two other components of attention: Selectivity and capacity |
| Digit span-DGS (number span forward and backward) [59] | Attention processes and working memory | Digit span (DGS) is a measure of verbal short term and working memory. DGS can be used in two formats, forward digit span and reverse digit span. This is a verbal task, with stimuli presented auditorily, and responses spoken by the participant and scored automatically by the software. Participants are presented with a random series of digits and are asked to repeat them in either the order presented (forward span) or in reverse order (backward span) |
| Letter fluency test (LFT) [60, 61] | Phonemic verbal fluency | Verbal fluency tests are brief assessment tools with relatively simple administration and scoring procedures. Semantic and phonemic fluency are measures of non-motor processing speed, language production, and executive functions. In particular, the phonemic verbal fluency test was shown to be sensitive for assessment of functional communication skills, commonly used for aphasic patients. Letter fluency is also referred to as phonemic test fluency |
| Category fluency test (CFT) [62, 63] | Semantic/category fluency | Verbal fluency tests are a kind of psychological test in which participants have to produce as many words as possible from a category in a given time (usually 60 s). This category can be semantic, including objects such as animals or fruits, or phonemic, including words beginning with a specified letter, such as p, for example. The semantic fluency test is sometimes described as the category fluency test or simply as “free listing” |

(continued)

**Table 3.1** (continued)

| Test/scale | Domain | Short description |
|---|---|---|
| Boston naming test (BNT) [64, 65] | Denomination<br>Language abilities | The Boston naming test (BNT) is a widely used neuropsychological assessment tool to measure confrontational word retrieval in individuals with aphasia or other language disturbance caused by stroke, Alzheimer's disease, or other dementing disorder. The BNT contains 60 line drawings graded in difficulty. Patients with anomia often have greater difficulties with the naming of not only difficult and low frequency objects but also easy and high frequency objects |
| Rey-osterrieth complex figure test (R-OCFT) [66] | Visuospatial abilities, memory, attention, planning, working memory, and executive functions | The Rey-Osterrieth complex figure (ROCF) is a neuropsychological assessment in which examinees are asked to reproduce a complicated line drawing, first by copying it freehand (recognition), and then drawing from memory (recall) |
| Wechsler memory scale-IV (WMS-IV) [67] | Memory functions | The Wechsler memory scale (WMS) is a neuropsychological test designed to measure different memory functions. A person's performance is reported as five index scores: Auditory memory, visual memory, visual working memory, immediate memory, and delayed memory |
| Stroop colour word (SCW) [68] | Executive functions | The Stroop colour and word test (SCWT) is a neuropsychological test extensively used to assess the ability to inhibit cognitive interference that occurs when the processing of a specific stimulus feature impedes the simultaneous processing of a second stimulus attribute, well-known as the Stroop effect |
| Tower of London (TOL) [69] | Executive functions | The tower of London test is a test used in applied clinical neuropsychology for the assessment of executive functioning specifically to detect deficits in planning, which may occur due to a variety of medical and neuropsychiatric conditions |
| Wisconsin card sorting test (WCST) [70] | Executive functions | The Wisconsin card sorting test (WCST) is a neuropsychological test that is frequently used to measure such higher level cognitive processes as attention, perseverance, working memory, abstract thinking, category fluency, and set shifting. The WCST consists of two card packs having four stimulus cards and 64 response cards in each |

(continued)

**Table 3.1** (continued)

| Test/scale | Domain | Short description |
|---|---|---|
| Frontal assessment battery (FAB) [71] | Frontal abilities | The frontal assessment battery (FAB) is a cognitive test that incorporates several clinical assessments, including S-word generation, similarities, Luria's test, grasp reflex, and the Go-No-Go test (inhibitory control) |
| Hamilton depression rating scale (HDRS) [72] | Mood/depression symptoms | HDRS also called the Hamilton depression rating scale (HDRS), sometimes also abbreviated as HAM-D, is a multiple-item questionnaire used to provide an indication of depression, and as a guide to evaluate recovery |
| Hamilton anxiety rating scale (HAM-A) [73] | Anxiety symptoms | The HAM-A was one of the first rating scales developed to measure the severity of anxiety symptoms. It is a psychological questionnaire used by clinicians to rate the severity of a patient's anxiety |
| Agitated behaviour scale (ABS) [74] | Behaviour agitation | Observational measure of agitation. 14 items. Four aggression items |
| Neurobehavioural functioning inventory (NFI) [75] | Neurobehavioural symptoms | Patient or informant-report (carer) of a range of behaviours/symptoms. 66–76 items depending on version. Nine aggression items |
| Neuropsychiatric inventory (NPI) [76] | Psychiatric symptoms | Patient and informant-report (carer). 12 domains such as depression, anxiety irritability, and disinhibition, 7–9 items in each. Seven aggression items |
| Behavioural assessment screening tool (BASTβ) [77] | Behaviour | Patient self-report of behavioural problems/ emotional symptoms, coping strategies, and major life events. 67 items |

These tests are designed to assess specific attentional functions and can help identify the nature and extent of attentional deficits in patients with acquired brain injury. In addition to these standardised tests, the evaluation may also include clinical observations of the patient's behaviour and performance in real-life situations. This can help provide a more comprehensive understanding of the patient's attentional deficits and their impact on daily functioning.

3. **Language**: Cognitive evaluation of language deficits in patients with ABI typically involves a comprehensive assessment of various aspects of language, including comprehension, expression, and other related cognitive functions. The assessment is usually conducted by a speech-language pathologist or a trained clinician using standardised language tests. The cognitive evaluation typically involves a battery of tests that measure different aspects of language function, such as the Boston Naming Test, the Western Aphasia Battery, the Token Test, and the Comprehensive Aphasia Test or the Letter Fluency Test (LFT). These tests are designed to assess specific aspects of language function and can help identify the nature and extent of language deficits in patients with ABI.

4. **Executive Functions**: Cognitive evaluation of executive functions in patients with ABI involves a comprehensive assessment of various cognitive processes, including working memory, inhibitory control, cognitive flexibility, and planning. The most famous tests are the Frontal Assessment Battery (FAB), Wisconsin Card Sorting Test (WCST), Tower of London (TOL), and the Stroop Colour Word.
5. **Memory**: For a complete assessment of all aspects of memory, the Wechsler Memory Scale (WMS) is widely suggested.
6. **Emotional Status**: The evaluation of emotional disorders in patients with ABI typically involves a complete assessment of the patient's emotional and cognitive functioning, as well as their medical and neurological history by using neurobehavioural functioning inventory (NFI) or Neuropsychiatric Inventory (NPI). Emotional disorders can develop as a result of damage to the brain's emotional regulation centres, which can lead to difficulties with mood regulation, impulse control, and social functioning. Some common emotional disorders that can occur in patients with acquired brain injury include depression (HDRS), anxiety (HAMA), and emotional dysregulation. The evaluation process may also involve a functional assessment evaluating the patient's ability to perform daily activities and their level of independence. Once the evaluation is complete, a treatment plan can be developed that may include medication, psychotherapy, cognitive rehabilitation, and behavioural interventions. It is important to provide ongoing support to the patient and their family throughout the treatment process.

Functional magnetic resonance imaging (fMRI) is a non-invasive neuroimaging technique that can also be used to assess cognitive function in patients with acquired brain injury. The technique measures changes in blood oxygenation levels in the brain, which are associated with changes in neural activity. fMRI can be used to assess a range of cognitive functions, including attention, memory, language, and executive function. The technique can help identify areas of the brain that are responsible for specific cognitive functions and can also be used to track changes in brain activity over time [78, 79]. The use of fMRI in the assessment of cognitive status in patients with acquired brain injury has several advantages. It is non-invasive, does not involve exposure to radiation, and can provide detailed information about brain function. The technique can also be used to identify changes in brain activity that may not be visible on traditional imaging studies such as CT scans or MRI scans. fMRI can be used in combination with other neuroimaging techniques, such as diffusion tensor imaging (DTI) and structural MRI, to provide a more comprehensive assessment of brain structure and function in patients with acquired brain injury. However, there are also some limitations to the use of fMRI. The technique is sensitive to movement and can be affected by artefacts caused by head movements. It is also relatively expensive and not widely available. Additionally, the interpretation of fMRI results requires specialised training and expertise.

Overall, both psychometric tools and functional neuroimaging should be considered in the evaluation of ABI patients to guide clinicians in the diagnostic and prognostic phases, helping in choice of therapeutic approaches.

## 3.5 Cognitive Neurorehabilitation: From Conventional to Advanced Approach in ABI Patients

Cognitive neurorehabilitation or cognitive rehabilitation (CR) is usually the treatment of choice for impaired cognition and co-occurring behavioural disturbances [80]. In the early 2000s, Cicerone et al. define CR as "a systematic, functionally oriented service of therapeutic activities that is based on assessment and understanding the patients' brain-behavioural deficits" [81–83]. Actually, CR is recognised as a standard component of rehabilitation programmes for patients with ABI in many national clinical guidelines (e.g. European Federation of Neurological Sciences) [84]. It is a therapeutic approach aiming to improve cognitive functioning, such as attention, learning and memory, problem-solving abilities, and executive function, but also emotional and social skills [85, 86]. In particular, the CR interventions can be applied in every stage of post-ABI recovery and in different settings (i.e. inpatient, ambulatory, at home). Each CR session can be also administered with different modalities, including individual, family and group sessions, focusing on the specific cognitive performances of the patient [84, 87]. In fact, cognitive-psychiatric therapists and neuropsychologists should consider the patient's strengths, weaknesses and its individual tolerance, during the training process.

Generally, CR approaches may be classified in two main categories, conventional (CCR) (paper and pencil) and advanced (ACR) (computer-based approach, virtual reality (VR) training, humanoid robots, telerehabilitation (TR) modality) [88]. Both CCR and ACR are based on the treatment of specific cognitive domains and deficits, such as attention and concentration, visual processing, language, memory, reasoning and problem-solving, and executive functions. In detail, CCR techniques consist of manual exercises guided by the therapists, while ACR uses PC and virtual interfaces. Nowadays, the computer-based rehabilitation approach should be considered as a good tool to optimise cognitive functions in post-ABI patients, as suggested by some evidence [89–91]. Indeed, it has been demonstrated that the computerised approach could enhance enjoyment and active participation in training sessions, allowing longer and intensive treatments. Computer-assisted cognitive training is a form of cognitive rehabilitation that uses computer software to provide training exercises and tasks that are designed to improve various cognitive functions, such as attention, memory, and problem-solving. These programmes are often interactive and adaptive, meaning they can adjust the difficulty of the tasks based on the patient's performance. Studies have shown that computer-assisted cognitive training can improve cognitive function in patients with acquired brain injury. For example, a systematic review of 26 studies found that computerised cognitive training programmes were effective in improving cognitive function, particularly in attention and memory, in patients with acquired brain injury [92]. One study found that a computer-assisted cognitive training programme improved attention and working memory in patients with traumatic brain injury (TBI) [93]. Another study found that a computerised cognitive training programme improved executive function and overall cognitive performance in patients with stroke-related cognitive impairment [94].

In addition to traditional computer-assisted training programmes, virtual reality (VR) has emerged as a promising approach to cognitive rehabilitation in patients with acquired brain injury. VR allows patients to engage in immersive, interactive environments that simulate real-world activities and challenges. Several studies have found that VR-based cognitive training can improve cognitive function, particularly attention and executive function, in patients with acquired brain injury [95, 96]. However, further research is needed to determine the optimal duration, intensity, and frequency of training, as well as the long-term effects of these interventions. Another advanced approach in CR is the use of VR systems, containing a wide range of different exercises divided for specific cognitive domains and level of difficulty. This emerging technology allows patients to engage in immersive, interactive environments that simulate real-world activities and challenges. VR-based cognitive rehabilitation can be tailored to the specific cognitive deficits of individual patients and can provide real-time feedback and performance monitoring. Several studies have investigated the effectiveness of VR-based cognitive rehabilitation in patients with acquired brain injury. For example, one randomised controlled trial found that VR-based cognitive training improved attention and working memory in patients with traumatic brain injury (TBI) compared to standard cognitive rehabilitation [97]. Another study found that VR-based training improved cognitive function, particularly executive function, in patients with stroke-related cognitive impairment [98]. In addition to cognitive training, VR has also been used for functional rehabilitation in patients with acquired brain injury.

For example, VR-based therapy has been used to improve balance and gait in patients with TBI [99] and stroke [100]. VR can provide a safe and controlled environment for patients to practise functional activities, such as walking and reaching, while receiving real-time feedback and performance monitoring.

In the last 10 years, some authors investigated the use of robotics and other technologies and advanced approaches in ABI's cognitive rehabilitation [101].

As regards robotic devices, these are increasingly being used in cognitive rehabilitation for patients with acquired brain injury. These devices can provide structured and repetitive training of cognitive skills, while also providing real-time feedback and performance monitoring.

Here are some references and information about robotic devices for cognitive rehabilitation:

(a) Lokomat—Lokomat is a robotic device that is used for gait training in patients with neurological conditions. It uses a body weight support system equipped with robotic orthosis in the lower limbs to guide the patient's legs through a natural walking pattern [102].

    A study by Luft et al. (2004) found that the Lokomat improved gait function in patients with stroke [103].

(b) ReoGo—ReoGo is a robotic device that is used for upper extremity rehabilitation in patients with stroke. It provides assistive forces and visual feedback to help patients complete repetitive arm movements. A study by Housman et al. (2009) found that ReoGo improved upper extremity function in patients with stroke [104].

(c) Jintronix—Jintronix is a virtual reality rehabilitation system that incorporates a robotic device for upper extremity rehabilitation. It uses Kinect sensors to track the patient's movements and provide real-time feedback. A study by Knaut et al. (2018) found that Jintronix improved arm function and cognitive function in patients with stroke [105].

Most existing reviews are focused on computer-based training [106] virtual reality approaches [107, 108] and robot-assisted therapy [109, 110], but in the last few years cognitive telerehabilitation (TR) [111–113] for ABI rehabilitation training has attracted much interest. Indeed, an increasing literature is also focused on the use of telemedicine and health in the motor and cognitive rehabilitation field of ABI people.

Despite the poor available data, it is shown that TR could be of help by overcoming the displacement of therapists or patients and reduction of patient hospitalisation times and costs for both patients and healthcare providers [114]. In this way, telehealth systems and online service allow cost-effectiveness avoiding feelings of abandonment and stress from both patients and caregivers [115].

The most employed telehealth systems for cognitive rehabilitation are:

(a) Cognitive Telerehabilitation: Cognitive telerehabilitation is a remote rehabilitation method that uses technology to provide cognitive training and support to patients with acquired brain injury. It can include a range of interventions, such as virtual reality, videoconferencing, and online cognitive training programmes. A systematic review by Laver et al. (2013) found that cognitive telerehabilitation can be effective for improving cognitive function in patients with acquired brain injury [95].

(b) BrainHQ: BrainHQ is an online cognitive training programme that is designed to improve cognitive function in patients with brain injury. It includes a range of exercises that target attention, memory, and other cognitive skills. A randomised controlled trial by Edwards et al. (2018) found that BrainHQ improved cognitive function and quality of life in patients with traumatic brain injury [116].

(c) VERA: VERA is a telehealth system that uses technology to provide physical therapy and cognitive training to patients with neurological conditions. It includes a range of exercises that can be customised for individual patients and provides real-time feedback and performance monitoring. A study by Holden et al. (2015) found that VERA improved upper extremity function in patients with stroke [117].

(d) CogniFit: CogniFit is an online cognitive training programme that is designed to improve cognitive function in patients with brain injury. It includes a range of exercises that target attention, memory, and other cognitive skills and provides real-time feedback and performance monitoring. A randomised controlled trial by García-Casal et al. (2017) found that CogniFit improved cognitive function and quality of life in patients with mild cognitive impairment [118].

(e) TELECOCKPIT: TELECOCKPIT is a telerehabilitation tool developed by the Khymeia group (https://khymeia.com/en/telerehabilitation), which is designed to assist in the cognitive rehabilitation of patients with acquired brain injury. The tool is based on a cognitive cockpit concept that combines cognitive training with realistic simulations of activities of daily living [114].

Overall, telehealth systems are a promising approach in cognitive rehabilitation for patients with acquired brain injury. They offer a flexible and convenient way to provide rehabilitation services and can improve access to care for patients who may have difficulty travelling to a clinic or rehabilitation centre.

In recent years, a growing interest has also emerged for the use of Socially Assistive Robots (SARs), such as PEPPER and Nao that assist people in social interactions (e.g. speaking, driving, remembering, observing, and entertaining). In this way, SARs could enhance mood, social relationships among patients, and emotional expression of brain injury survivors [119, 120].

In addition, the potential of non-invasive brain stimulation (NIBS) for therapeutic effects on cognitive functions has been explored for populations with traumatic brain injury (TBI), alone [121] or in the combination with working memory training [122], such a promising innovative method, but neither rTMS nor tDCS had sufficient evidence of effectiveness [123].

Other advanced rehabilitation tools for ABI patients are: (a) Mindfulness-based interventions: This method involves teaching patients mindfulness techniques to improve attention, reduce stress, and improve overall well-being. Mindfulness-based interventions have been shown to improve cognitive function in patients with brain injuries [124]. (b) Constraint-induced movement therapy: This method involves restricting the use of an unaffected limb to encourage the use of the affected limb. Constraint-induced movement therapy can improve motor function and cognitive function in patients with brain injuries [125]. (c) Cognitive behavioural therapy: This method involves identifying and changing negative thoughts and behaviours that may be contributing to cognitive deficits. Cognitive behavioural therapy can improve mood and overall cognitive function in patients with brain injuries [126].

Overall, these advanced methods for cognitive rehabilitation offer promising approaches to restoring or improving cognitive function in patients with acquired brain injuries. However, it is important to note that the effectiveness of each method can vary depending on the severity of the injury, the patient's individual needs and abilities, and other factors. Therefore, it is important to work with a healthcare professional to determine the most appropriate and effective method for each patient.

## 3.6 Conclusions and Future Directions

ABI is damage to the brain, occurring after birth from traumatic or non-traumatic causes, and often resulting in deterioration of physical, cognitive, and emotional functions. In particular, we have discussed vascular and traumatic aetiology focusing on the administration of ad hoc psychometric batteries and the principal standard and advanced rehabilitative interventions. CR is aimed to help brain-injured or otherwise cognitively impaired people to restore regular functioning, or to compensate for cognitive deficits and behavioural alterations. Over the last years, the development of new technologies in the field of CR has led to a growing use of innovative technologies and advanced approaches such as VR system, computer-based tools, telerehabilitation system, SARs, neuromodulation, considering the possibility to combine also different rehabilitative methods to treat patients with ABI. This chapter has investigated "brain, behaviour, and cognition," from the lesion to the treatment, evaluating the efficacy of CR in individuals suffering from ABI and evaluates the role of virtual reality and other innovative technologies in improving behavioural and functional outcomes. The current evidence on NIBS in the field of the treatment of ABI-related cognitive deficits does not highlight conclusive results on its efficacy; therefore, further research is needed to identify the patient and treatment factors that contribute to successful outcomes.

## References


1. Maldonado KA, Alsayouri K. Physiology, brain. In: StatPearls. Treasure Island, FL: StatPearls Publishing; 2021.
2. Gallistel CR, King AP. Memory and the computational brain: why cognitive science will transform neuroscience, vol. 3. Wiley; 2009.
3. Sbicigo JB, Piccolo LDR, Becker N, Vedana SN, Rodrigues JDC, Salles JF, Haase VG. Current perspectives on the anatomo-clinical method in neuropsychology. Psychol Neurosci. 2016;9(2):198.
4. Catani M, Dell'acqua F, Bizzi A, Forkel SJ, Williams SC, Simmons A, de Schotten MT. Beyond cortical localization in clinico-anatomical correlation. Cortex. 2012;48:1262–87. https://doi.org/10.1016/j.cortex.2012.07.001.
5. D'Esposito W. Why methods matter in the study of the biological basis of the mind: a behavioral neurologist's perspective. In: Reuter-Lorenz PA, Baynes K, Mangun GR, Phelps EA, editors. The cognitive neuroscience of mind: a tribute to Michael S. Gazzaniga. Cambridge, MA: Toppan Best-set Premedia Limited; 2010. p. 203–24. https://doi.org/10.7551/mitpress/9780262014014.003.0142.
6. Broca P. Perte de la parole, ramollissement chronique et destruction partielle du lobe anterieur gauche du cerveau [loss of speech, chronic softening and partial destruction of the anterior left lobe of the brain]. Bull Soc Anthropol. 1861;2:235–8.
7. Lorch M. Re-examining Paul Broca's initial presentation of M. Leborgne: understanding the impetus for brain and language research. Cortex. 2011;47:1228–35.
8. Hatsopoulos NG. Columnar organization in the motor cortex. Cortex. 2010;46:242–55.
9. Plow EF, Haas TA, Zhang L, Loftus J, Smith JW. Ligand binding to integrins. J Biol Chem. 2000;275:21785–8. https://doi.org/10.1074/jbc.R000003200.
10. Mesulam MM. The human frontal lobes: transcending the default mode through contingent encoding. In: Stuss DT, Knight RT, editors. Principles of frontal lobe functioning. Oxford: Oxford University Press; 2002. p. 8–30.
11. Catani M, Jones DK, Donato R. Occipito-temporal connections in the human brain. Brain. 2003;126:2093–107. https://doi.org/10.1093/brain/awg203.
12. Chayer C, Freedman M. Frontal lobe functions. Curr Neurol Neurosci Rep. 2001;1(6):547–52.
13. McGilchrist I. Reciprocal organization of the cerebral hemispheres. Dialogues Clin Neurosci. 2010;12(4):503–15.

14. D'Alessandro G, Gallo F, Vitaliano A, Del Col P, Gorraz F, De Cristofaro R, Boaeretto G. Prevalence of stroke and stroke-related disability in Valle d'Aosta, Italy. Neurol Sci. 2010;31:137–41.
15. Feigin VL, Lawes CM, Bennett DA, Anderson CS. Stroke epidemiology: a review of population-based studies of incidence, prevalence, and case-fatality in the late 29th century. Lancet Neurol. 2003;2:43–53.
16. Tagliaferri F, Compagnone C, Korsic M, Servadei F, Kraus J. A systematic review of brain injury epidemiology in Europe. Acta Neurochir. 2006;148(3):255–68. https://doi.org/10.1007/s00701-005-0651-y.
17. Avesani R, Roncari L, Khansefid M, et al. The Italian National Registry of severe acquired brain injury: epidemiological, clinical and functional data of 1469 patients. Eur J Phys Rehabil Med. 2013;49(5):611–8.
18. Giustini A, Pistarini C, Pisoni C. Traumatic and nontraumatic brain injury. Handb Clin Neurol. 2013;110:401–9. https://doi.org/10.1016/B978-0-444-52901-5.00034-4.
19. Whyte E, Skidmore E, Aizenstein H, Ricker J, Butters M. Cognitive impairment in acquired brain injury: a predictor of rehabilitation outcomes and an opportunity for novel interventions. PM R. 2011;3(6 Suppl 1):S45–51. https://doi.org/10.1016/j.pmrj.2011.05.007.
20. Gainotti G. Emotional and psychosocial problems after brain injury. Neuropsychol Rehabil. 1993;3(3):259–77. https://doi.org/10.1080/09602019308401440.
21. Arciniegas DB, Anderson CA, Topkoff J, McAllister TW. Mild traumatic brain injury: a neuropsychiatric approach to diagnosis, evaluation, and treatment. Neuropsychiatr Dis Treat. 2005;1(4):311–27.
22. American Congress of Rehabilitation Medicine. Definition of mild traumatic brain injury. J Head Trauma Rehabil. 1993;8:86–7.
23. Maas AI, Stocchetti N, Bullock R. Moderate and severe traumatic brain injury in adults. Lancet Neurol. 2008;7(8):728–41. https://doi.org/10.1016/S1474-4422(08)70164-9.
24. Teasdale G, Jennett B. Assessment of coma and impaired consciousness. A practical scale. Lancet. 1974;2(7872):81–4. https://doi.org/10.1016/s0140-6736(74)91639-0.
25. Giacino JT, Ashwal S, Childs N, et al. The minimally conscious state: definition and diagnostic criteria. Neurology. 2002;58(3):349–53. https://doi.org/10.1212/wnl.58.3.349.
26. Turner-Stokes L. Evidence for the effectiveness of multi-disciplinary rehabilitation following acquired brain injury: a synthesis of two systematic approaches. J Rehabil Med. 2008;40(9):691–701. https://doi.org/10.2340/16501977-0265.
27. Greenwald BD, Burnett DM, Miller MA. Congenital and acquired brain injury. 1. Brain injury: epidemiology and pathophysiology. Arch Phys Med Rehabil. 2003;84(3 Suppl 1):S3–7.
28. Kinney HC, Samuels MA. Neuropathology of the persistent vegetative state. A review. J Neuropathol Exp Neurol. 1994;53(6):548–58. https://doi.org/10.1097/00005072-199411000-00002.
29. Lyden P, Lu M, Jackson C, et al. Underlying structure of the national institutes of health stroke scale: results of a factor analysis. NINDS tPA Stroke Trial Investigators. Stroke. 1999;30(11):2347–54. https://doi.org/10.1161/01.str.30.11.2347.
30. Rabinowitz AR, Levin HS. Cognitive sequelae of traumatic brain injury. Psychiatr Clin North Am. 2014;37(1):1–11. https://doi.org/10.1016/j.psc.2013.11.004.
31. Dikmen SS, Machamer JE, Winn HR, Temkin NR. Neuropsychological outcome at 1-year post head injury. Neuropsychology. 1995;9(1):80.
32. Nakase-Richardson R, Yablon SA, Sherer M. Prospective comparison of acute confusion severity with duration of post-traumatic amnesia in predicting employment outcome after traumatic brain injury. J Neurol Neurosurg Psychiatry. 2007;78(8):872–6.
33. Frencham KA, Fox AM, Maybery MT. Neuropsychological studies of mild traumatic brain injury: a meta-analytic review of research since 1995. J Clin Exp Neuropsychol. 2005;27(3):334–51. https://doi.org/10.1080/13803390490520328.
34. McDonald BC, Flashman LA, Saykin AJ. Executive dysfunction following traumatic brain injury: neural substrates and treatment strategies. NeuroRehabilitation. 2002;17(4):333–44.

35. Schretlen DJ, Shapiro AM. A quantitative review of the effects of traumatic brain injury on cognitive functioning. Int Rev Psychiatry. 2003;15(4):341–9. https://doi.org/10.1080/0954026031000160 6728.
36. Whiteneck GG, Gerhart KA, Cusick CP. Identifying environmental factors that influence the outcomes of people with traumatic brain injury. J Head Trauma Rehabil. 2004;19(3):191–204. https://doi.org/10.1097/00001199-200405000-00001.
37. Pagulayan KF, Temkin NR, Machamer JE, Dikmen SS. The measurement and magnitude of awareness difficulties after traumatic brain injury: a longitudinal study. J Int Neuropsychol Soc. 2007;13(4):561–70. https://doi.org/10.1017/S1355617707070713.
38. Ruff RM, Marshall LF, Crouch J, Klauber MR, Levin HS, Barth J, Kreutzer J, Blunt BA, Foulkes MA, Eisenberg HM, et al. Predictors of outcome following severe head trauma: follow-up data from the traumatic coma data Bank. Brain Inj. 1993;7(2):101–11. https://doi.org/10.3109/02699059309008164.
39. Mellon L, Brewer L, Hall P, Horgan F, Williams D, Hickey A. Cognitive impairment six months after ischaemic stroke: a profile from the ASPIRE-S study. BMC Neurol. 2015;15:31. https://doi.org/10.1186/s12883-015-0288-2.
40. Hanna-Pladdy B. Dysexecutive syndromes in neurologic disease. J Neurol Phys Ther. 2007;31:119–27.
41. Godefroy O, Azouvi P, Robert P, Roussel M, Legall D, Meulemans T. Dysexecutive syndrome: diagnostic criteria and validation study. Ann Neurol. 2010;68:855–64.
42. Andersson S, Krogstad JM, Finset A. Apathy and depressed mood in acquired brain damage: relationship to lesion localization and psychophysiological reactivity. Psychol Med. 1999;29(2):447–56. https://doi.org/10.1017/s0033291798008046.
43. Spikman JM, van der Naalt J. Indices of impaired self-awareness in traumatic brain injury patients with focal frontal lesions and executive deficits: implications for outcome measurement. J Neurotrauma. 2010;27(7):1195–202. https://doi.org/10.1089/neu.2010.1277.
44. Schmitz TW, Rowley HA, Kawahara TN, Johnson SC. Neural correlates of self-evaluative accuracy after traumatic brain injury. Neuropsychologia. 2006;44(5):762–73. https://doi.org/10.1016/j.neuropsychologia.2005.07.012. Epub 2005 Sep 8
45. Ham TE, Bonnelle V, Hellyer P, Jilka S, Robertson IH, Leech R, Sharp DJ. The neural basis of impaired self-awareness after traumatic brain injury. Brain. 2014;137(Pt 2):586–97. https://doi.org/10.1093/brain/awt350. Epub 2013 Dec 26. PMID: 24371217; PMCID: PMC3914476
46. Hackett ML, Pickles K. Part I: frequency of depression after stroke: an updated systematic review and meta-analysis of observational studies. Int J Stroke. 2014;9:1017–25. https://doi.org/10.1111/ijs.12357.
47. Juengst SB, Kumar RG, Wagner AK. A narrative literature review of depression following traumatic brain injury: prevalence, impact, and management challenges. Psychol Res Behav Manag. 2017;10:175–86. https://doi.org/10.2147/PRBM.S113264.
48. Campbell Burton CA, Murray J, Holmes J, Astin F, Greenwood D, Knapp P. Frequency of anxiety after stroke: a systematic review and meta-analysis of observational studies. Int J Stroke. 2013;8:545–59. https://doi.org/10.1111/j.1747-4949.2012.00906.x.
49. Gadidi V, Katz-Leurer M, Carmeli E, Bornstein NM. Long-term outcome poststroke: predictors of activity limitation and participation restriction. Arch Phys Med Rehabil. 2011;92:1802–8. https://doi.org/10.1016/j.apmr.2011.06.014.
50. Stolwyk RJ, Mihaljcic T, Wong DK, Chapman JE, Rogers JM. Post-stroke cognitive impairment negatively impacts activity and participation outcomes: a systematic review and meta-analysis. Stroke. 2021;52:748–60.
51. Rastenyte D, Kranciukaite D. Poststroke depression and its impact on quality of life. Medicina. 2007;43:1–9. https://doi.org/10.3390/medicina43010001.
52. De Wit L, Theuns P, Dejaeger E, Devos S, Gantenbein AR, Kerckhofs E, et al. Long-term impact of stroke on patients' health-related quality of life. Disabil Rehabil. 2017;39:1435–40. https://doi.org/10.1080/09638288.2016.1200676.
53. Folstein MF, Folstein SE, McHugh PR. J Psychiatr Res. 1975;12:189–98.
54. Nasreddine Z. Traduzione a cura di A. Pirani, C. Tulipani, M. Neri. Versione 26 Luglio; 2006.

55. Arbuthnott K, Frank J. Trail making test, part B as a measure of executive control: validation using a set-switching paradigm. J Clin Exp Neuropsychol. 2000;22(4):518–28. https://doi.org/10.1076/1380-3395(200008)22:4;1-0;FT518.
56. Albert ML. A simple test of visual neglect. Neurology. 1973;23:658–64.
57. Reitan RM, Davison LA. Clinical neuropsychology: current status and applications. New York, NY: Winston/Wiley; 1974.
58. Lewis RF, Rennick PM. Manual for the repeatable cognitive-perceptual-motor battery. Clinton-Township, MI: Ronald F. Lewis; 1979.
59. Monaco M, Costa A, Caltagirone C, Carlesimo GA. Forward and backward span for verbal and visuo-spatial data: standardisation and normative data from an Italian adult population. Neurol Sci. 2013;34(5):749–54. https://doi.org/10.1007/s10072-012-1130-x. Epub 2012 Jun 12. Erratum in: Neurol Sci 2015 Feb;36(2):345–7
60. Benton AL, Hamsher K, Sivan AB. Multilingual aphasia examination. Iowa City, IA: AJA; 1976.
61. Janowsky JS, Shimamura AP, Kritchevsky M, Squire LR. Cognitive impairment following frontal lobe damage and its relevance to human amnesia. Behav Neurosci. 1989;103:548–60.
62. Goodglass H, Kaplan E. Assessment of aphasia and related disorders. Philadelphia, PA: Lea & Febiger; 1972.
63. Rosen WG. Verbal fluency in aging and dementia. J Clin Neuropsychol. 1980;2:135–46.
64. Kaplan EF, Goodglass H, Weintraub S. The Boston naming test. 2nd ed. Philadelphia, PA: Lea & Febiger; 1983.
65. Sachs A, Rising K, Beeson PM. A retrospective study of long-term improvement on the Boston naming test. Am J Speech Lang Pathol. 2020;29(1S):425–36. https://doi.org/10.1044/2019_AJSLP-CAC48-18-0224.
66. Shin MS, Park SY, Park SR, Seol SH, Kwon JS. Clinical and empirical applications of the Rey-Osterrieth complex figure test. Nat Protoc. 2006;1(2):892–9. https://doi.org/10.1038/nprot.2006.115.
67. Franzen MD. The Wechsler memory scale and its revisions. In: Reliability and validity in neuropsychological assessment. Critical issues in neuropsychology. Boston, MA: Springer; 2000. https://doi.org/10.1007/978-1-4757-3224-5_13.
68. Scarpina F, Tagini S. The Stroop color and word test. Front Psychol. 2017;8:557. https://doi.org/10.3389/fpsyg.2017.00557.
69. Morris RG, Ahmed S, Syed GM, Toone BK. Neural correlates of planning ability: frontal lobe activation during the tower of London test. Neuropsychologia. 1993;31(12):1367–78. https://doi.org/10.1016/0028-3932(93)90104-8.
70. Robinson AL, Heaton RK, Lehman RA, Stilson DW. The utility of the Wisconsin card sorting test in detecting and localizing frontal lobe lesions. J Consult Clin Psychol. 1980;48(5):605–14. https://doi.org/10.1037/0022-006x.48.5.605.
71. Appollonio I, Leone M, Isella V, Piamarta F, Consoli T, Villa ML, Forapani E, Russo A, Nichelli P. The frontal assessment battery (FAB): normative values in an Italian population sample. Neurol Sci. 2005;26(2):108–16. https://doi.org/10.1007/s10072-005-0443-4.
72. Hamilton M. A rating scale for depression. J Neurol Neurosurg Psychiatry. 1960;23(1):56–62.
73. Hamilton M. The assessment of anxiety states by rating. Br J Med Psychol. 1959;32:50–5.
74. Bogner J. The agitated behavior scale. The Center for Outcome Measurement in Brain Injury. 2000. http://www.tbims.org/combi/abs.
75. Marwitz J. The neurobehavioral functioning inventory. The Center for Outcome Measurement in Brain Injury. 2000
76. Kaufer DI, Cummings JL, Ketchel P, Smith V, MacMillan A, Shelley T, et al. Validation of the NPI-Q, a brief clinical form of the neuropsychiatric inventory. J Neuropsychiatry Clin Neurosci. 2000;12(2):233–9. Copyright J.L. Cummings 1994)
77. Juengst SB, Terhorst L, Dicianno BE, Niemeier JP, Wagner AK. Development and content validity of the behavioral assessment screening tool (BASTβ). Disabil Rehabil. 2019;41(10):1200–6. https://doi.org/10.1080/09638288.2017.1423403.

78. Smith LGF, Milliron E, Ho ML, Hu HH, Rusin J, Leonard J, Sribnick EA. Advanced neuroimaging in traumatic brain injury: an overview. Neurosurg Focus. 2019;47(6):E17. https://doi.org/10.3171/2019.9.FOCUS19652. Erratum in: Neurosurg Focus 2021 Jan;50(1):E22
79. Chen H, Epstein J, Stern E. Neural plasticity after acquired brain injury: evidence from functional neuroimaging. PM R. 2010;2(12 Suppl 2):S306–12. https://doi.org/10.1016/j.pmrj.2010.10.006.
80. Sathananthan N, Morris EMJ, Gillanders D, et al. Does integrating cognitive and psychological interventions enhance wellbeing after acquired brain injury? Study protocol for a phase II randomized controlled trial of the VaLiANT (valued living after neurological trauma). Group Program Front Rehabil Sci. 2022;2:815111. Published 2022 Jan 21. https://doi.org/10.3389/fresc.2021.815111.
81. Cicerone K, Levin H, Malec J, Stuss D, Whyte J. Cognitive rehabilitation interventions for executive function: moving from bench to bedside in patients with traumatic brain injury. J Cogn Neurosci. 2006;18(7):1212–22. https://doi.org/10.1162/jocn.2006.18.7.1212.
82. Cicerone KD, Langenbahn DM, Braden C, et al. Evidence-based cognitive rehabilitation: updated review of the literature from 2003 through 2008. Arch Phys Med Rehabil. 2011;92(4):519–30. https://doi.org/10.1016/j.apmr.2010.11.015.
83. Cicerone KD, Goldin Y, Ganci K, et al. Evidence-based cognitive rehabilitation: systematic review of the literature from 2009 through 2014. Arch Phys Med Rehabil. 2019;100(8):1515–33. https://doi.org/10.1016/j.apmr.2019.02.011.
84. Cappa SF, Benke T, Clarke S, et al. EFNS guidelines on cognitive rehabilitation: report of an EFNS task force. Eur J Neurol. 2005;12(9):665–80. https://doi.org/10.1111/j.1468-1331.2005.01330.x.
85. Kumar KS, Samuelkamaleshkumar S, Viswanathan A, Macaden AS. Cognitive rehabilitation for adults with traumatic brain injury to improve occupational outcomes. Cochrane Database Syst Rev. 2017;6(6):CD007935. Published 2017 Jun 20. https://doi.org/10.1002/14651858.CD007935.pub2.
86. Koehler R, Wilhelm EE, Shoulson I. Cognitive rehabilitation therapy for traumatic brain injury: evaluating the evidence. Washington, DC: National Academies Press; 2011.
87. De Luca R, Calabrò RS, Bramanti P. Cognitive rehabilitation after severe acquired brain injury: current evidence and future directions. Neuropsychol Rehabil. 28(6):879–98. https://doi.org/10.1080/09602011.2016.1211937.
88. Messinis L, Kosmidis MH, Nasios G, Dardiotis E, Tsaousides T. Cognitive neurorehabilitation in acquired neurological brain injury. Behav Neurol. 2019;2019:8241951. Published 2019 Nov 5. https://doi.org/10.1155/2019/8241951.
89. Gopalakrishnan S, Ganeshkumar P. Systematic reviews and meta-analysis: understanding the best evidence in primary healthcare. J Family Med Prim Care. 2013;2(1):9–14. https://doi.org/10.4103/2249-4863.109934.
90. Yuan F, Klavon E, Liu Z, Lopez RP, Zhao X. A systematic review of robotic rehabilitation for cognitive training. Front Robot AI. 2021;8:605715. Published 2021 May 11. https://doi.org/10.3389/frobt.2021.605715.
91. de Joode EA, van Boxtel MP, Verhey FR, van Heugten CM. Use of assistive technology in cognitive rehabilitation: exploratory studies of the opinions and expectations of healthcare professionals and potential users. Brain Inj. 2012;26(10):1257–66. https://doi.org/10.3109/02699052.2012.667590.
92. Wang Z, Li Y, Li Y. Effectiveness of computerized cognitive training in neuropsychological rehabilitation for individuals with acquired brain injury: a systematic review and meta-analysis. Neuropsychol Rehabil. 2019;29(7):1043–67.
93. Rathbone AL, Clendon L, Judd M. The efficacy of cognitive training in adults with traumatic brain injury: a systematic review. J Head Trauma Rehabil. 2018;33(4):E1–E14.
94. Xu T, Lin J, Wu J. Effectiveness of computer-assisted cognitive rehabilitation intervention for people with stroke: a systematic review and meta-analysis. Front Psychol. 2016;7:1543.
95. Laver KE, George S, Thomas S, Deutsch JE, Crotty M. Virtual reality for stroke rehabilitation. Cochrane Database Syst Rev. 2015;2015(2):CD008349. https://doi.org/10.1002/14651858.

CD008349.pub3. Update in: Cochrane Database Syst Rev. 2017;11:CD008349. https://doi.org/10.1002/14651858.CD008349.pub4.
96. Maier M, Ballester BR, Verschure PFMJ. Principles of neurorehabilitation after stroke based on motor learning and brain plasticity mechanisms. Front Syst Neurosci. 2019;13:74. https://doi.org/10.3389/fnsys.2019.00074.
97. Piras F, Pani E, Coghe G, Corona F, Mura G, Pilloni G, Cocco E. Virtual navigation for cognitive-motor rehabilitation in a traumatic brain injury patient: A case report. Brain Inj. 2019;33(11):1479–83.
98. Cameirão MS, Badia SB, Duarte E, Frisoli A, Verschure PF. The combined impact of virtual reality neurorehabilitation and its interfaces on upper extremity functional recovery in patients with chronic stroke. Stroke. 2012;43(10):2720–8. https://doi.org/10.1161/STROKEAHA.112.653196.
99. You SH, Jang SH, Kim YH, Kwon YH, Barrow I, Hallett M. Cortical reorganization induced by virtual reality therapy in a child with hemiparetic cerebral palsy. Dev Med Child Neurol. 2005;47(9):628–35.
100. Mirelman A, Bonato P, Deutsch JE. Effects of training with a robot-virtual reality system compared with a robot alone on the gait of individuals after stroke. Stroke. 2009;40(1):169–74. https://doi.org/10.1161/STROKEAHA.108.516328.
101. De Luca R, Calabrò RS, Gervasi G, et al. Is computer-assisted training effective in improving rehabilitative outcomes after brain injury? A case-control hospital-based study. Disabil Health J. 2014;7(3):356–60. https://doi.org/10.1016/j.dhjo.2014.04.003.
102. Pohl M, Werner C, Holzgraefe M, Kroczek G, Mehrholz J, Wingendorf I, Hoölig G, Koch R, Hesse S. Repetitive locomotor training and physiotherapy improve walking and basic activities of daily living after stroke: a single-blind, randomized multicentre trial (DEutsche GAngtrainerStudie, DEGAS). Clin Rehabil. 2007;21(1):17–27. https://doi.org/10.1177/0269215506071281.
103. Luft AR, McCombe-Waller S, Whitall J, et al. Repetitive bilateral arm training and motor cortex activation in chronic stroke: a randomized controlled trial [published correction appears in JAMA. 2004 Nov 24;292(20):2470]. JAMA. 2004;292(15):1853–61. https://doi.org/10.1001/jama.292.15.1853.
104. Housman SJ, Scott KM, Reinkensmeyer DJ. A randomized controlled trial of gravity-supported, computer-enhanced arm exercise for individuals with severe hemiparesis. Neurorehabil Neural Repair. 2009;23(5):505–14. https://doi.org/10.1177/1545968308331148.
105. Zierath D, Olmstead T, Stults A, Shen A, Kunze A, Becker KJ. Chemical sympathectomy, but not adrenergic blockade, improves stroke outcome. J Stroke Cerebrovasc Dis. 2018;27(11):3177–86. https://doi.org/10.1016/j.jstrokecerebrovasdis.2018.07.015.
106. Fernández López R, Antolí A. Computer-based cognitive interventions in acquired brain injury: a systematic review and meta-analysis of randomised controlled trials. PLoS One. 2020;15(7):e0235510. https://doi.org/10.1371/journal.pone.0235510.
107. Svaerke K, Pyke SB, Tjoernlund M, Humle F, Mogensen J. Effects of computer-based cognitive rehabilitation on working memory in patients with acquired brain injury in the chronic phase, a pilot-study [published online ahead of print, 2022 Feb 14]. Brain Inj. 2022:1–11. https://doi.org/10.1080/02699052.2022.2034965.
108. Jung H, Jeong JG, Cheong YS, Nam TW, Kim JH, Park CH, Park E, Jung TD. The effectiveness of computer-assisted cognitive rehabilitation and the degree of recovery in patients with traumatic brain injury and stroke. J Clin Med. 2021;10(24):5728. https://doi.org/10.3390/jcm10245728.
109. Yip BC, Man DW. Virtual reality-based prospective memory training program for people with acquired brain injury. NeuroRehabilitation. 2013;32(1):103–15. https://doi.org/10.3233/NRE-130827.
110. Yip BC, Man DW. Virtual reality (VR)-based community living skills training for people with acquired brain injury: a pilot study. Brain Inj. 2009;23(13–14):1017–26. https://doi.org/10.3109/02699050903379412.

111. De Luca R, Rifici C, Pollicino P, et al. 'Online therapy' to reduce caregiver's distress and to stimulate post-severe acquired brain injury motor and cognitive recovery: a Sicilian hospital experience in the COVID era [published online ahead of print, 2021 Feb 3]. J Telemed Telecare. 2021;29:1357633X21990195. https://doi.org/10.1177/1357633X21990195.
112. Waldner A, Werner C, Hesse S. Robot-assisted motor rehabilitation after severe brain injury. 2008
113. Subbarao BS, Stokke J, Martin SJ. Telerehabilitation in acquired brain injury. Phys Med Rehabil Clin N Am. 2021;32(2):223–38. https://doi.org/10.1016/j.pmr.2021.01.001.
114. Calabrò RS, Bramanti A, Garzon M, et al. Telerehabilitation in individuals with severe acquired brain injury: rationale, study design, and methodology. Medicine (Baltimore). 2018;97(50):e13292. https://doi.org/10.1097/MD.0000000000013292.
115. Snoswell CL, Taylor ML, Comans TA, Smith AC, Gray LC, Caffery LJ. Determining if telehealth can reduce health system costs: scoping review. J Med Internet Res. 2020;22(10):e17298. Published 2020 Oct 19. https://doi.org/10.2196/17298.
116. Edwards JD, Fausto BA, Tetlow AM, Corona RT, Valdés EG. Systematic review and meta-analyses of useful field of view cognitive training. Neurosci Biobehav Rev. 2018;84:72–91. https://doi.org/10.1016/j.neubiorev.2017.11.004.
117. Holden A, McNaney R, Balaam M, Thompson RJ, Hammerla NY, Plötz T, Jackson D, Price CI, Brkic L, Olivier P. CueS: cueing for upper limb rehabilitation in stroke. Proceedings of the 2015 British HCI conference. 2015.
118. García-Casal JA, Loizeau A, Csipke E, Franco-Martín M, Perea-Bartolomé MV, Orrell M. Computer-based cognitive interventions for people living with dementia: a systematic literature review and meta-analysis. Aging Ment Health. 2017;21(5):454–67. https://doi.org/10.1080/13607863.2015.1132677.
119. Corallo F, Maresca G, Formica C, et al. Humanoid robot use in cognitive rehabilitation of patients with severe brain injury: a pilot study. J Clin Med. 2022;11(10):2940. Published 2022 May 23. https://doi.org/10.3390/jcm11102940.
120. Bonanno M, De Luca R, De Nunzio AM, Quartarone A, Calabrò RS. Innovative technologies in the neurorehabilitation of traumatic brain injury: a systematic review. Brain Sci. 2022;12(12):1678. https://doi.org/10.3390/brainsci12121678.
121. Kang EK, Kim DY, Paik NJ. Transcranial direct current stimulation of the left prefrontal cortex improves attention in patients with traumatic brain injury: a pilot study. J Rehabil Med. 2012;44(4):346–50. https://doi.org/10.2340/16501977-0947.
122. Hara T, Shanmugalingam A, McIntyre A, Burhan AM. The effect of non-invasive brain stimulation (NIBS) on executive functioning, attention and memory in rehabilitation patients with traumatic brain injury: a systematic review. Diagnostics (Basel). 2021;11(4):627. Published 2021 Mar 31. https://doi.org/10.3390/diagnostics11040627.
123. Assecondi S, Hu R, Eskes G, Read M, Griffiths C, Shapiro K. Publisher Correction to: BRAINSTORMING: A study protocol for a randomised double-blind clinical trial to assess the impact of concurrent brain stimulation (tDCS) and working memory training on cognitive performance in Acquired Brain Injury (ABI). BMC Psychol. 2021;9(1):17. Published 2021 Jan 28. https://doi.org/10.1186/s40359-021-00516-7.
124. Zhang D, Lee EKP, Mak ECW, Ho CY, Wong SYS. Mindfulness-based interventions: an overall review. Br Med Bull. 2021;138(1):41–57. https://doi.org/10.1093/bmb/ldab005.
125. Corbetta D, Sirtori V, Castellini G, Moja L, Gatti R. Constraint-induced movement therapy for upper extremities in people with stroke. Cochrane Database Syst Rev. 2015;2015(10):CD004433. https://doi.org/10.1002/14651858.CD004433.pub3.
126. Wheeler S, Davis D, Basch J, James G, Lehman B, Acord-Vira A. Cognitive behavioral therapy interventions for adults with traumatic brain injury (2013-2020). Am J Occup Ther. 2022;76(Suppl 2):7613393160. https://doi.org/10.5014/ajot.2022/76S2016.

# Advances in Neuropsychology and Speech and Language Pathology in Patients with Stroke

4

M. Gobbo and A. Marini

## 4.1 Introduction

Aphasia is an acquired disorder that affects the ability to understand and produce language in oral (e.g., auditory comprehension and speech production) and written (e.g., reading, writing) modalities [1]. Aphasic symptoms are heterogeneous and vary according to lesion site and extension, vascular territory infarcted (in case of stroke), time post-onset, impaired cognitive functions [1–5]. Persons with aphasia (PWA) might have impairments at both microlinguistic (i.e., phonological, lexical, grammatical) and macrolinguistic (i.e., pragmatic and discourse) levels of language processing [6, 7] with degrees of illness ranging from mild (e.g., selective difficulty in word finding) to severe (e.g., difficulty in both auditory comprehension and speech production). Aphasia can compromise people's ability to communicate, take part into a social interaction, express feelings, and conduct everyday activities such as working or phone calling. It directly influences quality of life, the chance to return to work, and functional outcomes [8, 9]. At least one-third of stroke survivors experiences aphasic symptoms [3, 10]. These still persist in 61% of affected individuals 1 year after the insult [11] and continue after this period in 10–20% of cases [12]. Aphasia may also affect people suffering from Traumatic Brain Injury (TBI; e.g., [13]), brain tumors [14], and neurodegenerative disorders (e.g., Primary Progressive Aphasia [15]).

M. Gobbo (✉)
Department of Life Sciences, University of Trieste, Trieste, Italy

Cognitive Neuroscience Laboratory, University of Udine, Udine, Italy
e-mail: marika.gobbo@phd.units.it

A. Marini
Cognitive Neuroscience Laboratory, University of Udine, Udine, Italy
e-mail: andrea.marini@uniud.it


R. S. Calabrò (ed.), *Translational Neurorehabilitation*,
https://doi.org/10.1007/978-3-031-63604-2_4

Typically, recovery from vascular aphasia follows mechanisms of reorganization that vary according to the time from stroke onset: acute (from the first days to the first 4 weeks after stroke), post-acute (up to 3 months post-stroke), subacute (up to the 6 months post-stroke), and chronic (over 6 months post-stroke) phases of illness [16]. Notably, spontaneous neuroplastic modifications generally occur within the first year post-stroke [17, 18] with particular enhancement within the first 3 months [19–22].

Over the past 50 years, clinicians and researchers dealing with PWA have focused their efforts on applying theories and findings about language structures (from linguistics), linguistic processing (from psycholinguistics), and language organization in the brain (from neurolinguistics) to the clinical setting. These efforts have led to the development of standardized batteries of tests developed to be sensitive (i.e., able to differentiate between healthy and pathological performance) and specific (i.e., including tasks that selectively investigate particular aspects of language function). The results from the administration of such tests allow clinicians to determine the effects of a stroke on language functioning and typically guide the rehabilitation procedures.

Obviously, psycholinguistic and neurolinguistic models of language processing (e.g., [23–26]) are essential not only for the construction of tests but also for the interpretation of their results. For example, according to Levelt and colleagues (e.g., [23, 24]) the production of lexical and/or grammatical errors suggests the disruption of specific processes during the production of a message: verbal paraphasias (e.g., "tree" instead of "cat") suggest difficulties in a prelinguistic phase of message conceptualization; semantic paraphasias (e.g., "tree" instead of "flower") suggest difficulties in a phase of lexical selection; morphological errors (e.g., "trees" instead of "tree") suggest difficulties in a phase of morphological coding; phonological errors (e.g., "tlee" instead of "tree") suggest difficulties in a phase of phonological coding; and so on. Growing evidence suggests that each phase of language processing is subserved by a complex neural network characterized by cortical, subcortical, and even cerebellar epicenters in both hemispheres connected through pathways (fasciculi). For example, according to Friederici and colleagues [27] some lexical and grammatical functions are supported by two (generally left-lateralized) dorsal and ventral streams: (1) the first dorsal way is supported by a section of the Superior Longitudinal Fasciculus, connects the Superior Temporal Gyrus and the Premotor Cortex and supports auditory-motor mapping; (2) the second dorsal way is supported by the long segment of the Arcuate Fasciculus, connects the Superior Temporal Gyrus and a portion of the left inferior frontal gyrus (BA 44 in Broca's area), and supports syntactic processes; the first ventral way is supported by the Inferior Fronto-Occipital Fasciculus, connects the Superior and Middle Temporal Gyrus with a portion of the left inferior frontal gyrus (BA 45 and BA 47), and supports semantic processes; a second ventral way is supported by the Uncinate Fasciculus, connects the Frontal operculum to the anterior portions of the Superior Temporal Gyrus, and supports combinatorial processes in the semantic domain (sentential semantics which depends on syntactic structures). Notably, accumulating evidence suggests that lexical and grammatical functioning recruit also right hemisphere areas (e.g., [28]).

These sets of fibers and their terminations in the cortex were also studied in awake neurosurgery through errors committed at direct cortico-subcortical stimulation [29]. These models concur in defining the structural and functional connectivity subserving the language function and its restoration. Specific connectivity patterns define the goodness of recovery in post-stroke aphasia [5, 30]. In the subacute stage of the illness, PWA with persistent linguistic difficulties show a decreased connectivity in the networks of the lesioned hemisphere and a potential increase of connectivity in homologous areas of the non-affected hemisphere (usually the right one) [5]. Similarly, in case of persistent language impairments in the chronic phase, the functional connectivity is mainly focused within the unimpaired hemisphere with a disruption of connections in both lesioned intrahemispheric and contralateral homotopic regions [5, 31]. Several studies reported linguistic improvements after behavioral (speech and language therapy) and/or neuromodulation treatments in chronic PWA [16, 32, 33]. Nonetheless, a crucial predictor for positive language outcomes is the intensity of the linguistic training. Unfortunately, several factors prevent PWA to equally access the recommended amount of treatment sessions. Therefore, to enhance the efficacy of the rehabilitation process, over the past 20–30 years innovative rehabilitation methods have been developed. These rehabilitative techniques will be described in the following paragraphs of this chapter. Starting from traditional approaches to more innovative methods (i.e., neuromodulation techniques, virtual reality technologies, and pharmacological modulation programs), pros and cons of post-stroke aphasia treatments will be discussed. Finally, future implementations and considerations upon the next steps in aphasia research will be provided.

## 4.2 Traditional Behavioral Approaches

In the acute and post-acute phases an initial spontaneous recovery can be observed in persons with post-stroke aphasia [22]. Interestingly, there is also evidence suggesting that behavioral interventions within Speech and Language Therapy (SLT) may exert a positive effect on language recovery by restoring language impairments, helping patients and their families face everyday communication challenges, and eventually maintaining the obtained language improvements over time [2, 11, 34]. Since PWA face difficulties at several levels of language processing [35], behavioral treatments need to be planned and implemented considering the specific characteristics of every single PWA [2, 11]. Indeed, planning a behavioral intervention is a complex procedure. It should take into account different factors related to (1) the patient's stroke profile (e.g., time post-onset, lesion size and location, cognitive or motor functions affected), (2) his/her individual characteristics (e.g., age, sex, handedness, socioeconomic status, family support, and cognitive reserve) [36, 37], and (3) type of aphasic syndrome (e.g., linguistic levels impaired, residual communication abilities) [11, 38]. Furthermore, an effective intervention should be planned also in terms of (4) approach (e.g., conversational, semantic/phonologic training), (5) target function (e.g., naming, auditory comprehension), (6) regimen (e.g., frequency, duration), (7) mode (e.g., face-to-face, self-directed), (8) provider (e.g., clinicians, trained family members), and (9) customization (e.g., exercise difficulty and relevance) [11, 38].

Behavioral treatments for PWA are numerous and multifaceted [39–43]. Their general efficacy is not under discussion [2] as they are useful in enhancing functional communication and avoiding the development of compensatory maladaptive mechanisms [16, 44]. Namely, SLT can be targeted to specific linguistic levels (e.g., *Phonological* or *Semantic therapy*), communicative efficacy in real-life daily activities (e.g., *Functional* and/or *Pragmatic therapy*), and the ability to convey meanings in conversational exchanges by employing both linguistic and multimodal abilities (*Conversational therapies*) [2, 11, 34]. Despite their documented efficacy, behavioral treatments in aphasia rehabilitation pose some problems [11]. In rehabilitation research, it is usual to describe interventions in terms of frequency (number of sessions per week), duration (time spent into the therapeutic program), intensity (number of hours weekly), and dosage (total of hours spent into therapeutic program) [35]. Indeed, a crucial predictor for positive language outcomes is the intensity of the linguistic training (5–10 h per week; e.g., [44–46]). Unfortunately, intensive therapy programs involving face-to-face patient–clinician interactions are usually limited to the first few months after stroke and several reasons (e.g., the limited economic resources of the national health systems, the high costs of private health care services) do not allow PWA to have access to the recommended quantity of training [2, 11, 47, 48].

For these reasons, to enhance the efficacy of the rehabilitation process innovative rehabilitation methods have been developed. These may act either as a substitute or an adjunct to traditional behavioral approaches. Three examples of such innovations include approaches using neuromodulation, telerehabilitation and Virtual Reality (VR), and drugs [22]. These will be the target of the rest of this chapter.

## 4.3 Neuromodulation Approaches

Advancements in the rehabilitation of language consider the mechanisms underlying cerebral post-stroke reorganization [16, 22, 49, 50]. Neuromodulation interventions exploit non-invasive brain stimulation (NIBS) techniques to modify neural excitability, guide functional reorganization of language networks, and possibly enhance post-treatment plasticity [51, 52]. In recent years, the use of NIBS in post-stroke aphasia has been incremental. Namely, two methods of stimulation are usually performed [17]: (a) inhibitory stimulation of contralateral homolog areas to reduce hemispheric compensatory activity; (b) excitatory stimulation of perilesional areas to facilitate activity of the damaged hemisphere.[1]

[1] For most patients, lexical and grammatical impairments are usually related to lesions to cortical epicenters in the left hemisphere. Indeed, growing evidence suggests that approximately 95% of

Clearly, the application of neuromodulation techniques to PWA depends also on theoretical issues related to the involvement of cortical, subcortical, and even cerebellar structures in language processing and recovery. As to this regard, a critical issue regards the role potentially played by the right hemisphere in language recovery [56, 57]. Increased post-stroke activation in the right hemisphere may be related to either adaptive or maladaptive processes [16, 58–61]. Several hypotheses have been formulated so far to guide the use of neuromodulation techniques to excite or inhibit perilesional and/or contralateral areas of the brain in language treatment. According to the *interhemispheric inhibition hypothesis* [56, 58, 62] in case of stroke, the unaffected hemisphere (usually the right) would inhibit the injured contralateral one (usually the left) to bypass the activation of the damaged language networks [63, 64]. On the contrary, according to the *right hemisphere involvement hypothesis* this hemisphere is actively involved in the reorganization of language networks [57, 65, 66]. Other possibilities have been explored. For example, some authors consider the recruitment of both hemispheres as the *conditio* sine qua non to achieve the best possible outcome [67–69] while others support an indirect involvement of the right hemisphere in language recovery [70, 71]. Besides, additional brain structures (e.g., cerebellum) may concur in language recovery [72–74]. These different hypotheses have been tested in experiments showing the effectiveness of the excitation or inhibition of specific portions of the brain of PWA [17, 75].

Transcranial magnetic stimulation (TMS) and transcranial direct current stimulation (tDCS) are the two most used techniques in neuromodulation studies. Both TMS and tDCS can facilitate or inhibit cortical excitability and generate a neural response that lasts from minutes to days after the stimulation [16, 76–78]. This makes NIBS techniques particularly useful when paired with behavioral language interventions [17, 50].

Trascranial magnetic stimulation employs an 8-shaped coil to create a fluctuating magnetic field that generates an electrical changing over specific brain areas (stimulation passes the scalp, the skull, and arrives focally to the cerebral cortex) [79, 80]. The influence of TMS on cortical activity depends on stimulation parameters such as frequency and number of pulses per session. The frequency of stimulation controls for inhibitory or excitatory effects over the cerebral cortex. Specifically, Low-Frequency TMS stimulation (LF, $\leq$1 Hz) decreases cortical excitability (with an inhibitory effect) whereas High-Frequency TMS stimulation (HF, $\geq$5 Hz) increases cortical activity (with an excitatory effect) [81–83]. Also, a single-pulse TMS is generally associated to short-term variation over time while repetitive-pulses (rTMS) usually trigger more long-lasting effects [84–86]. Rehabilitation studies on post-stroke recovery usually adopt repetitive TMS at low frequencies of stimulation [75, 87, 88].

---

right-handed people and 75% of left-handed people have a lateralization of lexical and grammatical functions in the left hemisphere (LH) [55–57] and that the remaining percentage includes people with a double (both left and right) or crossed (only the right hemisphere) linguistic dominance for such functions [55, 57].

Transcranial Direct Cortical Stimulation modulates cortical excitability by delivering a low-intensity (1–2 mA) current across surface electrodes attached to the scalp [63, 89, 90]. tDCS affects membrane potentials by increasing or decreasing cortical excitability. Specifically, anodal stimulation (A-tDCS) increases cortical excitability via neuronal depolarization, while cathodal stimulation (C-tDCS) decreases cortical excitability via neuronal hyperpolarization [63, 90, 91]. The effects of tDCS may last until 1 h after a single stimulation session, whereas multiple sessions may extend their effects for days [92]. Other methods modulate cortical membrane potentials through electrical stimulation. These are transcranial alternating current stimulation (tACS) and transcranial random noise stimulation (tRNS) [93, 94]. Little evidence is available on their neurophysiological effects and clinical potential [94]. If with tDCS cortical areas can be either inhibited or excited, tACS exploits a low-intensity sinusoidal current to influence brain oscillations [94, 95]. Recent investigations and reviews suggest that tACS stimulations can be useful in both healthy and clinical settings (e.g., [96–99]. tRNS involves the application of current impulses that vary repetitively in their frequencies (from 0.1 Hz to 640 Hz) [93, 94]. Promising results support the positive effects of tRNS application in motor function [100], whereas more controversial is the effects of tRNS in higher cognitive domains [101].

With regard to stimulation protocols (e.g., site, frequency, and clinical population) in NIBS research, several programs have been designed for both TMS and tDCS techniques [17, 75]. Inhibitory Low-Frequency rTMS (LF-rTMS) over areas of the right hemisphere (e.g., right Inferior Frontal Gyrus [rIFG]) resulted in positive linguistic gains in subacute [102–104] and chronic PWA [105–107]. Particularly, the LF-rTMS stimulation of the pars triangularis is a common protocol among TMS studies [17, 75, 108], with evidence of naming improvements in right-handed, chronic, non-fluent, PWA at their first ischemic stroke [109–111]. Facilitatory High-Frequency rTMS (HF-rTMS) on areas of the left hemisphere directly modulates the activity of perilesional language areas and represents a promising neurostimulation protocol for language enhancement [82, 112–114]. Stimulation over the left dorsolateral frontal cortex associates with decreased reaction times in picture naming by healthy individuals [115] and supports action naming in non-fluent progressive aphasia [116, 117]. In a person with a conduction aphasia, HF-rTMS over the left IFG and SLT triggered linguistic improvements detectable on a standardized battery of tests (i.e., the Western Aphasia Battery) and changes in post-treatment fMRI connectivity patterns [118]. Among transcranial magnetic methods of stimulation, the theta burst stimulation (TBS) can induce lasting effects in shorter times [75, 119]. In TBS protocols, trains of pulses are delivered in repeated series (intermittent TBS–iTBS) or uninterruptedly (continuous TBS–cTBS) to facilitate or inhibit cortical activation [120]. Behavioral language improvements were reported for iTBS on left fronto-temporal areas [75, 82, 121], and some results support the application of bilateral TBS on the Inferior Frontal Gyri (i.e., cTBS in the RH and iTBS in the LH) [83, 122]. In a randomized, sham-controlled trial, Chou et al. [83] report language improvements in non-fluent, chronic PWA, after either iTBS on ipsilesional or 1-Hz rTMS on contralesional Pars Triangularis in the IFG. Interestingly, in auditory comprehension tasks ipsilesional iTBS obtained more promising linguistic results. These preliminary findings lead to future studies on the implementation of TBS protocols in aphasia rehabilitation [120].

As for tDCS protocols, different approaches have been implemented according to the cortical activity, either perilesional or contralesional, they aim to modulate. Anodal tDCS over perilesional areas (usually in the LH) increases cortical excitability in the spared tissue of the lesioned hemisphere. Both single- and multiple A-tDCS sessions result in long-lasting linguistic gains in chronic post-stroke aphasia [123–126]. Possible sites of stimulation are the IFG, the dorsolateral prefrontal cortex (DLPFC), the motor cortex, and the posterior perisilvian regions [125–127]. Cathodal tDCS over contralesional areas (usually in the RH) decreases the inhibitory effect of the intact hemisphere over the affected one. Studies on C-tDCS over the right IFG report facilitation effects in post-stroke naming performance [128, 129]. Furthermore, the concomitant delivery of Anodal and Cathodal tDCS aims to decrease contralesional excitability while increasing perilesional activation. Bihemispheric two-cephalic tDCS stimulation (paired with SLT) determined language improvements in PWA at a chronic and subacute stage of stroke [130–133]. Furthermore, in recent years, a new high-definition method of stimulation (HD-tDCS) has been increasingly used in tDCS research [134–136]. HD-tDCS presents smaller electrodes placed in a ring configuration (i.e., one anode placed on the cortical area of interest surrounded by four cathodes) that allows for a more selective and dense delivery of the current [136, 137]. It controls the electric field spreading across the cortex and improves the spatial accuracy of tDCS. Also, distinct patterns of cortical activity were detected after stimulation with tDCS and HD-tDCS [136].

As for NIBS applicability in language rehabilitation, studies usually report evidence of linguistic improvements when performing neuromodulation protocols combined with speech and language therapy [17, 138–140]. Non-invasive techniques allow for modulating neural plasticity while managing behavioral changes [86, 141, 142]. The joint delivery of SLT with either tDCS [139, 140, 143–146] or TMS stimulations [86, 142, 147–149] are related to improvements in several linguistic tasks [125, 150]. Such improvements are generally reported for measures tapping microlinguistic skills: not only on naming tasks [141, 150–152], but also on verb production and retrieval, repetition, comprehension, fluency, and writing tasks [17, 86, 141, 153]. On the other hand, growing evidence suggests that an accurate language assessment should also include the analysis of macrolinguistic aspects such as those involved in pragmatic and discourse level processing (see [154–156] for recent reviews and [157] for the application of a general communicative assessment in the clinical practice). Of note, recent advances in Artificial Intelligence (AI) including generative AI (e.g., ChatGPT) and natural language processing (NLP) models applied to Aphasiology [158, 159] suggest the possibility to use multilevel procedures of discourse analysis in such studies [160]. Indeed, recently, through Artificial Intelligence (AI) it is becoming possible to have a semi-automated transcription of the speech samples produced by PWA and to derive critical measures reflecting their narrative production skills [160]. Marangolo and colleagues investigated the efficacy of A-tDCS stimulation by applying a Multilevel Analysis of Narrative Discourse Production [161]. Namely, they showed that pairing conversational therapy procedures with A-TDCS over the lesioned left IFG improved the informative levels [162] and linguistic cohesion [163] of the speech outputs of severely non-fluent chronic PWA.

Altogether NIBS research supports the use of neurostimulation in post-stroke aphasia [164]. However, these studies usually involve very few participants to generalize their results [164, 165]. In an interesting investigation, Fridrikkson et al. [125] applied both A-tDCS and speech and language therapy to a wide cohort of 74 patients who were randomized in a double-blind, prospective study. The authors reported that A-TDS applied to the left temporal lobe enhanced their patients' naming skills. The peculiarity of this investigation rests on the fact that A-TDS was delivered through an fMRI-guided protocol. Namely, the authors used functional connectivity to detect eligible areas of stimulation. A method originally implemented in TMS studies in which excitatory stimulation was delivered on perilesional sites activated during linguistic tasks at fMRI acquisition [82, 112, 114]. This fMRI connectivity-based method might be promising to probe neuroplasticity and network reorganization, and eventually customize tailored stimulation protocols [166].

Neuromodulation approaches represent an innovative method to modify plasticity at neurophysiological and neurofunctional levels [5, 17, 50]. Despite their differences in temporal resolution (from milliseconds to minutes in TMS and tDCS, respectively), spatial accuracy (TMS offers a more focal stimulation than tDCS), and portability (tDCS is easier to manage than TMS), the application of both NIBS techniques provided encouraging results in post-stroke aphasia recovery [17, 88, 145, 165]. Nevertheless, while TMS positively affects recovery in chronic and subacute stroke populations, the potential effects of tDCS in non-chronic PWA still needs to be clarified [88, 165]. Additionally, the role of subcortical structures in functional reorganization should not be underestimated. Preliminary results show linguistic improvements after right cerebellar C-tDCS stimulation paired with SLT in chronic PWA [72, 73]. This is a promising issue to explore in NIBS research. As an ultimate goal, NIBS studies should define stimulation protocols taking into account the personal and cognitive characteristics of PWA, develop accurate theories of cortical neuromodulation, functional measures of efficacy, and precise methods to detect sites of stimulation [88, 145, 167].

## 4.4 Approaches Using Telerehabilitation and Virtual Reality

The integration of new technologies in neurorehabilitation is an expanding field of research [22, 168]. *Telerehabilitation (TR)* is a therapeutic technique applied in several areas of neurorehabilitation (e.g., motor and cognitive recovery), including post-stroke aphasia [48, 169, 170]. TR provides rehabilitation services through communication technologies such as video calls, telemonitoring, games, tablets, and computers [48, 171–173]. Again, AI can help clinicians monitoring the patients' performance across different assessments. TR can be synchronous or asynchronous and used as a supplementary or an alternative tool to aphasia treatment [48, 173, 174]. Those PWA who receive TR are usually at a chronic stage of stroke and participate in therapy sessions that last from 30 to 120 min, with a frequency of 1–6 times per week, and a duration of 4–24 weeks, for a total amount of treatment hours ranging from 20 to 100 [173]. TR therapeutic outcomes are similar (sometimes better) to those obtained through traditional in-person approaches [48, 173]. Studies on TR typically implement traditional behavioral approaches in computerized/tablet app-based settings [22]. For instance, in Meltzer et al. [175] a group of chronic PWA was administered SLT via in-person interaction or simultaneous video calls with the therapist. Results showed general improvements in both language and communication skills. Only communicative confidence benefited more from the person-to-person interaction. Similar results on the effectiveness of telerehabilitation in aphasia were reported by Maresca et al. [176] which developed a linguistic tablet-based program of tasks to perform at home.

Conversely, very few works exploited the potential of technology to create immersive experiences for persons with aphasia [22, 168, 177]. *Virtual reality (VR)* is a computer-generated simulation of 3D environments that allow users to experience semi-immersive interactions [22, 177]. Recently, preliminary studies implemented VR in aphasia rehabilitation [178, 179]. Devane et al. [168] identified four types of VR interventions: (1) virtual therapists who guide the training sessions (e.g., a clinician's face displayed while showing tongue or lip movements that will be a model for the patient); (2) virtual games played simultaneously by two PWA who sit in the same room but interact with different monitors; (3) 3D virtual scenes of a common place (e.g., supermarket) where the PWA interacts with virtual avatars that can be implemented though AI; (4) virtual worlds in which PWA interact with other users connected in the same virtual environment [177, 178, 180]. In a recent study, Giachero et al. [177] performed video-based conversational therapy (twice a week for 6 months) implemented in a semi-immersive VR environment. Eighteen chronic non-fluent PWA formed the VR group, while a control group of eighteen PWA performed the same approach without VR. The authors intended to test whether VR increased therapeutic benefits in language skills, communicative effectiveness, and psychosocial aspects (e.g., self-esteem and quality of life). As for the VR group, a semi-immersive scenario (the patient was not wearing visors) was projected onto a curved screen. Patients were asked to look at the monitor and indicate to a therapist the action they wanted to perform within that virtual environment. Therapists modulated the interaction. Of particular interest were the 3D environments specifically designed for the study. These corresponded to typical communication situations (e.g., within a city as supermarkets or restaurants) in which participants had to actively interact to complete frequent (e.g., requesting information to people in the same scenario) or unexpected (e.g., handling a robbery) tasks. All situations included a series of exercises that addressed language, working memory, attention, and executive functions. The results support the usefulness of VR in aphasia rehabilitation, with greater cognitive-linguistic and psychological improvements in the experimental group. It is in line with the idea that VR may reduce the communication difficulties in PWA by practicing social life like situations [168, 178, 181].

When using VR, clinicians and researchers should consider key concepts from VR research such as *immersion* (i.e., how well the scenario depicts environments' sensory properties and interactions) and *presence* (i.e., the feeling of being mentally within the environment) when planning their therapeutic settings [182, 183]. As to this regard, they should consider those factors that influence both immersion (e.g., visual resolution, auditory and tactile information, and the realism of the interactions) and presence (e.g., emotional responses, possibility of participants to create their own narratives about the events in the virtual scenarios) [182, 183]. A third variable that may concur in the efficacy of VR concerns the notion of *agency* (i.e., the sense of control experienced by the subject involved in the virtual scenario). Agency seems to favor users' feelings of being immersed in a virtual space and may be responsible for users' behaviors during VR games [184]. Transferred to VR therapy, agency might generate behavioral modifications by stimulating patients' involvement [168]. Accordingly, VR implementation in aphasia therapy is supported by the recent extension of the *embodied cognition hypothesis* to the language function [177, 185, 186]. An immersive multimodal experience (e.g., visual and/or acoustic stimulations) paired with emotional involvement may increase language learning [187] and recovery [177]. Legault et al. [187] investigated adults' performance in learning Mandarin Chinese as a second language (L2) through immersive VR and traditional word associations. An accuracy effect was found for vocabularies learnt via immersive reality. Interestingly, participants with major difficulties in studying L2 words exhibited significant benefits in learning through immersive VR. This result strengthens the idea that perceptual and sensorimotor integration, first-person activities, and multimodal interaction enhance linguistic learning and can be a valuable asset for language rehabilitation.

Overall, TR and VR are promising techniques that should be implemented in aphasia rehabilitation. This consideration relies on a range of facts. First of all, they may help coping with the limited economic resources of the national health systems. Indeed, such limitations and the high costs of private health care services too often do not allow PWA to have access to the recommended amount of training. TR and VR can potentially allow PWA to have access to rehabilitation protocols anywhere anytime, reducing the amount of hours that they should spend in the hospital [48, 170, 179]. Obviously, this also grants many more PWA to have access to therapeutic protocols. Furthermore, the ability to simulate real-world environments and practice every day like interactions, and the possibility to create contexts that stimulate expression of personal feelings and opinions [168, 181] further confirm the utility of VR in therapy. Indeed, if TR approaches are mostly limited to the digital implementation of classic language trainings [188], VR approaches can offer interesting therapeutic applications if considered in relation to immersive environments [168, 177, 181]. Nonetheless, even if highly promising, the application of VR to aphasia rehabilitation is still a new approach. This means that just few studies have adopted it so far. Because of its potentiality, there is urgent need of studies with a randomized-controlled design including large cohort of PWA. Only such studies may provide the necessary information for the routinary implementation of VR in health care systems.

## 4.5 Pharmacological Approaches

The stroke induces metabolic changes in brain connectivity (diaschisis) and revascularization (penumbra areas) [189–192]. Further developments in the field of aphasia rehabilitation stem from pharmacological research. Indeed, over time specific drugs have been developed with the aim to reduce the long-term effects of the lesion and potentially enhance language recovery in post-stroke aphasia by influencing the connectivity between neural cells, enhancing neurotransmitters' activity, and moderating plasticity during functional reorganization [193]. In recent years, growing evidence suggests that the application of recombinant tissue plasminogen activator (rtPA) to perform intravenous thrombolysis after ischemic stroke may reduce its neural and cognitive effects. The procedure is used to facilitate reperfusion in PWA with a typical dosage of 9 mg/kg within 4.5 h after symptom onset [194, 195]. There is evidence suggesting that specific drugs can modulate post-stroke mechanisms and affect spontaneous synaptic neuroplasticity in the subacute and chronic stages of recovery (see [22, 193, 196] for a recent review). Investigations have mainly focused on the effects of drugs that affect GABA, Acetylcholine, and Glutamate Systems (i.e., Piracetam, Donepezil, Galantamine, and Memantine) and those controlling for monoaminergic systems (i.e., Levodopa, Bromocriptine, and Amphetamines). In this paragraph, we will focus only on those drugs for which studies have provided promising or preliminary effects on linguistic recovery.

*Piracetam's* effect on aphasia recovery was explored through several controlled trials [197–199]. Notably, experimental groups (vs. placebo groups) obtained improvements in different linguistic domains (e.g., repetition, spontaneous speech, and written language) when paired with speech therapy [197–199]. Instead, when investigated as a substitutive treatment, piracetam exerted a positive influence on auditory comprehension limited to the period of pharmacological assumption [200]. Studies on *donepezil* have been conducted on chronic patients, reporting inconsistent results [22, 201]. Still, further research should investigate donepezil's influence on the acute stage of stroke since it might enhance transmission between cortical regions in this phase of neural modification. SLT paired with *memantine* was examined in two placebo-controlled studies on chronic PWA [202, 203]. Improvements in several linguistic skills were reported, and these lasted also after 6 months at follow-up. Thus, memantine could operate on long-term plasticity and restorative synaptic connections [204]. *Levodopa*, a dopamine precursor, might also exert an influence on post-stroke recovery. Since dopamine can modulate cognitive functions in healthy individuals [54, 205], it may also modulate cognitive-linguistic performance after stroke. Nevertheless, the current contrasting results and paucity of well-designed studies do not permit to advance any conclusion on its potential role in language recovery [206–208].

Overall, research on the selection of drugs and their effectiveness in the recovery of aphasia is in its infancy. Several reasons support the proceed of pharmacotherapy in aphasia research. Drugs may accelerate the reorganization of neuroplasticity, enhance linguistic gains generated by SLT, extend recovery to other cognitive functions and maintain improvements over time [209].

## 4.6 A Final Remark on Individual Differences

The influence of individual characteristics on language processing has been widely demonstrated. For instance, psycholinguistic studies have shown that word frequency and age of acquisition facilitate lexical access and speed in word retrieval. Also, the inclusion of the idea of a "cognitive reserve" (e.g., personal education level, work and life experiences) in research studies that predict or modulate functional recovery in neurological diseases is growing. Nevertheless, some individual characteristics are still underrepresented or not considered at all in cognitive neuroscience. Take, for example, a potential gender-related difference in language production: the (mythological) greater talkativeness of girls compared to boys. If this were true, not considering girls' talkativeness in studies on word or narrative production would generate a bias in performance. Unfortunately, as highlighted by a recent overview on this topic [210], gender-related differences in the neural networks of language production were matter of interest for few studies with controversial results. Another example of negligence in individual differences is provided by gender representation in neuroscience research. Although gender is not restricted to the biological sex of birth, neuroscience studies on human cognition generally recruit male and female participants based on this biological distinction rather than including sexual orientation as an influencing variable [211]. Furthermore, potential differences in brain structure and function exerted by socioeconomic status, ethnicity, and gender are almost never studied or considered in statistical analyses [212]. This leads to a gap in the representation of the population. How can results be generalized to a population that is not represented in a research sample? In other words, if a minority is not considered in the design of a study, its results cannot be extended to the entire population. This is a critical point for studies in the field of cognitive neuroscience. A question that researchers should address in their future investigations.

## 4.7 Conclusions

Language rehabilitation in PWA is the long-lasting target of an increasingly growing interest among aphasiologists. This renewed interest has been prompted, among other things, by the development of relatively new neuromodulation techniques that are promising in enhancing the efficacy of the rehabilitation process (i.e., tDCS and TMS), and new technologies that allow PWA to be immersed in virtual communicative scenarios that can be used not only in the clinical setting but even at home. The contribution of AI to language assessment and recovery is becoming increasingly more important. Nonetheless, to date, just few studies have begun to explore the potential usefulness of AI in the field of Aphasiology. Future investigations should aim at exploring the multifaceted applications of generative AI and NLP models to language assessment and rehabilitation. Pharmacological interventions may also boost neurophysiological mechanisms subserving neural plasticity and reperfusion. In our opinion, these are critical steps to provide patients tailored, intensive, effective programs of interventions. Indeed, the goal of aphasia rehabilitation should be the integration of techniques to reach the optimal approach for patient's needs.

**Acknowledgments** This work was supported by the Departmental Strategic Plan (PSD) of the University of Udine Interdepartmental Project on Artificial Intelligence (2020–2025).

## References

1. Basso A, Forbes M, Boller F. Chapter 27—Rehabilitation of aphasia. In: Barnes MP, editor. good DCBT-H of CN Neurological rehabilitation. Elsevier; 2013. p. 325–34.
2. Brady MC, Kelly H, Godwin J, Enderby P, Campbell P. Speech and language therapy for aphasia following stroke. Cochrane Database Syst Rev. 2016;2016(6):CD000425. https://doi.org/10.1002/14651858.CD000425.pub4.
3. Engelter ST, Gostynski M, Papa S, Frei M, Born C, Ajdacic-Gross V, Gutzwiller F, Lyrer PA. Epidemiology of aphasia attributable to first ischemic stroke: incidence, severity, fluency, etiology, and thrombolysis. Stroke. 2006;37:1379–84.
4. Bunker LD, Hillis AE. Chapter 3—Vascular syndromes: revisiting classification of poststroke aphasia. In: Hillis AE, editor. Fridriksson JBT-H of CN Aphasia. Elsevier; 2022. p. 37–55.
5. Meier EL. Chapter 7—The role of disrupted functional connectivity in aphasia. In: Hillis AE, editor. Fridriksson JBT-H of CN Aphasia. Elsevier; 2022. p. 99–119.
6. Andreetta S, Cantagallo A, Marini A. Narrative discourse in anomic aphasia. Neuropsychologia. 2012;50:1787–93.
7. Andreetta S, Marini A. The effect of lexical deficits on narrative disturbances in fluent aphasia. Aphasiology. 2015;29:705–23.
8. Worrall L, Holland A. Editorial: quality of life in aphasia. Aphasiology. 2003;17:329–32.
9. Rohde A, Worrall L, Le Dorze G. Systematic review of the quality of clinical guidelines for aphasia in stroke management. J Eval Clin Pract. 2013;19:994–1003.
10. Flowers HL, Skoretz SA, Silver FL, Rochon E, Fang J, Flamand-Roze C, Martino R. Poststroke aphasia frequency, recovery, and outcomes: a systematic review and meta-analysis. Arch Phys Med Rehabil. 2016;97:2188–2201.e8.
11. Brady MC, Ali M, VandenBerg K, et al. Complex speech-language therapy interventions for stroke-related aphasia: the RELEASE study incorporating a systematic review and individual participant data network meta-analysis. Southampton: National Institute for Health and Care Research; 2022.
12. Ali M, Lyden P, Brady M. Aphasia and dysarthria in acute stroke: recovery and functional outcome. Int J Stroke. 2013;10:400–6.
13. Heilman KM, Safran A, Geschwind N. Closed head trauma and aphasia. J Neurol Neurosurg Psychiatry. 1971;34:265–9.
14. Åke S, Hartelius L, Jakola AS, Antonsson M. Experiences of language and communication after brain-tumour treatment: a long-term follow-up after glioma surgery. Neuropsychol Rehabil. 2023;33(7):1225–61.
15. Marshall CR, Hardy CJD, Volkmer A, et al. Primary progressive aphasia: a clinical approach. J Neurol. 2018;265:1474–90.

16. Fama ME, Turkeltaub PE. Treatment of poststroke aphasia: current practice and new directions. Semin Neurol. 2014;34:504–13.
17. Harvey DY, Hamilton R. Chapter 15—Noninvasive brain stimulation to augment language therapy for poststroke aphasia. In: Hillis AE, Fridriksson J, editors. Handbook of clinical neurology. Elsevier; 2022. p. 241–50.
18. Wilson SM, Entrup JL, Schneck SM, et al. Recovery from aphasia in the first year after stroke. Brain. 2023;146(3):1021–39.
19. Kertesz A, Mccabe P. Recovery patterns and prognosis in aphasia. Brain. 1977;100:1–18.
20. Pedersen PM, Jørgensen HS, Nakayama H, Raaschou HO, Olsen TS. Aphasia in acute stroke: incidence, determinants, and recovery. Ann Neurol. 1995;38:659–66.
21. Laska AC, Hellblom A, Murray V, Kahan T, Von Arbin M. Aphasia in acute stroke and relation to outcome. J Intern Med. 2001;249:413–22.
22. Picano C, Quadrini A, Pisano F, Marangolo P. Adjunctive approaches to aphasia rehabilitation: a review on efficacy and safety. Brain Sci. 2021;11:41.
23. Levelt WJ, Roelofs A, Meyer AS. A theory of lexical access in speech production. Behav Brain Sci. 1999;22:1–38; discussion 38–75
24. Indefrey P, Levelt WJM. The spatial and temporal signatures of word production components. Cognition. 2004;92:101–44.
25. Hickok G. The functional neuroanatomy of language. Phys Life Rev. 2009;6:121–43.
26. Matchin W, Hickok G. The cortical organization of syntax. Cereb Cortex. 2020;30:1481–98.
27. Friederici AD, Chomsky N, Berwick RC, Moro A, Bolhuis JJ. Language, mind and brain. Nat Hum Behav. 2017;1:713–22.
28. Piervincenzi C, Petrilli A, Marini A, Caulo M, Committeri G, Sestieri C. Multimodal assessment of hemispheric lateralization for language and its relevance for behavior. NeuroImage. 2016;142:351–70.
29. Duffau H, Moritz-Gasser S, Mandonnet E. A re-examination of neural basis of language processing: proposal of a dynamic hodotopical model from data provided by brain stimulation mapping during picture naming. Brain Lang. 2014;131:1–10.
30. Cramer SC. Repairing the human brain after stroke: I. Mechanisms of spontaneous recovery. Ann Neurol. 2008;63:272–87.
31. Walenski M, Chen Y, Litcofsky KA, Caplan D, Kiran S, Rapp B, Parrish TB, Thompson CK. Perilesional perfusion in chronic stroke-induced aphasia and its response to behavioral treatment interventions. Neurobiol Lang. 2022;3:345–63.
32. Fridriksson J, Richardson JD, Fillmore P, Cai B. Left hemisphere plasticity and aphasia recovery. NeuroImage. 2012;60:854–63.
33. Marangolo P, Fiori V, Calpagnano MA, Campana S, Razzano C, Caltagirone C, Marini A. tDCS over the left inferior frontal cortex improves speech production in aphasia. Front Hum Neurosci. 2013;7:539.
34. Basso A. Aphasia and its therapy. Oxford University Press; 2003.
35. Cherney LR, Carpenter J. Chapter 1–Behavioral interventions for poststroke aphasia. In: Hillis AE, Fridriksson J, editors. Handbook of clinical neurology. Elsevier; 2022. p. 197–220.
36. Nucci M, Mapelli D, Mondini S. Cognitive reserve index questionnaire (CRIq): a new instrument for measuring cognitive reserve. Aging Clin Exp Res. 2012;24:218–26.
37. Montemurro S, Mondini S, Crovace C, Jarema G. Cognitive reserve and its effect in older adults on retrieval of proper names, logo names and common nouns. Front Commun. 2019;4:14.
38. Brady MC, Ali M, VandenBerg K, et al. RELEASE: a protocol for a systematic review based, individual participant data, meta- and network meta-analysis, of complex speech-language therapy interventions for stroke-related aphasia. Aphasiology. 2020;34:137–57.
39. Davis GA. PACE revisited. Aphasiology. 2005;19:21–38.
40. Boyle M. Semantic feature analysis treatment for aphasic word retrieval impairments: what's in a name? Top Stroke Rehabil. 2010;17:411–22.
41. Maddy KM, Capilouto GJ, McComas KL. The effectiveness of semantic feature analysis: an evidence-based systematic review. Ann Phys Rehabil Med. 2014;57:254–67.

42. Mack JE, Thompson CK. Recovery of online sentence processing in aphasia: eye movement changes resulting from treatment of underlying forms. J Speech Lang Hear Res. 2017;60:1299–315.
43. Kendall DL, Moldestad MO, Allen W, Torrence J, Nadeau SE. Phonomotor versus semantic feature analysis treatment for anomia in 58 persons with aphasia: a randomized controlled trial. J Speech Lang Hear Res. 2019;62:4464–82.
44. Menahemi-Falkov M, Breitenstein C, Pierce JE, Hill AJ, O'Halloran R, Rose ML. A systematic review of maintenance following intensive therapy programs in chronic post-stroke aphasia: importance of individual response analysis. Disabil Rehabil. 2022;44:5811–26.
45. Bhogal SK, Teasell R, Speechley M. Intensity of aphasia therapy, impact on recovery. Stroke. 2003;34:987–93.
46. Basso A. How intensive/prolonged should an intensive/prolonged treatment be? Aphasiology. 2005;19:975–84.
47. Palmer R, Witts H, Chater T. What speech and language therapy do community dwelling stroke survivors with aphasia receive in the UK? PLoS One. 2018;13:e0200096.
48. Cacciante L, Kiper P, Garzon M, Baldan F, Federico S, Turolla A, Agostini M. Telerehabilitation for people with aphasia: a systematic review and meta-analysis. J Commun Disord. 2021;92:106111.
49. Taub E, Uswatte G, Elbert T. New treatments in neurorehabilitation founded on basic research. Nat Rev Neurosci. 2002;3:228–36.
50. Fridriksson J, Hillis AE. Current approaches to the treatment of post-stroke aphasia. J Stroke. 2021;23:183–201.
51. De Ridder D, Stöckl T, To WT, Langguth B, Vanneste S. Noninvasive transcranial magnetic and electrical stimulation: working mechanisms. In: Rhythmic stimulation procedures in neuromodulation. 2017, pp 193–223.
52. Kiran S, Thompson CK. Neuroplasticity of language networks in aphasia: advances, updates, and future challenges. Front Neurol. 2019;10:295. https://doi.org/10.3389/fneur.2019.00295.
53. Pujol J, Deus J, Losilla JM, Capdevila A. Cerebral lateralization of language in normal left-handed people studies by functional MRI. Neurology. 1999;52:1038–43.
54. Knecht S, Deppe M, Dräger B, Bobe L, Lohmann H, Ringelstein E, Henningsen H. Language lateralization in healthy right-handers. Brain. 2000;123(Pt 1):74–81.
55. Szaflarski JP, Binder JR, Possing ET, McKiernan KA, Ward BD, Hammeke TA. Language lateralization in left-handed and ambidextrous people: fMRI data. Neurology. 2002;59:238–44.
56. Gainotti G. Contrasting opinions on the role of the right hemisphere in the recovery of language. A critical survey. Aphasiology. 2015;29:1020–37.
57. Marini A, Galetto V, Tatu K, Duca S, Geminiani G, Sacco K, Zettin M. Recovering two languages with the right hemisphere. Brain Lang. 2016;159:35–44.
58. Schlaug G, Marchina S, Wan CY. The use of non-invasive brain stimulation techniques to facilitate recovery from post-stroke aphasia. Neuropsychol Rev. 2011;21:288.
59. Turkeltaub PE, Messing S, Norise C, Hamilton RH. Are networks for residual language function and recovery consistent across aphasic patients? Neurology. 2011;76:1726–34.
60. Turkeltaub PE, Coslett HB, Thomas AL, Faseyitan O, Benson J, Norise C, Hamilton RH. The right hemisphere is not unitary in its role in aphasia recovery. Cortex. 2012;48:1179–86.
61. Torres J, Drebing D, Hamilton R. TMS and tDCS in post-stroke aphasia: integrating novel treatment approaches with mechanisms of plasticity. Restor Neurol Neurosci. 2013;31:501–15.
62. Postman-Caucheteux WA, Birn RM, Pursley RH, Butman JA, Solomon JM, Picchioni D, McArdle J, Braun AR. Single-trial fMRI shows contralesional activity linked to overt naming errors in chronic aphasic patients. J Cogn Neurosci. 2010;22:1299–318.
63. Lang N, Nitsche MA, Paulus W, Rothwell JC, Lemon RN. Effects of transcranial direct current stimulation over the human motor cortex on corticospinal and transcallosal excitability. Exp Brain Res. 2004;156:439–43.
64. Fregni F, Pascual-Leone A. Technology insight: noninvasive brain stimulation in neurology—perspectives on the therapeutic potential of rTMS and tDCS. Nat Rev Neurol. 2007;3:383–93.

65. Hartwigsen G, Saur D, Price CJ, Ulmer S, Baumgaertner A, Siebner HR. Perturbation of the left inferior frontal gyrus triggers adaptive plasticity in the right homologous area during speech production. Proc Natl Acad Sci. 2013;110:16402–7.
66. Chieffo R, Ferrari F, Battista P, et al. Excitatory deep transcranial magnetic stimulation with H-coil over the right homologous Broca's region improves naming in chronic post-stroke aphasia. Neurorehabil Neural Repair. 2014;28:291–8.
67. Heiss W-D, Thiel A. A proposed regional hierarchy in recovery of post-stroke aphasia. Brain Lang. 2006;98:118–23.
68. Hamilton RH, Chrysikou EG, Coslett B. Mechanisms of aphasia recovery after stroke and the role of noninvasive brain stimulation. Brain Lang. 2011;118:40–50.
69. Stockert A, Wawrzyniak M, Klingbeil J, Wrede K, Kümmerer D, Hartwigsen G, Kaller CP, Weiller C, Saur D. Dynamics of language reorganization after left temporo-parietal and frontal stroke. Brain. 2020;143:844–61.
70. van Oers CAMM, Vink M, van Zandvoort MJE, van der Worp HB, de Haan EHF, Kappelle LJ, Ramsey NF, Dijkhuizen RM. Contribution of the left and right inferior frontal gyrus in recovery from aphasia. A functional MRI study in stroke patients with preserved hemodynamic responsiveness. NeuroImage. 2010;49:885–93.
71. Geranmayeh F, Brownsett SLE, Wise RJS. Task-induced brain activity in aphasic stroke patients: what is driving recovery? Brain. 2014;137:2632–48.
72. Marangolo P, Fiori V, Caltagirone C, Pisano F, Priori A. Transcranial cerebellar direct current stimulation enhances verb generation but not verb naming in poststroke aphasia. J Cogn Neurosci. 2018;30:188–99.
73. Sebastian R, Kim JH, Brenowitz R, Tippett DC, Desmond JE, Celnik PA, Hillis AE. Cerebellar neuromodulation improves naming in post-stroke aphasia. Brain Communic. 2020;2:fcaa179.
74. Manto M, Argyropoulos GPD, Bocci T, et al. Consensus paper: novel directions and next steps of non-invasive brain stimulation of the cerebellum in health and disease. Cerebellum. 2022;21:1092–122.
75. Breining BL, Sebastian R. Neuromodulation in post-stroke aphasia treatment. Curr Phys Med Rehabil Rep. 2020;8:44–56.
76. Hallett M. Transcranial magnetic stimulation and the human brain. Nature. 2000;406:147–50.
77. Nitsche MA, Paulus W. Sustained excitability elevations induced by transcranial DC motor cortex stimulation in humans. Neurology. 2001;57:1899–901.
78. Zaghi S, Acar M, Hultgren B, Boggio PS, Fregni F. Noninvasive brain stimulation with low-intensity electrical currents: putative mechanisms of action for direct and alternating current stimulation. Neuroscientist. 2010;16:285–307.
79. Rothwell JC. Techniques and mechanisms of action of transcranial stimulation of the human motor cortex. J Neurosci Methods. 1997;74:113–22.
80. Hartwigsen G. The neurophysiology of language: insights from non-invasive brain stimulation in the healthy human brain. Brain Lang. 2015;148:81–94.
81. Abo M, Kakuda W, Watanabe M, Morooka A, Kawakami K, Senoo A. Effectiveness of Low-frequency rTMS and intensive speech therapy in poststroke patients with aphasia: a pilot study based on evaluation by fMRI in relation to type of aphasia. ENE. 2012;68:199–208.
82. Szaflarski JP, Griffis J, Vannest J, et al. A feasibility study of combined intermittent theta burst stimulation and modified constraint-induced aphasia therapy in chronic post-stroke aphasia. Restor Neurol Neurosci. 2018;36:503–18.
83. Chou T-Y, Wang J-C, Lin M-Y, Tsai P-Y. Low-frequency vs. theta burst transcranial magnetic stimulation for the treatment of chronic non-fluent aphasia in stroke: a proof-of-concept study. Front Aging Neurosci. 2021;13:800377.
84. Chen R, Classen J, Gerloff C, Celnik P, Wassermann EM, Hallett M, Cohen LG. Depression of motor cortex excitability by low-frequency transcranial magnetic stimulation. Neurology. 1997;48:1398–403.
85. Barwood CHS, Murdoch BE, Riek S, O'Sullivan JD, Wong A, Lloyd D, Coulthard A. Long term language recovery subsequent to low frequency rTMS in chronic non-fluent aphasia. NeuroRehabilitation. 2013;32:915–28.

86. Gholami M, Pourbaghi N, Taghvatalab S. Evaluation of rTMS in patients with poststroke aphasia: a systematic review and focused meta-analysis. Neurol Sci. 2022;43:4685–94.
87. Dionísio A, Duarte IC, Patrício M, Castelo-Branco M. The use of repetitive transcranial magnetic stimulation for stroke rehabilitation: a systematic review. J Stroke Cerebrovasc Dis. 2018;27:1–31.
88. Kielar A, Patterson D, Chou Y. Efficacy of repetitive transcranial magnetic stimulation in treating stroke aphasia: systematic review and meta-analysis. Clin Neurophysiol. 2022;140:196–227.
89. Chrysikou EG, Hamilton RH. Noninvasive brain stimulation in the treatment of aphasia: exploring interhemispheric relationships and their implications for neurorehabilitation. Restor Neurol Neurosci. 2011;29:375–94.
90. Fregni F, El-Hagrassy MM, Pacheco-Barrios K, et al. Evidence-based guidelines and secondary meta-analysis for the use of transcranial direct current stimulation in neurological and psychiatric disorders. Int J Neuropsychopharmacol. 2021;24:256–313.
91. Floel A, Cohen LG. Recovery of function in humans: cortical stimulation and pharmacological treatments after stroke. Neurobiol Dis. 2010;37:243–51.
92. Reis J, Schambra HM, Cohen LG, Buch ER, Fritsch B, Zarahn E, Celnik PA, Krakauer JW. Noninvasive cortical stimulation enhances motor skill acquisition over multiple days through an effect on consolidation. Proc Natl Acad Sci. 2009;106:1590–5.
93. Antal A, Herrmann CS. Transcranial alternating current and random noise stimulation: possible mechanisms. Neural Plast. 2016;2016:3616807. https://doi.org/10.1155/2016/3616807.
94. Woods AJ, Antal A, Bikson M, et al. A technical guide to tDCS, and related non-invasive brain stimulation tools. Clin Neurophysiol. 2016;127:1031–48.
95. Wischnewski M, Alekseichuk I, Opitz A. Neurocognitive, physiological, and biophysical effects of transcranial alternating current stimulation. Trends Cogn Sci. 2023;27:189–205.
96. Antonenko D, Faxel M, Grittner U, Lavidor M, Floël A. Effects of transcranial alternating current stimulation on cognitive functions in healthy young and older adults. Neural Plast. 2016;2016:4274127. https://doi.org/10.1155/2016/4274127.
97. Elyamany O, Leicht G, Herrmann CS, Mulert C. Transcranial alternating current stimulation (tACS): from basic mechanisms towards first applications in psychiatry. Eur Arch Psychiatry Clin Neurosci. 2021;271:135–56.
98. Booth SJ, Taylor JR, Brown LJE, Pobric G. The effects of transcranial alternating current stimulation on memory performance in healthy adults: a systematic review. Cortex. 2022;147:112–39.
99. Takeuchi N. Pain control based on oscillatory brain activity using transcranial alternating current stimulation: an integrative review. Front Hum Neurosci. 2023;17:941979.
100. Brancucci A, Rivolta D, Nitsche MA, Manippa V. The effects of transcranial random noise stimulation on motor function: a comprehensive review of the literature. Physiol Behav. 2023;261:114073.
101. van der Groen O, Potok W, Wenderoth N, Edwards G, Mattingley JB, Edwards D. Using noise for the better: the effects of transcranial random noise stimulation on the brain and behavior. Neurosci Biobehav Rev. 2022;138:104702.
102. Thiel A, Hartmann A, Rubi-Fessen I, Anglade C, Kracht L, Weiduschat N, Kessler J, Rommel T, Heiss W-D. Effects of noninvasive brain stimulation on language networks and recovery in early poststroke aphasia. Stroke. 2013;44:2240–6.
103. Rubi-Fessen I, Hartmann A, Huber W, Fimm B, Rommel T, Thiel A, Heiss W-D. Add-on effects of repetitive transcranial magnetic stimulation on subacute aphasia therapy: enhanced improvement of functional communication and basic linguistic skills. A randomized controlled study. Arch Phys Med Rehabil. 2015;96:1935–1944.e2.
104. Ren C, Zhang G, Xu X, Hao J, Fang H, Chen P, Li Z, Ji Y, Cai Q, Gao F. The effect of rTMS over the different targets on language recovery in stroke patients with global aphasia: a randomized sham-controlled study. Biomed Res Int. 2019;2019:e4589056.

105. Hara T, Abo M, Kobayashi K, Watanabe M, Kakuda W, Senoo A. Effects of low-frequency repetitive transcranial magnetic stimulation combined with intensive speech therapy on cerebral blood flow in post-stroke aphasia. Transl Stroke Res. 2015;6:365–74.
106. Harvey DY, Podell J, Turkeltaub PE, Faseyitan O, Coslett HB, Hamilton RH. Functional reorganization of right prefrontal cortex underlies sustained naming improvements in chronic aphasia via repetitive transcranial magnetic stimulation. Cogn Behav Neurol. 2017;30:133.
107. Rossetti A, Malfitano C, Malloggi C, Banco E, Rota V, Tesio L. Phonemic fluency improved after inhibitory transcranial magnetic stimulation in a case of chronic aphasia. Int J Rehabil Res. 2019;42:92.
108. Sebastianelli L, Versace V, Martignago S, Brigo F, Trinka E, Saltuari L, Nardone R. Low-frequency rTMS of the unaffected hemisphere in stroke patients: a systematic review. Acta Neurol Scand. 2017;136:585–605.
109. Kindler J, Schumacher R, Cazzoli D, Gutbrod K, Koenig M, Nyffeler T, Dierks T, Müri RM. Theta burst stimulation over the right broca's homologue induces improvement of naming in aphasic patients. Stroke. 2012;43:2175–9.
110. Lefaucheur J-P, André-Obadia N, Antal A, et al. Evidence-based guidelines on the therapeutic use of repetitive transcranial magnetic stimulation (rTMS). Clin Neurophysiol. 2014;125:2150–206.
111. Georgiou A, Konstantinou N, Phinikettos I, Kambanaros M. Neuronavigated theta burst stimulation for chronic aphasia: two exploratory case studies. Clin Linguist Phon. 2019;33:532–46.
112. Szaflarski JP, Vannest J, Wu SW, DiFrancesco MW, Banks C, Gilbert DL. Excitatory repetitive transcranial magnetic stimulation induces improvements in chronic post-stroke aphasia. Med Sci Monit. 2011;17:CR132–9.
113. Dammekens E, Vanneste S, Ost J, de Ridder D. Neural correlates of high frequency repetitive transcranial magnetic stimulation improvement in post-stroke non-fluent aphasia: a case study. Neurocase. 2014;20:1–9.
114. Griffis JC, Nenert R, Allendorfer JB, Szaflarski JP. Interhemispheric plasticity following intermittent theta burst stimulation in chronic poststroke aphasia. Neural Plast. 2016;2016:e4796906.
115. Cotelli M, Manenti R, Rosini S, Calabria M, Brambilla M, Bisiacchi PS, Zanetti O, Miniussi C. Action and object naming in physiological aging: an rTMS study. Front Aging Neurosci. 2010;2:151.
116. Cotelli M, Manenti R, Alberici A, Brambilla M, Cosseddu M, Zanetti O, Miozzo A, Padovani A, Miniussi C, Borroni B. Prefrontal cortex rTMS enhances action naming in progressive non-fluent aphasia. Eur J Neurol. 2012;19:1404–12.
117. Huang Y, Tan Y, Hao H, et al. Treatment of primary progressive aphasia by repetitive transcranial magnetic stimulation: a randomized, double-blind, placebo-controlled study. J Neural Transm (Vienna). 2023;130(2):111–23. https://doi.org/10.1007/s00702-023-02594-w.
118. Zhang H, Chen Y, Hu R, Yang L, Wang M, Zhang J, Lu H, Wu Y, Du X. rTMS treatments combined with speech training for a conduction aphasia patient. Medicine (Baltimore). 2017;96:e7399.
119. Suppa A, Huang Y-Z, Funke K, Ridding MC, Cheeran B, Di Lazzaro V, Ziemann U, Rothwell JC. Ten years of theta burst stimulation in humans: established knowledge, unknowns and prospects. Brain Stimul. 2016;9:323–35.
120. Rounis E, Huang Y-Z. Theta burst stimulation in humans: a need for better understanding effects of brain stimulation in health and disease. Exp Brain Res. 2020;238:1707–14.
121. Versace V, Schwenker K, Langthaler PB, Golaszewski S, Sebastianelli L, Brigo F, Pucks-Faes E, Saltuari L, Nardone R. Facilitation of auditory comprehension after theta burst stimulation of Wernicke's area in stroke patients: a pilot study. Front Neurol. 2020;10:1319. https://doi.org/10.3389/fneur.2019.01319.
122. Vuksanović J, Jelić MB, Milanović SD, Kačar K, Konstantinović L, Filipović SR. Improvement of language functions in a chronic non-fluent post-stroke aphasic patient following bilateral sequential theta burst magnetic stimulation. Neurocase. 2015;21:244–50.

123. Baker JM, Rorden C, Fridriksson J. Using transcranial direct-current stimulation to treat stroke patients with aphasia. Stroke. 2010;41:1229–36.
124. Campana S, Caltagirone C, Marangolo P. Combining voxel-based lesion-symptom mapping (VLSM) with A-tDCS language treatment: predicting outcome of recovery in nonfluent chronic aphasia. Brain Stimul. 2015;8:769–76.
125. Fridriksson J, Rorden C, Elm J, Sen S, George MS, Bonilha L. Transcranial direct current stimulation vs sham stimulation to treat aphasia after stroke: a randomized clinical trial. JAMA Neurol. 2018;75:1470–6.
126. Pestalozzi MI, Di Pietro M, Martins Gaytanidis C, Spierer L, Schnider A, Chouiter L, Colombo F, Annoni J-M, Jost LB. Effects of prefrontal transcranial direct current stimulation on lexical access in chronic poststroke aphasia. Neurorehabil Neural Repair. 2018;32:913–23.
127. Meinzer M, Darkow R, Lindenberg R, Flöel A. Electrical stimulation of the motor cortex enhances treatment outcome in post-stroke aphasia. Brain. 2016;139:1152–63.
128. Kang EK, Kim YK, Sohn HM, Cohen LG, Paik N-J. Improved picture naming in aphasia patients treated with cathodal tDCS to inhibit the right Broca's homologue area. Restor Neurol Neurosci. 2011;29:141–52.
129. Wu D, Wang J, Yuan Y. Effects of transcranial direct current stimulation on naming and cortical excitability in stroke patients with aphasia. Neurosci Lett. 2015;589:115–20.
130. Ciavarro M, Ambrosini E, Tosoni A, Committeri G, Fattori P, Galletti C. rTMS of medial parieto-occipital cortex interferes with attentional reorienting during attention and reaching tasks. J Cogn Neurosci. 2013;25:1453–62.
131. Marangolo P, Fiori V, Cipollari S, Campana S, Razzano C, Di Paola M, Koch G, Caltagirone C. Bihemispheric stimulation over left and right inferior frontal region enhances recovery from apraxia of speech in chronic aphasia. Eur J Neurosci. 2013;38:3370–7.
132. Marangolo P, Fiori V, Sabatini U, De Pasquale G, Razzano C, Caltagirone C, Gili T. Bilateral transcranial direct current stimulation language treatment enhances functional connectivity in the left hemisphere: preliminary data from aphasia. J Cogn Neurosci. 2016;28:724–38.
133. Feil S, Eisenhut P, Strakeljahn F, et al. Left shifting of language related activity induced by bihemispheric tDCS in postacute aphasia following stroke. Front Neurosci. 2019;13:295.
134. Edwards D, Cortes M, Datta A, Minhas P, Wassermann EM, Bikson M. Physiological and modeling evidence for focal transcranial electrical brain stimulation in humans: a basis for high-definition tDCS. NeuroImage. 2013;74:266–75.
135. Fiori V, Nitsche MA, Cucuzza G, Caltagirone C, Marangolo P. High-definition transcranial direct current stimulation improves verb recovery in aphasic patients depending on current intensity. Neuroscience. 2019;406:159–66.
136. Masina F, Arcara G, Galletti E, Cinque I, Gamberini L, Mapelli D. Neurophysiological and behavioural effects of conventional and high definition tDCS. Sci Rep. 2021;11:7659.
137. Miniussi C, Harris JA, Ruzzoli M. Modelling non-invasive brain stimulation in cognitive neuroscience. Neurosci Biobehav Rev. 2013;37:1702–12.
138. Saxena S, Hillis AE. An update on medications and noninvasive brain stimulation to augment language rehabilitation in post-stroke aphasia. Expert Rev Neurother. 2017;17:1091–107.
139. Marangolo P. The potential effects of transcranial direct current stimulation (tDCS) on language functioning: combining neuromodulation and behavioral intervention in aphasia. Neurosci Lett. 2020;719:133329.
140. Elsner B, Kugler J, Pohl M, Mehrholz J. Transcranial direct current stimulation (tDCS) for improving aphasia in adults with aphasia after stroke. Cochrane Database Syst Rev. 2019; https://doi.org/10.1002/14651858.CD009760.pub4.
141. Elsner B, Kugler J, Mehrholz J. Transcranial direct current stimulation (tDCS) for improving aphasia after stroke: a systematic review with network meta-analysis of randomized controlled trials. J Neuroeng Rehabil. 2020;17:88.
142. Arheix-Parras S, Barrios C, Python G, Cogné M, Sibon I, Engelhardt M, Dehail P, Cassoudesalle H, Moucheboeuf G, Glize B. A systematic review of repetitive transcranial magnetic stimulation in aphasia rehabilitation: leads for future studies. Neurosci Biobehav Rev. 2021;127:212–41.

143. Fiori V, Cipollari S, Di Paola M, Razzano C, Caltagirone C, Marangolo P. tDCS stimulation segregates words in the brain: evidence from aphasia. Front Hum Neurosci. 2013;7:269.
144. de Aguiar V, Paolazzi CL, Miceli G. tDCS in post-stroke aphasia: the role of stimulation parameters, behavioral treatment and patient characteristics. Cortex. 2015;63:296–316.
145. Zettin M, Bondesan C, Nada G, Varini M, Dimitri D. Transcranial direct-current stimulation and behavioral training, a promising tool for a tailor-made post-stroke aphasia rehabilitation: a review. Front Hum Neurosci. 2021;15:742136.
146. Zhao Q, Wang J, Li Z, Song L, Li X. Effect of anodic transcranial direct current stimulation combined with speech language therapy on nonfluent poststroke aphasia. Neuromodulation. 2021;24:923–9.
147. Martin PI, Naeser MA, Theoret H, Tormos JM, Nicholas M, Kurland J, Fregni F, Seekins H, Doron K, Pascual-Leone A. Transcranial magnetic stimulation as a complementary treatment for aphasia. Semin Speech Lang. 2004;25:181–91.
148. Hara T, Abo M, Kakita K, Mori Y, Yoshida M, Sasaki N. The effect of selective transcranial magnetic stimulation with functional near-infrared spectroscopy and intensive speech therapy on individuals with post-stroke aphasia. Eur Neurol. 2017;77:186–94.
149. Low TA, Lindland K, Kirton A, Carlson HL, Harris AD, Goodyear BG, Monchi O, Hill MD, Dukelow SP. Repetitive transcranial magnetic stimulation (rTMS) combined with multimodality aphasia therapy for chronic post-stroke non-fluent aphasia: a pilot randomized sham-controlled trial. Brain Lang. 2023;236:105216. https://doi.org/10.1016/j.bandl.2022.105216.
150. Naeser MA, Martin PI, Nicholas M, et al. Improved picture naming in chronic aphasia after TMS to part of right Broca's area: an open-protocol study. Brain Lang. 2005;93:95–105.
151. Martin PI, Naeser MA, Ho M, et al. Overt naming fMRI pre- and post-TMS: two nonfluent aphasia patients, with and without improved naming post-TMS. Brain Lang. 2009;111:20–35.
152. Harvey DY, Mass JA, Shah-Basak PP, Wurzman R, Faseyitan O, Sacchetti DL, DeLoretta L, Hamilton RH. Continuous theta burst stimulation over right pars triangularis facilitates naming abilities in chronic post-stroke aphasia by enhancing phonological access. Brain Lang. 2019;192:25–34.
153. Ntemou E, Svaldi C, Jonkers R, Picht T, Rofes A. Verb and sentence processing with TMS: a systematic review and meta-analysis. Cortex. 2023;162:38–55. https://doi.org/10.1016/j.cortex.2023.01.005.
154. Bryant L, Ferguson A, Spencer E. Linguistic analysis of discourse in aphasia: a review of the literature. Clin Linguist Phon. 2016;30:489–518.
155. Pritchard M, Hilari K, Cocks N, Dipper L. Reviewing the quality of discourse information measures in aphasia. Int J Lang Commun Disord. 2017;52:689–732.
156. Doedens WJ, Meteyard L. Measures of functional, real-world communication for aphasia: a critical review. Aphasiology. 2020;34:492–514.
157. Stark BC, Dutta M, Murray LL, Fromm D, Bryant L, Harmon TG, Ramage AE, Roberts AC. Spoken discourse assessment and analysis in aphasia: an international survey of current practices. J Speech Lang Hear Res. 2021;64:4366–89.
158. Privitera AJ, Ng SHS, Kong AP-H., Weekes BS. AI and Aphasia in the Digital Age: A Critical Review. 2024. https://doi.org/10.3390/brainsci14040383.
159. Rezaii N et al. Artificial intelligence classifies primary progressive aphasia from connected speech. 2024. https://doi.org/10.1093/brain/awae196.
160. Brunato D, Cimino A, Dell'Orletta F, Venturi G, Montemagni S. Profiling-UD: a Tool for Linguistic Profiling of Texts. 2020. https://aclanthology.org/2020.lrec-1.883.
161. Marini A, Andreetta S, del Tin S, Carlomagno S. A multi-level approach to the analysis of narrative language in aphasia. Aphasiology. 2011;25:1372–92.
162. Marangolo P. tDCS over the left inferior frontal cortex improves speech production in aphasia. Front Hum Neurosci. 2013;7:539.
163. Marangolo P, Fiori V, Campana S, Calpagnano MA, Razzano C, Caltagirone C, Marini A. Something to talk about: enhancement of linguistic cohesion through tdCS in chronic non fluent aphasia. Neuropsychologia. 2014;53:246–56.

164. Kidwai J, Sharma S, Peper M, Brumberg J. Investigating NIBS for language rehabilitation in aphasia. Aphasiology. 2022;37(8):1285–314. https://doi.org/10.1080/02687038.2022.2089972.
165. Shah-Basak PP, Wurzman R, Purcell JB, Gervits F, Hamilton R. Fields or flows? A comparative metaanalysis of transcranial magnetic and direct current stimulation to treat post-stroke aphasia. Restor Neurol Neurosci. 2016;34:537–58.
166. Hartwigsen G, Volz LJ. Probing rapid network reorganization of motor and language functions via neuromodulation and neuroimaging. NeuroImage. 2021;224:117449.
167. Crosson B, Rodriguez AD, Copland D, Fridriksson J, Krishnamurthy LC, Meinzer M, Raymer AM, Krishnamurthy V, Leff AP. Neuroplasticity and aphasia treatments: new approaches for an old problem. J Neurol Neurosurg Psychiatry. 2019;90:1147–55.
168. Devane N, Behn N, Marshall J, Ramachandran A, Wilson S, Hilari K. The use of virtual reality in the rehabilitation of aphasia: a systematic review. Disabil Rehabil. 2023;45(23):3803–22.
169. Cherney LR, Van Vuuren S. Telerehabilitation, virtual therapists, and acquired neurologic speech and language disorders. Semin Speech Lang. 2012;33:243–57.
170. Hall N, Boisvert M, Steele R. Telepractice in the assessment and treatment of individuals with aphasia: a systematic review. Int J Telerehabil. 2013;5:27–38.
171. Winters JM. Telerehabilitation research: emerging opportunities. Annu Rev Biomed Eng. 2002;4:287–320.
172. Gruska M, Aigner G, Altenberger J, et al. Recommendations on the utilization of telemedicine in cardiology. Wien Klin Wochenschr. 2020;132:782–800.
173. Nikolaev VA, Nikolaev AA. Recent trends in telerehabilitation of stroke patients: a narrative review. NeuroRehabilitation. 2022;51:1–22.
174. Morse H, Biggart L, Pomeroy V, Rossit S. Exploring perspectives from stroke survivors, carers and clinicians on virtual reality as a precursor to using telerehabilitation for spatial neglect post-stroke. Neuropsychol Rehabil. 2022;32:767–91.
175. Meltzer JA, Baird AJ, Steele RD, Harvey SJ. Computer-based treatment of poststroke language disorders: a non-inferiority study of telerehabilitation compared to in-person service delivery. Aphasiology. 2018;32:290–311.
176. Maresca G, Maggio MG, Latella D, et al. Toward improving poststroke aphasia: a pilot study on the growing use of telerehabilitation for the continuity of care. J Stroke Cerebrovasc Dis. 2019;28(10):104303. https://doi.org/10.1016/j.jstrokecerebrovasdis.2019.104303.
177. Giachero A, Calati M, Pia L, La Vista L, Molo M, Rugiero C, Fornaro C, Marangolo P. Conversational therapy through semi-immersive virtual reality environments for language recovery and psychological Well-being in post stroke aphasia. Behav Neurol. 2020;2020:e2846046.
178. Marshall J, Booth T, Devane N, Galliers J, Greenwood H, Hilari K, Talbot R, Wilson S, Woolf C. Evaluating the benefits of aphasia intervention delivered in virtual reality: results of a quasi-randomised study. PLoS One. 2016;11:e0160381.
179. Grechuta K, Rubio Ballester B, Espín Munne R, Usabiaga Bernal T, Molina Hervás B, Mohr B, Pulvermüller F, San Segundo R, Verschure P. Augmented dyadic therapy boosts recovery of language function in patients with nonfluent aphasia. Stroke. 2019;50:1270–4.
180. Grechuta K, Bellaster BR, Munne RE, Bernal TU, Hervas BM, Segundo RS, Verschure PFMJ. The effects of silent visuomotor cueing on word retrieval in Broca's aphasies: a pilot study. IEEE Int Conf Rehabil Robot. 2017;2017:193–9.
181. Carragher M, Talbot R, Devane N, Rose M, Marshall J. Delivering storytelling intervention in the virtual world of EVA Park. Aphasiology. 2018;32:37–9.
182. Riches S, Elghany S, Garety P, Rus-Calafell M, Valmaggia L. Factors affecting sense of presence in a virtual reality social environment: a qualitative study. Cyberpsychol Behav Soc Netw. 2019;22:288–92.
183. Smith SA, Mulligan NW. Immersion, presence, and episodic memory in virtual reality environments. Memory. 2021;29:983–1005.
184. Cole T, Gillies M. Thinking and doing: challenge, agency, and the eudaimonic experience in video games. Games and Culture. 2021;16:187–207.

185. Barsalou LW, Kyle Simmons W, Barbey AK, Wilson CD. Grounding conceptual knowledge in modality-specific systems. Trends Cogn Sci. 2003;7:84–91.
186. Marangolo P, Bonifazi S, Tomaiuolo F, Craighero L, Coccia M, Altoè G, Provinciali L, Cantagallo A. Improving language without words: first evidence from aphasia. Neuropsychologia. 2010;48:3824–33.
187. Legault J, Zhao J, Chi Y-A, Chen W, Klippel A, Li P. Immersive virtual reality as an effective tool for second language vocabulary learning. Languages. 2019;4:13.
188. Choi MJ, Kim H, Nah H-W, Kang D-W. Digital therapeutics: emerging new therapy for neurologic deficits after stroke. J Stroke. 2019;21:242–58.
189. Carrera E, Tononi G. Diaschisis: past, present, future. Brain. 2014;137:2408–22.
190. Hillis AE. Pharmacological, surgical, and neurovascular interventions to augment acute aphasia recovery. Am J Phys Med Rehabil. 2007;86:426.
191. Zeiler SR, Krakauer JW. The interaction between training and plasticity in the poststroke brain. Curr Opin Neurol. 2013;26:609.
192. Wawrzyniak M, Schneider HR, Klingbeil J, Stockert A, Hartwigsen G, Weiller C, Saur D. Resolution of diaschisis contributes to early recovery from post-stroke aphasia. NeuroImage. 2022;251:119001.
193. Keser Z, Francisco GE. Neuropharmacology of Poststroke motor and speech recovery. Phys Med Rehabil Clin N Am. 2015;26:671–89.
194. Furlanis G, Ridolfi M, Polverino P, Menichelli A, Caruso P, Naccarato M, Sartori A, Torelli L, Pesavento V, Manganotti P. Early recovery of aphasia through thrombolysis: the significance of spontaneous speech. J Stroke Cerebrovasc Dis. 2018;27:1937–48.
195. Menichelli A, Furlanis G, Sartori A, Ridolfi M, Naccarato M, Caruso P, Pesavento V, Manganotti P. Thrombolysis' benefits on early post-stroke language recovery in aphasia patients. J Clin Neurosci. 2019;70:92–5.
196. Stockbridge MD. Chapter 17—Better language through chemistry: augmenting speech-language therapy with pharmacotherapy in the treatment of aphasia. In: Hillis AE, Fridriksson J, editors. Handbook of clinical neurology. Elsevier; 2022. p. 261–72.
197. Enderby P, Broeckx J, Hospers W, Schildermans F, Deberdt W. Effect of Piracetam on recovery and rehabilitation after stroke: a double-blind, placebo-controlled study. Clin Neuropharmacol. 1994;17:320.
198. Huber W, Willmes K, Poeck K, Vleymen BV, Deberdt W. Piracetam as an adjuvant to language therapy for aphasia: a randomized double-blind placebo-controlled pilot study. Arch Phys Med Rehabil. 1997;78:245–50.
199. Kessler J, Thiel A, Karbe H, Heiss WD. Piracetam improves activated blood flow and facilitates rehabilitation of poststroke aphasic patients. Stroke. 2000;31:2112–6.
200. Güngör L, Terzi M, Onar MK. Does long term use of piracetam improve speech disturbances due to ischemic cerebrovascular diseases? Brain Lang. 2011;117:23–7.
201. Sarter M, Hasselmo ME, Bruno JP, Givens B. Unraveling the attentional functions of cortical cholinergic inputs: interactions between signal-driven and cognitive modulation of signal detection. Brain Res Rev. 2005;48:98–111.
202. Berthier ML, Green C, Lara JP, Higueras C, Barbancho MA, Dávila G, Pulvermüller F. Memantine and constraint-induced aphasia therapy in chronic poststroke aphasia. Ann Neurol. 2009;65:577–85.
203. Barbancho MA, Berthier ML, Navas-Sánchez P, et al. Bilateral brain reorganization with memantine and constraint-induced aphasia therapy in chronic post-stroke aphasia: an ERP study. Brain Lang. 2015;145–146:1–10.
204. Parsons CG, Stöffler A, Danysz W. Memantine: a NMDA receptor antagonist that improves memory by restoration of homeostasis in the glutamatergic system—too little activation is bad, too much is even worse. Neuropharmacology. 2007;53:699–723.
205. Kandel ER. The molecular biology of memory storage: a dialogue between genes and synapses. Science. 2001;294:1030–8.

206. Seniów J, Litwin M, Litwin T, Leśniak M, Członkowska A. New approach to the rehabilitation of post-stroke focal cognitive syndrome: effect of levodopa combined with speech and language therapy on functional recovery from aphasia. J Neurol Sci. 2009;283:214–8.
207. Leemann B, Laganaro M, Chetelat-Mabillard D, Schnider A. Crossover trial of subacute computerized aphasia therapy for anomia with the addition of either levodopa or placebo. Neurorehabil Neural Repair. 2011;25:43–7.
208. Breitenstein C, Korsukewitz C, Baumgärtner A, Flöel A, Zwitserlood P, Dobel C, Knecht S. L-dopa does not add to the success of high-intensity language training in aphasia. Restor Neurol Neurosci. 2015;33:115–20.
209. Berthier ML. Ten key reasons for continuing research on pharmacotherapy for post-stroke aphasia. Aphasiology. 2021;35:824–58.
210. Marini A. The beauty of diversity in cognitive neuroscience: the case of sex-related effects in language production networks. J Neurosci Res. 2022;00:1–10.
211. Edmiston EK, Juster R-P. Refining research and representation of sexual and gender diversity in neuroscience. Biol Psychiatry Cogn Neurosci Neuroimaging. 2022;7:1251–7.
212. Dotson VM, Duarte A. The importance of diversity in cognitive neuroscience. Ann N Y Acad Sci. 2020;1464:181–91.

# Dynamic Postural Stability in Patients with Neurological Disorders

5

Marco Tramontano, Giuseppe Vannozzi, Irene Ciancarelli, and Giovanni Morone

## 5.1 Dynamic Postural Stability

The integration of multisensory information (visual, vestibular, and somatosensory) could be impaired in patients with neurological diseases. Several studies have focused on the identification of fall risk factors in patients with cerebrovascular event outcomes to help in the prevention and treatment of stroke survivors at high risk of falling [1–4]. Among the various risk factors identified, [5] the presence of gait impairment has been recognized as one of the most important predictors of

M. Tramontano
Department of Biomedical and Neuromotor Sciences, Alma Mater Studiorum University of Bologna, Bologna, Italy

Unit of Occupational Medicine, IRCCS Azienda Ospedaliero-Universitaria di Bologna, Bologna, Italy
e-mail: marco.tramontano@unibo.it

G. Vannozzi
Santa Lucia Foundation, IRCCS, Rome, Italy

Department of Movement, Human and Health Sciences, University of Rome "Foro Italico", Rome, Italy
e-mail: giuseppe.vannozzi@uniroma4.it

I. Ciancarelli
Department of Life, Health and Environmental Sciences, University of L'Aquila, L'Aquila, Italy
e-mail: irene.ciancarelli@univaq.it

G. Morone (✉)
Department of Life, Health and Environmental Sciences, University of L'Aquila, L'Aquila, Italy

San Raffaele Institute of Sulmona, Sulmona, Italy
e-mail: giovanni.morone@univaq.it


R. S. Calabrò (ed.), *Translational Neurorehabilitation*,
https://doi.org/10.1007/978-3-031-63604-2_5

falling [5, 6]. Indeed, it has been shown that individuals with reduced mobility are 1.65 times more likely to suffer a fall and that up to 70% of falls occur while walking [3]. Ninety per cent of patients diagnosed with Multiple Sclerosis report balance as a primary motor problem, and more than 50% report a moderate to severe impact on quality of life [7]. Gait disturbance represents one of the most characteristic and disabling symptoms of PD [8, 9]. Spatio-temporal gait parameters are often altered in PD patients, showing a decrease in foot height, step length, and gait speed [8]. Furthermore, Rinne et al. [10] showed that in people with TBI despite a good recovery of other motor functions, there is still an impairment of balance and agility compared to healthy controls. Several studies [11] showed that people with TBI walked slower than healthy controls, mainly due to reduced stride length and the impairment could be related to the severity level [12]. Given the importance of walking as the most frequent dynamic activity in daily life [6], the recovery of stability during walking represents one of the most important goals of neurorehabilitation [13, 14]. Dynamic postural stability can be defined as the ability to minimize oscillations, progressively, from lower to higher levels of the human body, and thus maintain upright balance during walking [15]. An adequate and timely assessment of balance and gait disorders can help in identifying patients at high risk of falling and in setting up an individual rehabilitation project that takes into account the individual's abilities and needs as much as possible to prevent possible falls [13]. In this context, it is crucial to quantify gait and balance deficits and monitor patients over time [16]. Several tests and rating scales are used in the clinic to assess the gait and balance disorders in patients with neurological disorders; however, these scales and questionnaires may sometimes lack reliability and specificity-interoperability [17, 18]. Combining a clinical and instrumental assessment may be useful to obtain an objective quantitative approach to support the clinical staff in better design personalized treatment [19, 20]. The use of wearable wireless Inertial Measurement Units (IMUs) within tests could represent a great advantage in the assessment of balance and gait disorders in patients with neurological diseases. Indeed, these sensors allow clinicians to obtain valid and reliable measurements, especially in screening studies and therapeutic interventions [21]. Upper-body accelerations measured with a multi-sensor approach can provide information on the relationship between the trunk and head during dynamic balance tasks. This approach, already adopted in gait assessment, has proven feasible in the clinical context, supporting the planning of balance rehabilitation strategies and determining the effectiveness of therapeutic procedures [20]. Furthermore, the use of wearable inertial sensors makes it possible to step out of the laboratory context and assess patients in a more ecological environment by evaluating the patient while performing tasks that are therefore as close as possible to the daily life activities performed by the patient [13, 20]. Thus, the instrumental analysis aims to collect information directly in the field and to obtain concise quantitative metrics of global gait quality (gait quality indices), which can assess the motor capacity of patients at the person level, and which take into account characteristics such as balance maintenance during walking and gait symmetry. Specifically, the analysis with inertial sensors makes it possible to assess not only spatio-temporal gait parameters (gait speed, gait duration, gait frequency) but also qualitative indices relating to different domains of interest, such as dynamic postural stability. Several indices pertaining to the different domains have already been used in neurological patients and have been shown not only to discriminate between the pathological and control group but also to highlight differences between different levels of motor skills within groups with the same pathology [12, 22, 23].

## 5.2 Gait and Postural Stability Assessment Through IMUs in Patients with Neurological Disorders: State of the Art

Thanks to the introduction of integrated circuits and the evolution of silicon micro-electro-mechanical systems (MEMS) technology, the first cumbersome and expensive inertial sensors dating back to the 1950s have been replaced by newer devices. This has made it possible to cut production costs and significantly reduce the size of the device so that these sensors can be implemented in everyday practice (air-bags, smartphones, games consoles, tablets, but also helicopters, planes, ships, cars, etc.). Inertial sensors have several fields of application, indeed a progression in the use of gait analysis in the neurological population [24], as well as runners [25] was registered. In addition, inertial sensors also find applications in the rehabilitation field, where they are used to enhance standard rehabilitation programs, as in the Euleria Lab (built on the legacy device Riablo). Euleria Lab is a class IIa medical device according to the MDR, an adaptive system composed of several IMUs that allows the performance of exercises to be improved through visual feedback [26]. Among the most interesting and promising applications, still in the experimental phase, inertial sensors have been used both in healthy people, such as athletes [25] and in para-athletes [27]. In fact, the sensors are able to derive spatio-temporal gait parameters [28], stability indices [29] as well as provide information on the symmetry and fluidity of movement, useful for example to quantify the risk of falling [30]. In the case of athletes, they made it possible to observe athletic gestures, providing important information in the prevention of injuries and/or for the development of new materials and equipment, as well as providing feedback on performance that could help both athletes and coaches to improve performance. In the clinical field, they have proved particularly useful in the assessment of tremor, providing objective information in terms of frequency and intensity of oscillations [31, 32]; in the assessment of people at risk of falling [30]; and in orthopedics, where they have been used to assess the change in walking pattern in people with hip and knee replacements; and in the study of neurological disorders such as stroke outcomes [33], Parkinson's Disease (PD) [13], cerebellar ataxia, in Multiple Sclerosis (MS), in patients with foot drop [32] or in cases of infantile cerebral palsy [32]. Several studies have already demonstrated the potential of using IMUs in patients with MS, as they provide quantitative information on the main gait alterations and allow discrimination between different walking abilities [34]. Recently, several studies reported the usefulness of gait and balance assessment through wearable inertial sensors in patients with MS [22]. The articles showed that different types of tests and/or clinical scales were administered when assessing walking using inertial sensors in patients with MS. Specifically, short-duration tests to assess walking in a typical manner (through spatio-temporal parameters of gait) such as the Timed 25 Foot Walk (T25FW); long duration tests, mainly to assess the impact of fatigue; tests for functional abilities such as the 6 Min Walk Test (6MWT) and tests to assess the level of mobility such as the Time Up and Go (TUG) [35]. Most studies have focused on the main spatio-temporal parameters (speed, cadence, single and double stance phase time, swing phase time, stride length and time), while only a few have considered complementary criteria such as intensity, stability, frequency, symmetry, and fluidity of movement (smoothness). In particular, indicators such as the Root Mean Square (RMS), the Harmonic Ratio (HR), and autocorrelation indices were used to assess the latter. Also different is the number and type of sensors that were used: in most of the studies, the number of sensors was no more than three. The primary outcome was mostly derived from the correlation of the InertiaLocoGraphy (ILG) [35], which is the quantification of gait using IMUs, with the EDSS scores. Secondary outcomes, on the other hand, are derived by comparing the data collected by the IMUs with the scores of the scales or clinical tests used (T25FW, 6MWT, TUG). These studies have shown that not only are IMUSs able to detect gait alterations in patients with overt difficulties, but also patients with low disability, often already present in the early stages of the disease. An interesting study [36] compared three different neurological diseases (MS, PD, stroke) and healthy participants using a sternum-level inertial sensor in a motor task that consisted of climbing 10 steps. Significant, clinically undetected differences between the different disease groups were underlined by this study. As recently reported in a narrative scoping review [37] different approaches and fundamentals were used, however, despite the results obtained, IMUs have not yet been adopted in routine clinical practice; probably, the protocols used are too heterogeneous, both in terms of the tests and/or scales used and the number of sensors. In the following section, the main application of IMUs for the evaluation of gait and balance in patients with neurological disorders is reported, highlighting when this type of approach was effective.

### 5.2.1 Stroke

The application of IMUs in patients with stroke has been predominantly related to the assessment of gait and postural stability [13]; to the symmetry of trunk movements [38], dynamic stability [39], particularly of the upper body [38], and to highlight major gait alterations [40]. In general, the study of the walking pattern of the hemiparetic patient using sensors has shown a marked loss of symmetry in the stance phase; whereas it would appear that the swing phase occurs more slowly and is characterized by a significant reduction in commuting on the affected side, high energy consumption, and altered clearance. Depending on the study, different protocols were applied both for the sensors (type and number) and for the scales and/or tests administered. Among the tests, the most frequently used was the 10 Meter Walk Test (10MWT), which was performed in different ways. The number and type of sensors, on the other hand, changed according to the study; from a minimum of

1 sensor to a maximum of 5. With regard to the location of the sensors, the lumbar spine (L4-L5) and the lateral malleoli were the segments most frequently considered. The main reference indicator used in the assessment of the walking pattern of the stroke patient was the root mean-square (RMS), which is an indicator of the standard deviation of the signal along one direction and represents the effective value of the signal considered. Only a few studies have also used other indicators such as attenuation coefficients (CV), the Harmonic Ratio (HR), and the Gait Ratio (GR).

### 5.2.2 Traumatic Brain Injury

Gait disturbances are a common consequence of head injuries. Studies have shown that TBI patients walk slower than healthy participants; they have shorter strides and present balance alterations with marked postural oscillations. There are several studies in the literature that have used an integrated approach of sensors and clinical scales for gait assessment. The main applications are related to the evaluation of certain elements such as balance and postural oscillations [41]; the impact of curvilinear gait on walking disorders [16], in terms of an alteration of spatio-temporal parameters, stability and symmetry and alterations during dual-task activities [42], and quantifying the characteristics of walking patterns [43]. Depending on the focus to be investigated, the scales and/or tests used were: the Berg Balance Scale (BBS), the Community Balance and Mobility Scale (CB&M), the Dynamic Gait Index (DGI), the TUG, the 10MWT, the Figure-of-8 Walk Test (F8WT), the Fukuda Stepping Test (FST); as well as non-specific tests to assess balance (standing station with open/closed eyes on a stable/unstable surface) [41]. Different sensors were used but the most common were IMUs. Concerning the body segments chosen for sensor location, again, the malleoli and the L4/L5 lumbar spine were the sites of choice. Among the most used indicators are the normalized root mean square (nRMS); attenuation coefficients; Harmonic Ratio (HR); and the improved Harmonic Ratio (iHR). All these studies revealed the potential of this integrated approach to objectively assess gait and balance changes during different tasks, showing significant differences both between healthy participants and patients and within patients themselves in terms of walking impairment severity [43]. Furthermore, IMUs were used also to evaluate the effectiveness of vestibular rehabilitation in patients with TBI [44].

### 5.2.3 Parkinson Disease

In this pathology, the use of sensors allowed to investigate different aspects for various purposes: diagnostic, to monitor both activities of daily living and the progression of the pathology [35, 45], to investigate specific walking disorders such as freezing [46], or to quantify the risk of falling [47]. Also in this case, as in the MS studies, the main data were derived from a comparison of the qualitative indices obtained from the sensors with the scores of the traditional clinical scales; in addition, in the case of falls, the correlation was also made with "fall" events that occurred 6 months after patient's discharge [47]. The different studies have adopted heterogeneous protocols both in terms of scales and number of sensors: from three sensors, in the case of fall assessment and dynamic stability [48] to a maximum of seven sensors; sometimes even in association with an optoelectronic device. The most common anatomical districts on which the sensors were located include the pelvis (L4/L5), sternum, and head. Common indicators include the Harmonic Ratio, attenuation and autocorrelation coefficients, and the RMS.

### 5.2.4 Cerebellar Ataxia

The main applications concerning the use of inertial sensors in patients with cerebellar ataxia have been to identify biological markers to monitor disease progression and measure the effectiveness of interventions. Several studies have shown that patients with cerebellar ataxia exhibit altered gait characteristics and that some of these may reflect disease progression [49, 50]. In particular, postural instability, as measured by triaxial accelerometers, has often been correlated with ataxia severity in patients with the degenerative cerebellar syndrome (DCS). A recent study by Shirai et al. [51] found that the mean amplitude of acceleration in the mid- to lateral direction, measured by accelerometers placed on the back during a linear gait test, could be considered a clinically relevant biological marker in patients with DCS. Recent studies have then exploited the potential of sensors to validate these biological markers in an ecological context, by monitoring patients in real life over an extended period [52]. Protocols, as in previous studies, were heterogeneous in terms of scales and/or tests (6MWT, 10MWT,) and sensors (number and type). Most studies used RMS as the main indicator, often in combination with CV.

### 5.2.5 Cerebral Palsy

They are the leading cause of disability in childhood and lead to various motor deficits including major neuromotor impairments (spasticity, weakness, lack of muscle control) and pathological movement patterns. These motor disabilities have a more important impact than in adults as they affect normal growth and development of musculoskeletal structures, with important consequences on functional activities including walking. Another characteristic is the variability with which cerebral palsies (CP) manifest themselves; including a wide range of pathological walking patterns that usually change over time. For all these reasons, in CP there is a constant need to monitor and assess the clinical condition of patients and, above all, there is a need for a dynamic assessment tool that can objectively quantify the child's motor functions and describe changes in their motor pattern. Several studies have therefore been carried out with inertial sensors with different applications: monitoring walking during activities of daily living [53], assessing spasticity [54], to identify the orthosis best suited to the needs of these patients [55], or to assess upper body movements [56]. Depending on the aspect to be investigated, a certain number of sensors and one or more types of scales were used. The main indicators include the Root Mean Square, attenuation coefficients, and specific algorithms.

## 5.3 General Discussion and Future Perspectives

Combining instrumental and clinical assessment during dynamic gait tasks could be a useful clinical approach that is increasingly supported by convincing evidence. Indeed, it is worth to underline the importance of obtaining an objective quantitative approach to inform the clinical decision-making process in treating balance disorders in patients with neurological disorders. From an applied biomechanics perspective, this recent literature trend promotes the adoption of integrated instrumented approaches to motor performance assessment. The added value of instrumented protocol is in the fact that we can go far beyond a global collective score (as it happens in the clinical tests), allowing the clinician to objectively examine subtle improvements in gait quality from different domain perspectives (such as gait stability and smoothness, for instance). As previously reported [21], the usefulness of the IMU-based approach could be perceived and confirmed also by the clinicians, indeed the sensor-based assessment seems to be user-friendly and useful in providing information for designing rehabilitation programs [21].

It is reasonable to hypothesize that rehabilitative programs focused on increasing dynamic postural adjustments might reduce unsmooth movements with changes that are not easily detected by traditional evaluation techniques. This body of evidence is promoting the opportunity to design rehabilitative programs focused on increasing dynamic postural stability [57–59].

From a neuro-rehabilitative point of view, integrating IMUs in the clinical assessment pathway could help the management of patients with neurological diseases for two important reasons: (i) gait and balance instrumental evaluation should be performed during more ecological and challenging contexts, including curvilinear walking pathways as a condition more representative of real-life scenarios; (ii) neurorehabilitation program could be tailored according to the results of instrumental assessments [21]. In this respect, the literature is progressively proposing the joint use of wearables to objectify rehabilitation-driven movement changes and the adoption of motor tasks involving rotations to challenge the musculoskeletal system in motion. This approach is preferable when aiming at identifying postural instability and could be also prospectively implemented into software applications that calculate IMU-based parameters in real time and with a low computational cost [60].

The assessment of the dynamic postural stability offers new clinical perspectives in obtaining biomarkers of gait deficits in neurological conditions with a procedure shared in real time with electronic medical records. Furthermore, this evaluation approach can support clinicians in estimating patients' independence in the ADL and offers the opportunity to continue clinical monitoring of the patients in a home-based setting representing a cost-effective solution for the National Health Care System.

## References


1. Bruijn SM, van Dieën JH. Control of human gait stability through foot placement. J R Soc Interface. 2018;15(143):20170816. https://doi.org/10.1098/rsif.2017.0816.
2. Tramontano M, Piermaria J, Morone G, et al. Postural changes during exteroceptive thin plantar stimulation: the effect of prolonged use and different plantar localizations. Front Syst Neurosci. 2019;13:49. https://doi.org/10.3389/fnsys.2019.00049.
3. Fletcher PC, Hirdes JP. Risk factors for falling among community-based seniors using home care services. J Gerontol A Biol Sci Med Sci. 2002;57(8):M504–10. https://doi.org/10.1093/gerona/57.8.m504.
4. Campbell GB, Matthews JT. An integrative review of factors associated with falls during post-stroke rehabilitation. J Nurs Scholarsh. 2010;42(4):395–404. https://doi.org/10.1111/j.1547-5069.2010.01369.x.
5. Masud T, Morris RO. Epidemiology of falls. Age Ageing. 2001;30(Suppl 4):3–7. https://doi.org/10.1093/ageing/30.suppl_4.3.
6. Hamacher D, Singh NB, Van Dieën JH, et al. Kinematic measures for assessing gait stability in elderly individuals: a systematic review. J R Soc Interface. 2011;8(65):1682–98. https://doi.org/10.1098/rsif.2011.0416.
7. Zackowski KM. Gait and balance assessment. Semin Neurol. 2016;36(5):474–8. https://doi.org/10.1055/s-0036-1584949.
8. Boonstra TA, van der Kooij H, Munneke M, Bloem BR. Gait disorders and balance disturbances in Parkinson's disease: clinical update and pathophysiology. Curr Opin Neurol. 2008;21(4):461–71. https://doi.org/10.1097/WCO.0b013e328305bdaf.
9. Mirelman A, Bonato P, Camicioli R, et al. Gait impairments in Parkinson's disease. Lancet Neurol. 2019;18(7):697–708. https://doi.org/10.1016/S1474-4422(19)30044-4.
10. Rinne MB, Pasanen ME, Vartiainen MV, et al. Motor performance in physically well-recovered men with traumatic brain injury. J Rehabil Med. 2006;38(4):224–9. https://doi.org/10.1080/16501970600582989.
11. Williams G, Galna B, Morris ME, et al. Spatiotemporal deficits and kinematic classification of gait following a traumatic brain injury: a systematic review. J Head Trauma Rehabil. 2010;25(5):366–74. https://doi.org/10.1097/HTR.0b013e3181cd3600.
12. Dever A, Powell D, Graham L, Mason R, Das J, Marshall SJ, Vitorio R, Godfrey A, Stuart S. Gait impairment in traumatic brain injury: a systematic review. Sensors (Basel). 2022;22(4):1480. https://doi.org/10.3390/s22041480.
13. Bergamini E, Iosa M, Belluscio V, Morone G, Tramontano M, Vannozzi G. Multi-sensor assessment of dynamic balance during gait in patients with subacute stroke. J Biomech. 2017;61:208–15. https://doi.org/10.1016/j.jbiomech.2017.07.034.
14. Morone G, Paolucci S, Iosa M. In what daily activities do patients achieve independence after stroke? J Stroke Cerebrovasc Dis. 2015;24(8):1931–7. https://doi.org/10.1016/j.jstrokecerebrovasdis.2015.05.006.
15. Cappozzo A. Analysis of the linear displacement of the head and trunk during walking at different speeds. J Biomech. 1981;14(6):411–25. https://doi.org/10.1016/0021-9290(81)90059-2.
16. Belluscio V, Bergamini E, Tramontano M, et al. Does curved walking sharpen the assessment of gait disorders? An instrumented approach based on wearable inertial sensors. Sensors (Basel). 2020;20(18):5244. https://doi.org/10.3390/s20185244.
17. Bloem BR, Marinus J, Almeida Q, et al. Measurement instruments to assess posture, gait, and balance in Parkinson's disease: critique and recommendations. Mov Disord. 2016;31(9):1342–55. https://doi.org/10.1002/mds.26572.
18. Carpinella I, Gervasoni E, Anastasi D, et al. Instrumentally assessed gait quality is more relevant than gait endurance and velocity to explain patient-reported walking ability in early-stage multiple sclerosis. Eur J Neurol. 2021;28(7):2259–68. https://doi.org/10.1111/ene.14866.
19. Nardone A, Schieppati M. The role of instrumental assessment of balance in clinical decision making. Eur J Phys Rehabil Med. 2010;46(2):221–37.

20. Belluscio V, Bergamini E, Iosa M, Tramontano M, Morone G, Vannozzi G. The iFST: an instrumented version of the Fukuda stepping test for balance assessment. Gait Posture. 2018;60:203–8. https://doi.org/10.1016/j.gaitpost.2017.12.010.
21. Iosa M, de Sanctis M, Summa A, et al. Usefulness of magnetoinertial wearable devices in neurorehabilitation of children with cerebral palsy. Appl Bionics Biomech. 2018;2018:5405680. https://doi.org/10.1155/2018/5405680.
22. Vienne-Jumeau A, Quijoux F, Vidal PP, Ricard D. Wearable inertial sensors provide reliable biomarkers of disease severity in multiple sclerosis: a systematic review and meta-analysis. Ann Phys Rehabil Med. 2020;63(2):138–47. https://doi.org/10.1016/j.rehab.2019.07.004.
23. Angelini L, Buckley E, Bonci T, Radford A, Sharrack B, Paling D, Nair KPS, Mazza C. A multifactorial model of multiple sclerosis gait and its changes across different disability levels. IEEE Trans Biomed Eng. 2021;11:3196–204. https://doi.org/10.1109/TBME.2021.3061998.
24. Celik Y, Stuart S, Woo WL, Godfrey A. Gait analysis in neurological populations: progression in the use of wearables. Med Eng Phys. 2021;87:9–29. https://doi.org/10.1016/j.medengphy.2020.11.005.
25. Mason R, Pearson LT, Barry G, Young F, Lennon O, Godfrey A, Stuart S. Wearables for running gait analysis: a systematic review. Sports Med. 2022;53(1):241–68. https://doi.org/10.1007/s40279-022-01760-6.
26. Leardini A, Chiari L, Della Croce U, et al. Human movement analysis using stereophotogrammetry. Part 3. Soft tissue artifact assessment and compensation. Gait Posture. 2005;21(2):212–25. https://doi.org/10.1016/j.gaitpost.2004.05.002.
27. Rum L, Sten O, Vendrame E, et al. Wearable sensors in sports for persons with disability: a systematic review. Sensors (Basel). 2021;21(5):1858. Published 2021 Mar 7. https://doi.org/10.3390/s21051858.
28. Kavanagh JJ, Menz HB. Accelerometry: a technique for quantifying movement patterns during walking. Gait Posture. 2008;28(1):1–15. https://doi.org/10.1016/j.gaitpost.2007.10.010.
29. Zago M, Sforza C, Pacifici I, et al. Gait evaluation using inertial measurement units in subjects with Parkinson's disease. J Electromyogr Kinesiol. 2018;42:44–8. https://doi.org/10.1016/j.jelekin.2018.06.009.
30. Shany T, Redmond SJ, Marschollek M, et al. Assessing fall risk using wearable sensors: a practical discussion. A review of the practicalities and challenges associated with the use of wearable sensors for quantification of fall risk in older people. Z Gerontol Geriatr. 2012;45(8):694–706. https://doi.org/10.1007/s00391-012-0407-2.
31. Picerno P, Iosa M, D'Souza C, Benedetti MG, Paolucci S, Morone G. Wearable inertial sensors for human movement analysis: a five-year update. Expert Rev Med Devices. 2021;18(sup1):79–94. https://doi.org/10.1080/17434440.2021.1988849.
32. Iosa M, Picerno P, Paolucci S, Morone G. Wearable inertial sensors for human movement analysis. Expert Rev Med Devices. 2016;13(7):641–59. https://doi.org/10.1080/17434440.2016.1198694.
33. Lowry KA, Smiley-Oyen AL, Carrel AJ, et al. Walking stability using harmonic ratios in Parkinson's disease. Mov Disord. 2009;24(2):261–7. https://doi.org/10.1002/mds.22352.
34. Angelini L, Carpinella I, Cattaneo D, et al. Is a wearable sensor-based characterisation of gait robust enough to overcome differences between measurement protocols? A multi-centric pragmatic study in patients with multiple sclerosis. Sensors (Basel). 2019;20(1):79. https://doi.org/10.3390/s20010079.
35. Witchel HJ, Oberndorfer C, Needham R, et al. Thigh-derived inertial sensor metrics to assess the sit-to-stand and stand-to-sit transitions in the timed up and go (TUG) task for quantifying mobility impairment in multiple sclerosis. Front Neurol. 2018;9:684. https://doi.org/10.3389/fneur.2018.00684.
36. Carpinella I, Gervasoni E, Anastasi D, et al. Instrumental assessment of stair ascent in people with multiple sclerosis, stroke, and Parkinson's disease: a wearable-sensor-based approach. IEEE Trans Neural Syst Rehabil Eng. 2018;26(12):2324–32. https://doi.org/10.1109/TNSRE.2018.2881324.

37. Celik R, Toruner EK. The effect of technology-based programmes on changing health behaviours of adolescents: systematic review. Compr Child Adolesc Nurs. 2020;43(2):92–110. https://doi.org/10.1080/24694193.2019.1599083.
38. Iosa M, Fusco A, Morone G, et al. Assessment of upper-body dynamic stability during walking in patients with subacute stroke. J Rehabil Res Dev. 2012;49(3):439–50. https://doi.org/10.1682/jrrd.2011.03.0057.
39. Iosa M, Bini F, Marinozzi F, et al. Stability and harmony of gait in patients with subacute stroke. J Med Biol Eng. 2016;36(5):635–43. https://doi.org/10.1007/s40846-016-0178-0.
40. Yang S, Zhang JT, Novak AC, et al. Estimation of spatio-temporal parameters for post-stroke hemiparetic gait using inertial sensors. Gait Posture. 2013;37(3):354–8. https://doi.org/10.1016/j.gaitpost.2012.07.032.
41. Gera G, Chesnutt J, Mancini M, et al. Inertial sensor-based assessment of central sensory integration for balance after mild traumatic brain injury [published correction appears in Mil Med. 2019 1;184(1–2):56]. Mil Med. 2018;183(suppl_1):327–32. https://doi.org/10.1093/milmed/usx162.
42. Newman MA, Hirsch MA, Peindl RD, et al. Use of an instrumented dual-task timed up and go test in children with traumatic brain injury. Gait Posture. 2020;76:193–7. https://doi.org/10.1016/j.gaitpost.2019.12.001.
43. Belluscio V, Bergamini E, Tramontano M, et al. Gait quality assessment in survivors from severe traumatic brain injury: an instrumented approach based on inertial sensors. Sensors (Basel). 2019;19(23):5315. https://doi.org/10.3390/s19235315.
44. Tramontano M, Russo V, Spitoni GF, et al. Efficacy of vestibular rehabilitation in patients with neurologic disorders: a systematic review. Arch Phys Med Rehabil. 2021;102(7):1379–89. https://doi.org/10.1016/j.apmr.2020.11.017.
45. Yang K, Xiong WX, Liu FT, et al. Objective and quantitative assessment of motor function in Parkinson's disease-from the perspective of practical applications. Ann Transl Med. 2016;4(5):90. https://doi.org/10.21037/atm.2016.03.09.
46. Rezvanian S, Lockhart TE. Towards real-time detection of freezing of gait using wavelet transform on wireless accelerometer data. Sensors (Basel). 2016;16(4):475. https://doi.org/10.3390/s16040475.
47. Weiss A, Herman T, Giladi N, et al. Objective assessment of fall risk in Parkinson's disease using a body-fixed sensor worn for 3 days. PLoS One. 2014;9(5):e96675. https://doi.org/10.1371/journal.pone.0096675.
48. Buckley C, Galna B, Rochester L, Mazzà C. Upper body accelerations as a biomarker of gait impairment in the early stages of Parkinson's disease. Gait Posture. 2019;71:289–95. https://doi.org/10.1016/j.gaitpost.2018.06.166.
49. Widener GL, Conley N, Whiteford S, et al. Changes in standing stability with balance-based torso-weighting with cerebellar ataxia: a pilot study. Physiother Res Int. 2020;25(1):e1814. https://doi.org/10.1002/pri.1814.
50. Hickey A, Gunn E, Alcock L, et al. Validity of a wearable accelerometer to quantify gait in spinocerebellar ataxia type 6. Physiol Meas. 2016;37(11):N105–17. https://doi.org/10.1088/0967-3334/37/11/N105.
51. Shirai S, Yabe I, Takahashi-Iwata I, et al. The responsiveness of triaxial accelerometer measurement of gait ataxia is higher than that of the scale for the assessment and rating of ataxia in the early stages of spinocerebellar degeneration. Cerebellum. 2019;18(4):721–30. https://doi.org/10.1007/s12311-019-01025-5.
52. Ilg W, Seemann J, Giese M, et al. Real-life gait assessment in degenerative cerebellar ataxia: toward ecologically valid biomarkers. Neurology. 2020;95(9):e1199–210. https://doi.org/10.1212/WNL.0000000000010176.
53. Carcreff L, Paraschiv-Ionescu A, Gerber CN, et al. A personalized approach to improve walking detection in real-life settings: application to children with cerebral palsy. Sensors (Basel). 2019;19(23):5316. https://doi.org/10.3390/s19235316.

54. Choi S, Shin YB, Kim SY, et al. A novel sensor-based assessment of lower limb spasticity in children with cerebral palsy. J Neuroeng Rehabil. 2018;15(1):45. https://doi.org/10.1186/s12984-018-0388-5.
55. Contini BG, Bergamini E, Alvini M, et al. A wearable gait analysis protocol to support the choice of the appropriate ankle-foot orthosis: a comparative assessment in children with cerebral palsy. Clin Biomech (Bristol, Avon). 2019;70:177–85. https://doi.org/10.1016/j.clinbiomech.2019.08.009.
56. Summa A, Vannozzi G, Bergamini E, et al. Multilevel upper body movement control during gait in children with cerebral palsy. PLoS One. 2016;11(3):e0151792. https://doi.org/10.1371/journal.pone.0151792.
57. Tramontano M, Belluscio V, Bergamini E, Allevi G, De Angelis S, Verdecchia G, Formisano R, Vannozzi G, Buzzi MG. Vestibular rehabilitation improves gait quality and activities of daily living in people with severe traumatic brain injury: a randomized clinical trial. Sensors (Basel). 2022;22(21):8553. https://doi.org/10.3390/s22218553.
58. Tramontano M, Morone G, Curcio A, et al. Maintaining gait stability during dual walking task: effects of age and neurological disorders. Eur J Phys Rehabil Med. 2017;53(1):7–13. https://doi.org/10.23736/S1973-9087.16.04203-9.
59. Tramontano M, Bergamini E, Iosa M, et al. Vestibular rehabilitation training in patients with subacute stroke: a preliminary randomized controlled trial. NeuroRehabilitation. 2018a;43(2):247–54. https://doi.org/10.3233/NRE-182427.
60. Lavikainen J, Vartiainen P, Stenroth L, Karjalainen PA. Open-source software library for real-time inertial measurement unit data-based inverse kinematics using OpenSim. PeerJ. 2023;11:e15097. https://doi.org/10.7717/peerj.15097. PMID: 37038471; PMCID: PMC10082569.

# Functional Neuroimaging and Rehabilitation

6

Elisa Tatti and Alberto Cacciola

## 6.1 Introduction

Neurorehabilitation is a multidisciplinary approach aimed at improving the functional outcomes and quality of life of individuals with neurological disorders, promoting recovery of function, preventing disability, and enhancing participation in daily activities [1]. The recovery of sensorimotor functioning after brain injury or disease is dependent on the activation of adaptive neuroplastic mechanisms. Neuroplasticity, the capacity of the nervous system to constantly modify its structure and function in response to changing conditions, is the building block of our capacity to learn and plays a key role in the development of new motor skills and the recovery of impaired ones [1–3]. Although neuroplasticity contributes to motor recovery after brain injury, research has also highlighted how neuroplastic changes may turn in a maladaptive direction after an injury or following intensive training. Indeed, studies have linked "maladaptive plasticity" to dystonia, phantom limb syndrome, and reduced functional recovery [4]. As such, the role of neurorehabilitation is to design tailored interventions aiming at both promoting adaptive plastic changes that support full recovery and assessing and targeting compensatory maladaptive plasticity that could hinder the effects of rehabilitation. For this purpose, neuroimaging techniques have become increasingly important tools in the field of neurorehabilitation [5–7] and have the potential to identify neural biomarkers of motor functioning and recovery, providing fundamental insights to guide and predict treatment response. Although studying the human brain in vivo and non-invasively is not

E. Tatti
Department of Molecular, Cellular and Biomedical Sciences, City University of New York (CUNY), School of Medicine, New York, NY, USA

A. Cacciola (✉)
Brain Mapping Lab, Department of Biomedical, Dental Sciences and Morphological and Functional Imaging, University of Messina, Messina, Italy


R. S. Calabrò (ed.), *Translational Neurorehabilitation*,
https://doi.org/10.1007/978-3-031-63604-2_6

trivial, the last years have been characterized by paradigmatic advances in brain imaging techniques which allow the investigation of structural and functional neural changes at different spatial and temporal resolutions [8, 9]. Functional imaging is a relatively novel field including different techniques that permit the disclosure of which areas of the brain are preferentially active during the rest condition or a particular task. However, functional neuroimaging has gained popularity in the field of neurorehabilitation as a tool for understanding brain plasticity and monitoring the effectiveness of rehabilitation interventions.

In this chapter, we will discuss three commonly used functional neuroimaging techniques: functional magnetic resonance imaging (fMRI), electroencephalography (EEG), and magnetoencephalography (MEG).

fMRI is a non-invasive imaging technique that measures changes in the cerebral blood flow [10]. When a specific area of the brain becomes more active, it requires more oxygen and nutrients, which leads to an increase in blood flow that can be detected by fMRI [11, 12]. During an fMRI scan, the participant lies inside the MRI scanner and is instructed to perform a specific task or to rest quietly while the scan is being performed. fMRI is a widely used technique in neuroscience research due to its relatively high spatial resolution. However, fMRI has a relatively poor temporal resolution, which means it cannot track changes in brain activity as they occur in real time [10], thus limiting its usefulness for studying rapid cognitive processes, including decision-making and language processing.

On the other hand, recent technological advances in the recording and analyses of electromagnetic signals have led to a renewed interest in the use of EEG and MEG for neurorehabilitation. Indeed, as highlighted by [13], only in the last decade, the number of scientific contributions on the use of EEG to assess, predict and guide therapeutic interventions has increased by 14.3%, thus suggesting how electromagnetic techniques are just beginning to unveil their potential in the field of rehabilitation.

Indeed, neuroimaging techniques, such as fMRI, EEG, and MEG, have revolutionized the field of neuroscience by allowing researchers to study human brain function in vivo and non-invasively. Together, these imaging modalities have provided invaluable insights into the structure and function of the human brain and continue to drive advances in our understanding of cognition, motor control, and diseases. In this chapter, we will explore new concepts of network organization underpinning brain function and provide a brief but comprehensive overview of the key functional neuroimaging techniques. We will focus on fMRI, EEG, and MEG because they are currently the most relevant and widely used in the fields of neuroscience and neurorehabilitation. We will examine their application in stroke and Parkinson's disease (PD) neurorehabilitation, and the identification of biomarkers to guide rehabilitation interventions. Finally, we will provide promising insights into the development of brain-computer interfaces (BCIs) to recover and/or compensate for sensorimotor dysfunction.

## 6.2 Theories of Brain Functions and Their Importance in Neurorehabilitation

Theories of brain function have evolved, reflecting advances in scientific knowledge, and changing philosophical perspectives. One of the key questions in the study of the brain is the extent to which different functions are localized to specific regions of the cortex, and what happens when a discrete lesion occurs in one of these regions. This section will explore three different theories of brain function—localizationist, holistic, and associationist—and consider their implications for understanding the consequences of brain diseases and neurorehabilitation strategies.

Localizationist theories of brain function suggest that specific regions of the brain are responsible for particular functions. This view has a long history, dating back at least to the work of phrenologists in the early nineteenth century, who believed that different mental faculties were associated with different bumps on the skull. The modern version of localizationism is more sophisticated and is based on the idea that different areas of the brain are specialized for different types of information processing. For example, Broca's area in the left hemisphere of the brain is thought to be responsible for language production, while Wernicke's area is responsible for language comprehension [14, 15]. The localizationist theory reached its acme, along with the beginning of its fall, in 1934 when Kleist provided an extensive cartography of brain functions as derived from brain lesion studies [16]. Hence, one of the main implications of localizationist theories is that a discrete cortical lesion can result in a specific functional deficit. This has been demonstrated in numerous studies of patients with brain damage, such as those who have suffered a stroke or traumatic brain injury. For example, a lesion in Broca's area can result in a condition known as Broca's aphasia, in which the patient has difficulty producing speech but can still understand language [14, 15]. Similarly, damage to the visual cortex can result in blindness in specific areas of the visual field.

Holistic theories of brain function, in contrast, suggest that the brain is a highly interconnected system in which all regions work together to produce behavior. This view emphasizes the importance of the whole brain, rather than specific regions, in understanding complex brain functions. Holistic theories have been influenced by the work of neuroscientists such as Karl Lashley, who conducted experiments in the 1940s and 1950s attempting to locate specific memory traces in the cortex by lesioning different areas of the brain [17]. Lashley found that even after extensive lesions to various parts of the cortex, rats were still able to perform certain tasks, suggesting that memories are distributed across the brain rather than localized to specific regions. The implication of holistic theories for the consequences of cortical lesions is that deficits may be more widespread and difficult to predict than in localizationist theories. For example, a lesion in one area of the cortex may disrupt the functioning of multiple regions that are interconnected with it, leading to more generalized deficits in cognitive function. This is consistent with the observation that patients with brain damage often have a range of cognitive and behavioral problems, rather than a single specific deficit.

On the other hand, associationist theories suggest that the brain is organized in parallel distributed networks around cortical epicenters and cognitive processes are the result of the interaction between different regions of the brain, whereas primary sensory and motor functions are localized. This view emphasizes the importance of both localized and distributed processing in understanding cognitive function. The associationist theory has been originally elaborated by Meynert [18], reformulated by Geschwind's neoassociationist school [19, 20], and recently supported by functional neuroimaging studies [21].

The implication of associationist theories for the consequences of cortical lesions is that the nature and extent of deficits may depend on the specific pattern of connectivity between different regions of the brain; therefore, a lesion in one region of the brain may disrupt the functioning of a network of interconnected regions, leading to deficits that are specific to that network. This is consistent with the observation that different types of brain damage can result in similar deficits, depending on the patterns of connectivity between regions.

Theories of brain function have evolved, reflecting advances in scientific knowledge, and changing neuroanatomical and pathophysiological perspectives. Localizationist, holistic, and associationist theories all offer different insights into how the brain works and how it responds to cortical lesions. Each theory has its implications for understanding the consequences of cortical lesions, with localizationist theories suggesting specific deficits, holistic theories suggesting more widespread deficits, and associationist theories suggesting that neurological deficits may depend on patterns of connectivity between regions.

As such, neurorehabilitation strategies aim to capitalize on brain plasticity by providing targeted and intensive treatments that are designed to lead to brain network reorganization. Indeed, clinical and neuroimaging research has shown that different rehabilitation programs can be effective in improving functional outcomes for individuals with brain injuries or neurological conditions [5, 6, 22, 23]. In conclusion, understanding the main theories of brain function is of major relevance in neurorehabilitation to guide the clinical assessment of patients with "dysconnectivity" syndromes and design effective rehabilitation programs tailored to the individual needs of patients.

## 6.3 Principles of fMRI

fMRI is a relatively novel neuroimaging technique that evaluates the fluctuation of the blood-oxygenation-level-dependent (BOLD) signal thus estimating the cerebral blood oxygenation changes in different brain areas. Figure 6.1 shows axial sections of morphological (T1-weighted) and functional images. The basic principle on which fMRI relies is that neuronal activity demands high levels of oxygen that cause a reduction of deoxygenated blood (paramagnetic) thus inducing magnetic properties changes in that same brain region. Such fluctuations of oxygenated (diamagnetic) and deoxygenated (paramagnetic) blood are therefore quantified by the

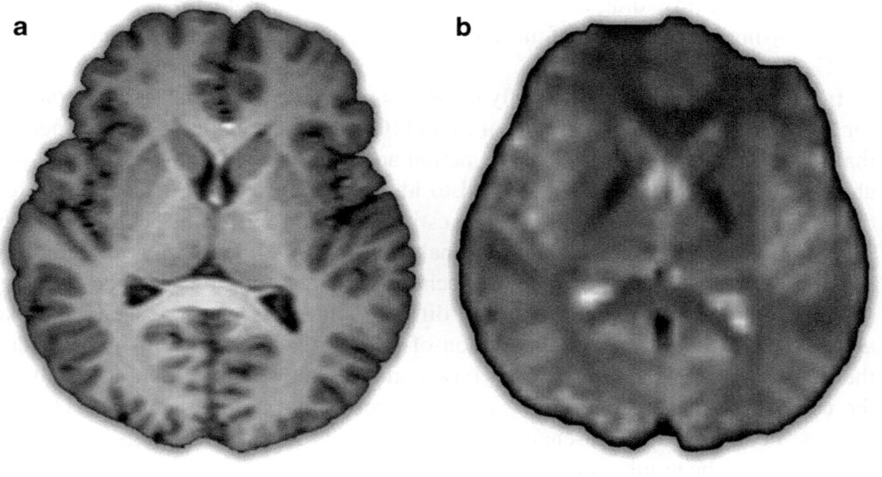


**Fig. 6.1** Axial sections of T1-weighted and fMRI images of the same subject show different spatial resolutions. (**a**) T1-weighted images provide a high spatial resolution permitting distinguishing the different brain structures and identifying the gray-matter/white matter interface. (**b**) The spatial resolution of fMRI is limited primarily by the signal-to-noise ratio due to the need for rapid acquisition of time series information. Typically, fMRI pixel size is 3–4 mm, although high-field magnet scanners (7 T) allow reaching a pixel size of 500 μm

MRI scanner—both in resting state (rs-fMRI) and during different tasks—and considered as a hemodynamic index of local brain activity [24, 25].

As for all neuroimaging techniques, fMRI data requires preprocessing and analysis before meaningful conclusions can be drawn. Here, we will provide a more detailed overview of some common fMRI preprocessing and analysis steps. As first steps, motion and slice-timing correction are usually performed. The former corrects for any head motion during the scan that can result in image artifacts or distortions. Such a step is therefore crucial as head motion can introduce noise and confounds in the fMRI data, which can in turn reduce the accuracy of estimated neural activity. Slice timing correction instead corrects for the small differences in acquisition time between slices in a multi-slice fMRI image. However, no consensus has been reached on how and if performing it [26]. Spatial and temporal filtering are further preprocessing steps needed in fMRI analysis. Spatial filtering or smoothing is the process of blurring the fMRI data, i.e., by averaging the values of one voxel with its neighbors, to improve the signal-to-noise ratio and reduce the variance. The temporal filtering, instead, reduces the temporal noise removing frequencies within the raw signal that are not of interest or are due to the noise of the scanner drifts as well as to that of the cardiac and respiratory cycles. Furthermore, the fMRI data need to be normalized to a standardized template to compare activity across individuals and studies. Spatial normalization allows researchers to investigate how neural activity is related to specific brain regions or networks, and it can help to identify brain regions that are consistently activated across individuals or populations. As providing an extensive and technically detailed overview of fMRI preprocessing is not the aim of this chapter, the readers can find further information in [27, 28].

In addition, we will herein briefly present some of the most common approaches for fMRI analysis. The general linear model (GLM) is a statistical analysis method that models the relationship between neural activity and experimental design variables [29, 30]. The GLM can be used to identify brain regions that are activated during a specific task or condition and to estimate the strength of the neural response. This method is widely used in fMRI analysis and is well-suited for investigating specific hypotheses about the neural mechanisms underlying for instance recovery or the effects of neurorehabilitation in different clinical scenarios, such as stroke and movement disorders patients. Region-of-interest (ROI) analysis involves instead the selection of specific brain regions based on prior knowledge or hypotheses about the neural mechanisms underlying a specific rehabilitation intervention [31]. This approach can increase the sensitivity and specificity of fMRI findings, and it can help to reduce the number of comparisons and statistical tests required to investigate brain activity. On the other hand, independent component analysis (ICA) is a data-driven method that identifies networks of brain regions that are co-activated during a specific task or condition. ICA can identify functionally relevant networks even when there is no a priori knowledge about the underlying neural mechanisms [32]. This method can be particularly useful for investigating the neural mechanisms underlying neurorehabilitation as it can identify complex and dynamic patterns of neural activity. Finally, connectivity analysis investigates the functional connectivity between brain regions at rest or during a specific task or condition [33]. Connectivity analysis can provide insights into the brain network-level changes that occur during neurorehabilitation and can help to identify specific brain regions, connectivity patterns, or systems that are affected by a specific rehabilitation intervention. Figure 6.2 shows a simplified workflow for fMRI data analysis.

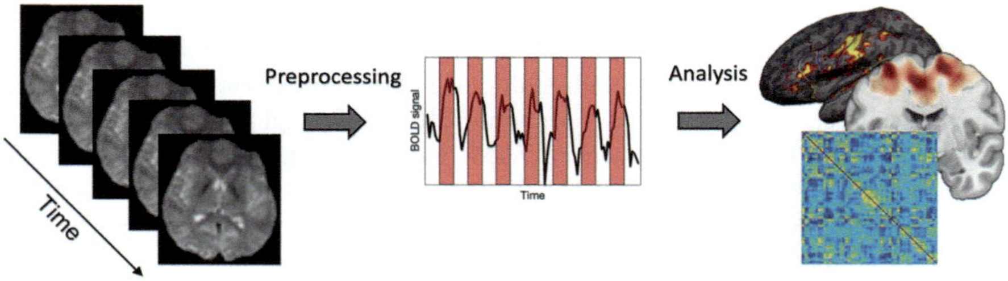


**Fig. 6.2** Typical fMRI workflow in brief. Researchers and clinicians need to understand the physical bases of fMRI and its intrinsic hemodynamic characteristics in order to properly design and perform neuroimaging studies in the neurorehabilitation field. Appropriate data acquisition is the first essential step. The acquired data then undergoes several quality control and preprocessing steps. The intended analysis methods should be implemented for resting-state or task-based fMRI and statistical inferences performed. Finally, results interpretation and translation into clinical practice should be made with extreme caution

## 6.4 The Role of fMRI in Neurorehabilitation

As stated above, fMRI can provide information on brain regions activation in resting conditions or during specific tasks or stimuli. This evidence can be used to investigate the neural mechanisms underlying neurological disorders and their recovery, as well as to assess the effectiveness of rehabilitation interventions. One of the key advantages of fMRI in neurorehabilitation is its ability to provide objective and quantitative measures of brain function. Traditional clinical assessments rely on subjective ratings and are often limited in their ability to capture subtle changes in brain function. fMRI can provide more sensitive measures of brain function and can help to identify changes in brain activity that may not be observable with traditional clinical assessments. In addition, it can also offer important information on the neural substrates of motor and cognitive functions that can be used to design targeted rehabilitation interventions tailored to patients' individual needs. In this regard, fMRI can be used to identify and target regions of the brain that are most affected by a given pathology with specific rehabilitation interventions, thus improving patients' functional outcomes. Another important use of fMRI in neurorehabilitation is in the development of neuroprosthetic devices. Indeed, fMRI can identify the neural activity patterns associated with specific motor or cognitive functions, which can be used to develop neuroprosthetic devices that can mimic or enhance these functions.

To summarize, fMRI-guided interventions have several potential benefits over traditional rehabilitation interventions. First, they can be tailored to the specific needs of the individual patient based on their unique patterns of brain activation. Second, they can be adapted in real time based on changes in brain activity that are detected during the intervention. Third, they can be employed for targeting specific brain structures or networks that are involved in the patient's deficits. Fourth, they can be used to monitor changes in brain activity that occur because of the intervention, which can provide valuable feedback to the patient and the therapist. Finally, fMRI-guided interventions represent the basis for future brain-computer interface devices aimed at enhancing brain plasticity and motor and non-motor function recovery.

## 6.5 Mapping Human Brain Function and Recovery with fMRI in Stroke

Stroke is a leading cause of long-term disability worldwide. The major stroke-related neuroanatomical and neurophysiological plasticity changes include functional alterations to the spared regions adjacent and remote to the lesioned area, as well as rerouting of intracortical pathways to novel territories [34]. These changes can be enhanced by the experience [35] and are linked to functional recovery. However, transposing animal experiments to humans has limitations, especially when exploring brain regions and functions that are not present in animals. Functional imaging methods have been employed to investigate the cerebral structures responsible for spontaneous recovery in humans following stroke, with changes observed specifically for motor recovery. These changes include differences in the site of activation foci relative to healthy controls and over time, changes in laterality of activation (typically from contralesional to ipsilesional), involvement of distributed sensorimotor networks, activation in the peri-infarct region [36], and increase of cortical tissue thickness in the vicinity of activation [37]. These changes have been documented in the acute and chronic phases of recovery and over time [38], and neuroplastic changes have been identified in relation to different functions, including movement [39], sensation [40], language [41], and cognition [42]. Evidence suggests that training-related experience and rehabilitation can impact recovery. Clinical improvement accompanied by cortical reorganization, even in chronic stages of recovery, is influenced by factors such as the type of intervention and lesion site (cortical vs subcortical) [43]. Proof-of-principle studies in humans indicate that somatosensory input, cortical stimulation, and interhemispheric interactions can modulate cortical plasticity in the post-stroke neurorehabilitation [44]. Furthermore, the potential for specific motor interventions to facilitate adaptive changes in the brain is expanding. To optimize recovery, it is essential to advance our understanding of how the neurobiology, systems physiology, and functional neuroanatomy of the brain are affected by injury and the recovery process, and how they may be manipulated to facilitate improved function.

The brain functions as a cohesive unit with distributed functional systems governing the behavior [45]. Motor learning, for example, involves various circuits beyond the motor cortex, which are determined by the level of skill and task-solving strategy [46, 47]. Simply knowing the location of a lesion is insufficient for predicting the functional impact of the injury or the potential for recovery and response to therapy. The arterial lesion pattern in stroke is distinct from the organization of functional brain systems, and indirect lesion effects such as cognitive impairments, depression, and altered processing speed and attention can also impact the action coding [48]. Therefore, it is essential to enhance our comprehension of how the brain's functional circuits adapt during recovery, both following the initial injury and under varying recovery conditions. Recovery of sensorimotor function after stroke likely involves dynamic bilateral interactions between connected gray matter, white matter tracts, and learning-dependent neuroplastic changes [44].

Recent studies have revealed the importance of functional systems and the "balance" of activation within these systems during the recovery process. For instance, changes in activation patterns have been observed in both hemispheres during motor recovery [49, 50], even in structurally intact perilesional brain tissue and remote brain regions [37, 51]. The postlesional reorganization has been shown to involve a network typically not active in sensorimotor activity, including the thalamus, visual association cortex, and bilateral cerebellum [52]. In addition, recent coherence analysis has demonstrated a shift of functional connectivity towards the contralesional side, with reduced corticocortical connections in the lesioned hemisphere but increased connections in the contralesional hemisphere [53], which is more likely to occur within the first week post-stroke [54, 55].

Changes in the balance of cortical excitability have also been observed in both the lesioned and non-lesioned hemisphere [56, 57], with persistent alterations in intracortical and interhemispheric interactions, despite complete functional recovery [50].

Ongoing debates exist regarding the functional significance of activity changes that occur following a stroke [58]. Some argue that the increase in contralesional activity could support neural processing of the lesioned hemisphere, while others suggest that it may result from transcallosal disinhibition, which could negatively affect neural processing in the lesioned hemisphere (Fig. 6.3). Computational models of connectivity, such as dynamic causal modeling (DCM), have clarified the role of a particular area in the entire network [59]. Applying DCM to fMRI data collected while stroke patients performed movements with their stroke-affected hand, it was found that the contralesional primary motor cortex (M1) exhibited an inhibitory influence on the activity of the ipsilesional M1. The strength of this inhibitory coupling correlated with the degree of motor impairment, with more impaired patients showing stronger inhibition [60]. These findings suggest a maladaptive role of contralesional M1 2–3 months post-stroke. However, longitudinal data revealed that in the first few

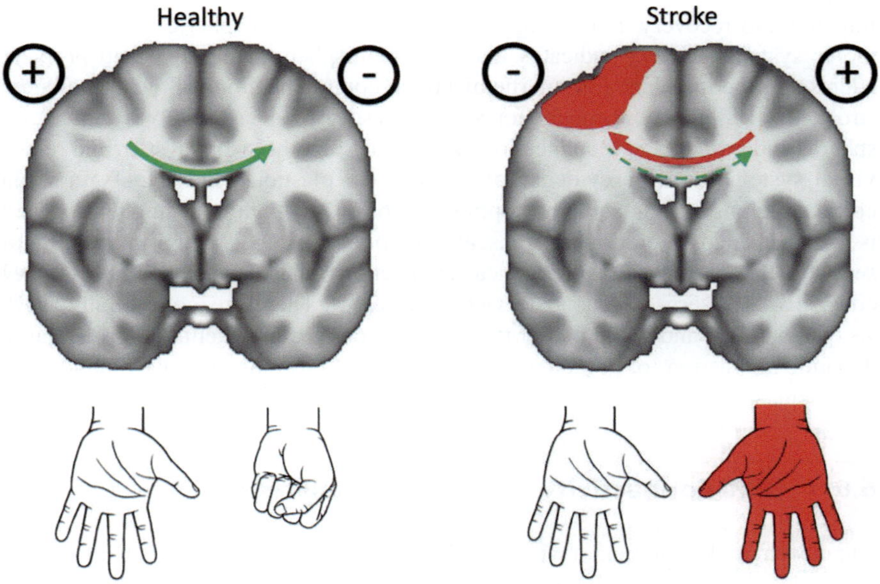


**Fig. 6.3** Maladaptive plasticity occurs in stroke patients. Interhemispheric inhibition Unilateral upper limb motor tasks in healthy subjects lead to the activation of the contralateral hemisphere which in turn inhibits the activation of the ipsilateral hemisphere. However, when the contralateral motor cortex is impacted by stroke (red areas), such an interhemispheric inhibition is lost (dashed green line), thus making the contralesional more activated. In addition, evidence suggests that the enhanced contralesional activation (thick red line) may further contribute to the interhemispheric imbalance by leading to increased inhibition of the ipsilesional hemisphere. The red hand indicates the paretic limb

days after a stroke, the contralesional M1 exerts a positive influence on the ipsilesional M1. Correlating DCM coupling changes over time with motor outcome revealed that patients with less favorable outcomes developed an inhibitory coupling, while higher increases in coupling from ipsilesional premotor areas to ipsilesional M1 were associated with good motor outcomes [54, 55]. The reinstatement of a network configuration lateralized to the ipsilesional hemisphere, similar to that observed in healthy subjects, is linked to good motor outcomes. Nevertheless, the contralesional hemisphere may also exert a supportive influence on well-recovered patients. For example, the contralesional dorsolateral premotor cortex has been shown to exert a positive influence on ipsilesional M1 activity in patients 1.5 years post-stroke the more demanding a motor task gets for the paretic hand [61]. This finding lacks in healthy subjects, suggesting that the dorsolateral premotor cortex may contribute to motor recovery after stroke. The analysis of fMRI activity and connectivity indicates that a focal ischemic lesion affects brain areas not only in the vicinity but also at remote locations in both hemispheres. Recovery from stroke-induced motor impairments is closely related to network changes, with contralesional areas also influencing the motor performance of the paretic hand.

Recently proposed models of recovery provide a framework for understanding the mechanisms underlying recovery and possible strategies that may influence function and recovery. For example, a model of interhemispheric interactions for motor system recovery indicates that both hemispheres are active and proposes addressing the imbalance in interhemispheric communication following unilateral stroke [62]. Similarly, a model for somatosensory recovery involves both hemispheres and proposes new connections within the somatosensory system and related visual system based on evidence of cross-modal plasticity and activity in visual cortical regions during tactile perception [52, 63]. Drawing on this model, recovery is likely linked to alterations in transcallosal connections between primary somatosensory (S1) regions, thalamocortical connections, and the development of new connectivity patterns to secondary visual areas [64]. Therefore, connectivity analysis of functional and white matter tract connections is fundamental in investigating dynamic interactions of new circuits and potentially devising effective rehabilitative treatments (see subsequent paragraphs).

## 6.6 fMRI Connectivity in Stroke Rehabilitation

Our current understanding of why some patients recover more fully than others after a brain lesion is limited. This is due to the unique expression of brain lesions, which may not align with functional units, and the fact that the functional deficit resulting from a brain lesion may not mirror the activation in healthy individuals performing the same task. Brain pathology can offer partial insight into brain reorganization following recovery, particularly as cortical brain lesions often only affect a portion of the functional representation. Interestingly, research has shown that individual finger movements are more closely associated with the proportion of surviving sensorimotor cortex in healthy individuals, based on fMRI activation maps, than with total infarct volume in patients with damage to the sensorimotor cortex. This suggests that incorporating a measure of functional map injury may better predict and characterize neurological deficit [65]. Indeed, specific sites and patterns of activation have been linked to the severity of impairment and recovery, and correlated activity in these areas may have a predictive role in recovery and tailored targeted interventions. In this regard, the use of fMRI in the early stages of stroke allowed for the prediction of changes in motor impairment within the following 3 months [66, 67]. A specific brain activation pattern was found to be a strong predictor of Fugl-Meyer score changes, independent of the initial stroke severity and lesion volume. Although the analysis method used did not allow for anatomical conclusions, it was clear that the function of the brain's response to injury provided information about future clinical progression beyond the injury's anatomy. Furthermore, upper limb function at 3 months was shown to correlate more strongly with intracortical excitability within the primary motor cortex, measured with transcranial magnetic stimulation (TMS), than with the structural integrity of the corticospinal tract. Such findings support the notion that motor improvement is not solely linked to corticospinal tract integrity but also to other characteristics of the motor cortex [68]. In addition, it was found a correlation between the improvement of motor performance induced by TMS and task-related activity in the ipsilesional primary motor cortex (M1) as measured by fMRI at the baseline [69]. Although this may appear to be a straightforward outcome, the utilization of residual functional and structural anatomy for such classification is seldom considered, even though it has the potential to enhance the trial design.

In general, fMRI studies performed on lower limb rehabilitation showed increased bilateral activation of cortical areas as well as functional recovery of the unaffected side along with improvement in locomotor function [70–73]. Two studies specifically emphasized the crucial role of certain brain structures, such as the thalamus [73], cerebellum, and midbrain [74], in the recovery of gait function. In fact, greater activation of these areas was associated with a faster walking speed [73, 74] and endurance [73].

In contrast, fMRI changes in response to upper limb rehabilitation are highly variable. Some studies have reported increased activation of sensorimotor areas on the ipsilesional sensorimotor cortical areas [75–79] while others have shown increased activation on the contralesional side [80–83] or both hemispheres [84]. Additionally, reduced activation of the ipsilesional hemisphere was demonstrated following rehabilitation and associated with clinical improvement [85–88].

Interpreting these heterogeneous and seemingly contradictory results is challenging, as many factors need to be considered, including the patient's level of disability, the type of rehabilitation approach used, the extent of tissue damage, and the involvement of cortical and subcortical regions. The type of rehabilitation approach adopted in the studies plays a crucial role in determining the functional activation and potential reorganization of brain networks. For example, constraint-induced movement therapy has been associated with the activation of sensorimotor areas and the cerebellum, while bilateral limb training is likely to facilitate the activation of bilateral areas. Action-observation therapy has been shown to increase mirror neuron system activation, and passive proprioceptive training triggers the activation of secondary somatosensory areas. On the other hand, rehabilitation strategies employing virtual reality, robotics, or brain-computer interfaces often induce the activation of motor regions but also of visuomotor or associative areas.

The use of rehabilitative treatments that involve challenging cognitive tasks can result in significant clinical improvement and activation of various functional brain areas compared to other approaches, as supported by several studies [77, 79, 89]. These treatments, which may involve visuomotor loop tasks, have been found to promote neuroplastic changes by engaging motor and other areas related to executive and visuospatial functions [38, 72, 90]. Some studies have also investigated the relationship between functional connectivity and mental practice, which involves cognitive rehearsal of movements or activities [85, 91, 92]. While it has been suggested that mental practice can promote brain reorganization and enhance the effects of physical therapy [91, 92], a recent Cochrane review found no evidence of efficacy for this kind of rehabilitation [93].

Another important unresolved question pertains to whether the improvements in clinical and functional changes achieved during rehabilitation persist over time or diminish in the subsequent months. This is a critical issue as it determines whether the skills and brain network reorganization acquired during rehabilitation are merely a response to the intensive physical activity or solid acquisitions that patients can apply in real-world settings. To the best of our knowledge, only a few studies have included longer follow-ups, and their results have been inconsistent. Indeed, some have reported clinical functional improvements paralleled by a more focused activation of the ipsilesional motor cortex, whereas others have observed maintenance of the gains immediately after training. However, partial or complete loss of the functional changes achieved through rehabilitation has been also reported [82, 94].

While most of the abovementioned studies have employed static connectivity approaches, the human brain is a dynamic system fluctuating at the time scale of milliseconds [95]. For this reason, many time-resolved approaches have been devised to explore the so-called *dynamic* functional connectivity (dFC) (see [96] for a review), outperforming *static* methods, notably to study cognition and neuropsychiatric neurological diseases [96, 97].

Several studies have also employed time-tailored methods to explore the neural correlates of stroke [98, 99] evidence in favor of a stroke-related dynamic reconfiguration of functional brain networks, despite their intrinsic limitations, including the relatively small sample sizes, a priori selection of regions of interest and time-windowed estimates [96]. Notwithstanding, the innovation-driven coactivation patterns (iCAP) framework [97, 100] has been shown to be able to overcome these limitations and assess whether spatial and temporal properties of large-scale brain networks following stroke correlate with anatomical damage and behavioral recovery. The iCAP method relies on a data-driven approach that employs single frames without a priori selection of regions of interest, along with the potential of identifying large-scale brain networks that can be spatially and temporally overlapping. While large-scale spatial patterns of brain networks were preserved after stroke, their durations were altered, with stroke subjects exhibiting a longer duration in the lateral precentral gyrus and anterior cingulum, and a shorter duration in the occipital lobe and the cerebellum, compared to healthy controls. In addition to showing significant correlations with recovery from attention, language, and spatial memory deficits, these functional network durations were associated with white matter damage in projection and association pathways.

This is of particular importance if we consider that white matter connections, when intact, form the anatomical scaffold that mediates the dynamic fluctuations between networks [8, 9, 101]. Disruptions in these connections may impact the functional brain organization, not only in the proximity of the lesion, but also between distant brain regions, thus altering the dynamics of network transitions and consequently their temporal characteristics. Therefore, further investigation of structural-functional connections and their predictive value is needed to better understand the dynamic interactions at a systems level and guide the selection of optimal rehabilitation strategies for individual patients. One of the issues concerning the existing studies is that structural and functional MRI data are generally processed separately, in general by employing the results of voxel-wise statistical analysis of fMRI as seeds for tractography or by correlating functional with structural white matter information, making the results indirect and bound to the specific experimental hypothesis. Integrating functional and structural information to investigate the function of white matter pathways thus remains a rare endeavor. In this regard, Nozais et al. have recently developed the Functionnectome, a method that can directly project resting-state and task-related fMRI on the white matter, thus enabling the evaluation of brain regions' interaction rather than their independent contribution during brain processes [9]. Furthermore, track-weighted dynamic functional connectivity (tw-dFC) is a recently developed technique that allows for a joint analysis of structural and functional connectivity data, by mapping time-windowed functional connectivity, sampled from resting-state functional MRI, back on the underlying white matter anatomy, reconstructed by tractography [101]. In this regard, independent component analysis (ICA) applied to the tw-dFC time series was shown to be suitable to identify highly reliable and biologically meaningful functional units within the human white matter of healthy subjects [8]: this feature makes it useful to investigate white matter structures in an entirely data-driven fashion, based on the fluctuations of functional connectivity at their endpoints located at the gray-white matter interface.

Therefore, dynamic functional connectivity approaches and integrating the functional signal and white matter pathways' anatomy may play a fundamental role in informing the therapeutic process in stroke patients. They may indeed identify post-stroke pathophysiology changes informing whether the rehabilitation-mediated changes are adaptive (mimicking the functional network organization of the healthy brain) or maladaptive. Furthermore, they may be employed for patients' outcome prediction and tailored rehabilitative interventions.

Overall, recent research provides further evidence of the potential of fMRI for investigating the neural mechanisms underlying stroke recovery and monitoring the effectiveness of rehabilitation interventions. The results suggest that rehabilitation interventions may promote neural plasticity, functional connectivity, and white matter reorganization, which may contribute to improved functional outcomes in stroke patients. However, further research is needed to determine the optimal rehabilitation interventions and to identify the patient characteristics that are most likely to benefit from these interventions.

## 6.7 fMRI in Parkinson's Disease Rehabilitation

Pharmacological and surgical interventions have been broadly utilized in treating PD, particularly for motor symptoms [102]. However, these methods become less effective over time and the use of antiparkinsonian drugs can cause various side effects leading to non-motor symptoms [103]. Therefore, there is a pressing need to identify non-pharmacological strategies that can reduce the overall burden of PD. Rehabilitation has emerged as a promising treatment for PD, as it has been shown to reduce functional disabilities and enhance patients' quality of life [104]. Physical exercise can mitigate motor symptoms, improve gait and balance skills, and ultimately result in reduced falling and increased mobility [105]. Additionally, PD animal models studies have shown that motor exercise can have a neuroprotective effect by inducing antioxidant activity, reducing inflammation, and promoting synaptogenesis [106, 107]. Hence, due to its effectiveness in addressing both motor and non-motor symptoms, physical exercise should be considered the foundation of any neuro-rehabilitative approach for PD [104]. A variety of physical exercise approaches have been employed in PD rehabilitation, including aerobic exercise, which aims to improve cardio-pulmonary functionality and cerebral perfusion, and goal-based exercises that instead target core domains, such as manual dexterity, balance, posture, and gait control [108]. Notably, both kinds of physical exercises involve cognitive and automatic components of motor control, which are known to be affected in PD [106]. Indeed, another key aspect of PD rehabilitation is cognitive engagement, which can help patients develop strategies to compensate for their motor impairment [108]. In the last decade, increasing evidence has suggested the role of exercise, physical activity, and environmental enrichment in improving not only motor performance per se but also both cognitive and automatic functions [109]. Furthermore, basic research reports suggest that exercise induces benefits on brain health by acting on blood flow, trophic factors, and the immune system to create the optimal milieu required for enhancing neuroplasticity [109].

Hence, it is essential to comprehend the neural basis responsible for the modifications caused by various exercise approaches in PD for effective and targeted rehabilitation strategies. A recent review summarized the outcomes of motor learning in PD, which revealed that movement automaticity in PD relies on sustained recruitment of attentional and sensorimotor networks as a compensatory mechanism, as observed through fMRI and EEG [110].

Advanced MRI techniques have significantly expanded our understanding of the pathophysiological mechanisms underlying the evolution of PD [111, 112], as well as monitoring the efficacy of pharmacological and non-pharmacological treatments [113]. As stated in previous paragraphs, MRI offers a variety of non-invasive and complementary techniques that can measure structural changes and functional connectivity. Functional changes have been consistently reported within the cortical-striatal pathway across resting-state fMRI studies in PD patients, with reduced connectivity in the posterior putamen being the most frequent finding [114]. Additionally, aberrant cerebellar connectivity, observed both in terms of increased connectivity [115] within the cerebellum and disrupted connectivity between the cerebellum and sensorimotor network [116], has been interpreted as a compensatory phenomenon.

A meta-analysis of motor task fMRI studies reported both reduced activation of the right posterior putamen and increased activation of the superior parietal lobule in PD patients OFF medication, while decreased recruitment of the M1 was observed during the ON medication state [117]. Moreover, the same study showed that levodopa administration partially restored the activity within the putamen during motor performances. Noteworthy, fMRI studies specifically investigating the effect of pharmacological treatment showed that levodopa partially restored connectivity within the cortico-striatal network and normalized the cerebellar hyperactivation [118]. Additionally, some studies have described a compensatory increase in fronto-parietal network activation to overcome gait and motor difficulties in PD [119, 120]. Finally, pathological aspects such as aberrant cerebral perfusion have been detected in various cortices in PD patients, with functional abnormalities potentially preceding overt structural modification.

Post-rehabilitation neuroplastic changes tend to compensate for altered connectivity of the cortico-striatal circuitry, primarily associated with motor symptoms, by enhancing activation of the same structures or recruiting compensatory areas. Exercise-induced neuroplasticity in PD was previously suggested to be associated with specific movement characteristics such as exercise complexity, frequency, specificity, and intensity [104, 106]. Interestingly, the efficacy of high-intensity aerobic exercise is likely to be based on restoration of function within the cortico-thalamic-striatal pathway, which represents a major target in PD and is also associated with the development of motor symptoms.

Furthermore, some functional mechanisms induced by aerobic exercise are the same as those induced by the administration of antiparkinsonian drugs that target dopaminergic pathways [121, 122]. These results not only support the efficacy of rehabilitation in relieving the motor symptoms of PD but also support the idea that neuroplastic changes involve the regulation and transmission of dopamine [106, 108, 122]. Moreover, a recent ex vivo study of autopsy hippocampal tissue showed that neurogenesis did not decrease over time, whereas angiogenesis was significantly lower in the elderly, in parallel with a concomitant decrease in neuroplasticity.

These results highlight the potential benefits of aerobic exercise in promoting not only functional connectivity but also vascular function, ultimately leading to increased neuroplasticity [123].

Goal-directed rehabilitative interventions serve as support for enhancing sensorimotor integration and connectivity in the premotor cortex, which is usually affected in the early stages of disease [124].

Finally, cognitively challenging exercises primarily affected frontoparietal networks, resulting in changes in functional connectivity associated with clinical and cognitive improvements, and positive effects lasting longer than controls [125]. Rehabilitation interventions that enhanced cognitively challenging tasks have been applied to other pathological entities with comparable results, with higher task complexity resulting in greater and more effective brain network reorganization [38, 79].

Interestingly, improved clinical outcomes are associated with increased [125] or decreased [126] activity in frontoparietal networks involved in cognitive and attentional processes.

This apparent discrepancy can be interpreted considering the concepts of positive adaptive and maladaptive brain plasticity, occurring in response to any type of stimulus, including pathological processes, that aim to adapt or compensate for new conditions. As there is only a very thin line between the adaptive and maladaptive processes, the clinical outcomes are the only factors that can define whether a functional reorganization is adaptive or maladaptive. Given that the frontoparietal network is involved in motor learning, in the above studies, increased activation may be related to compensatory mechanisms [125], and decreased activation may be interpreted as a restoration of automatism in motor control [126].

However, the small sample size of the studies, along with pronounced data heterogeneity, reduces the reproducibility and generalizability of results, preventing the identification of reliable MRI-derived biomarkers. Indeed, several studies have enrolled patients without specifying the disease phenotype, such as tremor dominant or akinetic-rigid, even though they involve distinct pathophysiological mechanisms. This heterogeneity may result in different treatment responses, prognosis, as well as distinct patterns of functional brain changes [127, 128]. Additionally, PD patients can be categorized based on the presence of freezing of gait (FoG), a complex symptomatology with unknown pathophysiology that is likely to be linked to various brain regions, including cortical and subcortical areas [125, 129]. Patients with FoG exhibit elevated activity in pre-frontal regions, mesencephalic locomotor region, and cerebellar locomotor region, likely due to impaired automatism of motor control. This issue translates into cognitive dysfunction and postural instability, among the other symptoms [129]. This altered functional connectivity pattern related to FoG may, in part, explain the varying response to cognitively enriched rehabilitation. In patients with FoG, this rehabilitation approach strengthens the connections between regions involved in motor learning and goal-directed processing, resulting in increased functional connectivity of the frontoparietal network, along with improved clinical outcomes [125]. On the other hand, patients without FoG showed rehabilitation-mediated decreased functional connectivity of the frontoparietal network, likely due to increased neuronal efficiency [126].

Overall, neurorehabilitation has positive effects on patients with PD from both clinical and MRI perspectives. Improved clinical outcomes are usually associated with brain reorganization at the functional level, such as restoration of the cortico-striatal pathway and activation of compensatory patterns such as the frontoparietal networks.

## 6.8 Challenges and Limitations of fMRI in Neurorehabilitation

Despite showing great potential in neurorehabilitation, fMRI also suffers from several drawbacks. One challenge is the issue of generalization. While fMRI can provide insights into the neural mechanisms underlying specific rehabilitation interventions, it may not always generalize to real-world functional outcomes. Indeed, improvements in brain activation patterns may not necessarily translate into improvements in activities of daily living. Therefore, combining fMRI with functional assessments is crucial to ensure that brain function improvements translate into better functional outcomes.

Another challenge is the issue of variability. Stroke patients, for instance, vary widely in terms of the location and extent of their brain lesions, which can affect the response to rehabilitation interventions. It is essential to consider the medication status of patients while undergoing MRI (as well as EEG and MEG), as pharmacological treatments can impact functional activity and represent a confounding factor. Most of the current studies exhibit heterogeneity in the assessment of both functional activity and clinical outcomes and do not provide details about the medication status of patients during the neuroimaging assessment or treatment. It is therefore fundamental to design rigorous inclusion criteria in terms of patient characteristics, such as disease phenotypes and clinical features, as these factors may have a significant impact on the functional brain response to rehabilitation. Additionally, there are variations in the clinical scales used to assess disability, rehabilitation duration, and frequency. At this stage, drawing definitive conclusions from fMRI studies that are based on small sample sizes or heterogeneous patient populations is not trivial and free from strains. It should be also considered and kept in mind that fMRI requires the patient to lie still for an extended period, which can be challenging for some patients.

Another limitation of fMRI is the issue of task-related confounds. Although fMRI studies use motor or cognitive tasks to investigate changes in brain activation patterns, these tasks may not accurately reflect real-world functional demands and may introduce task-related confounds, such as fatigue or attentional demands, that can affect the results.

Furthermore, fMRI data processing and analysis can be complex and time-consuming, and several errors or inconsistencies can occur. Therefore, ensuring rigorous and standardized methods for data preprocessing and analysis is of major relevance and still represents one of the hottest topics in the field of neuroimaging [130].

Additionally, while showing improved spatial resolution compared to other neuroimaging modalities (i.e., EEG), thus providing information about the location and extent of brain activation patterns at the macro- and meso-scale, fMRI cannot provide micro-scale information such as the underlying cellular or molecular changes occurring during neurorehabilitation.

Finally, cost and accessibility limit the use of fMRI in neurorehabilitation, as it is a relatively expensive technique that requires specialized equipment and trained personnel, which may constrain its availability in some clinical settings.

## 6.9 Principles of EEG/MEG

Differently from fMRI and PET, EEG and MEG directly measure the brain's activity, respectively, by detecting small changes in the electrical and magnetic fields with millisecond accuracy, a temporal resolution crucial to finely investigate the quick functional dynamic changes that occur in the brain.

EEG is recorded by placing electrodes directly on the scalp while MEG uses sensors called SQUIDs (superconducting quantum interference devices) that detect the

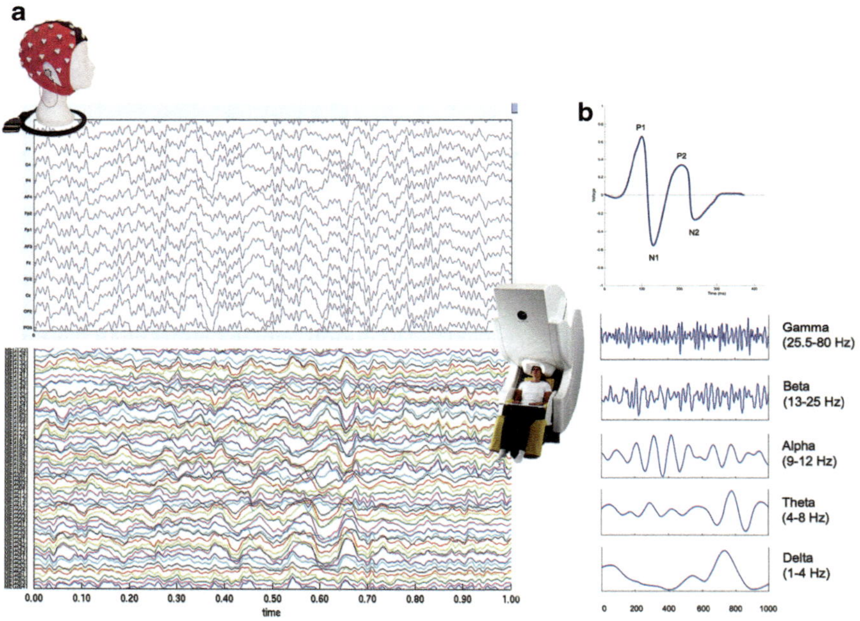


**Fig. 6.4** Principles of EEG/MEG. (**a**) Raw EEG (upper section) and MEG (bottom section) signals in sensor space. (**b**) Schematic representation of the event-related potential (ERP) components generated in the brain structures in response to specific events or stimuli (upper part). Components with positive (P100-P1 and P200-P2) and negative peaks (N100-N1 and N200-N2) are displayed. The term P100 indicates that this component is a positive going wave with a latency of ~100 ms after the stimulus onset, whereas the term P1 indicates that it is the first positive peak. The bottom part of the panel shows the main patterns of brain oscillatory activity, ranging from lower frequencies, such as delta (0.5–4 Hz) and theta waves (4–8 Hz), to higher frequencies, such as alpha (8–13 Hz), beta (13–30 Hz), and gamma waves (>30 Hz)

magnetic fields generated by the neuronal electrical activity (Fig. 6.4a). The primary source of both EEG and MEG signals is the synchronous activity of large populations of pyramidal neurons, which are perpendicularly oriented to the scalp and have cell bodies predominantly located in layers three and five of the cerebral cortex. The recorded signal represents the electromagnetic changes occurring in the extracellular space as a by-product of the sum of excitatory and inhibitory postsynaptic potentials (i.e., graded changes in the potential of a postsynaptic neuron caused by the binding of the neurotransmitter to its receptor), thus, EEG/MEG signal represents cyclic changes in the excitability of local neuronal assemblies [131]. These potential changes create electrical fields that are typically modeled as pairs of electric charges with opposite polarities, called *dipoles* (i.e., fields with positive—*source*- and negative—*sinks*- poles), and are typically characterized by their strength, location, and orientation (radial, tangential, oblique). Although the sources of the electromagnetic signal can be estimated from dipoles using mathematical models for both EEG and MEG, these two techniques present some differences in terms of spatial accuracy. Indeed, even with high-density EEG systems with large numbers of electrodes (>180), the spatial accuracy of EEG is limited by the distortion and attenuation of the signal through the brain, skull, and skin. While MEG is subject to the same signal attenuation of EEG with distance, the tissues and bone do not distort the magnetic field, thus resulting in greater spatial resolution (2–3 mm compared to 7–10 mm for EEG) [132].

Despite being more affected by spatial smearing than MEG, EEG has some advantages that make it ideal for clinical and rehabilitation settings; specifically, EEG's cost-effectiveness, portability, and variety of applications (e.g., BCI, Neurofeedback, clinical diagnosis) make it an inexpensive and versatile technique to investigate cognitive, sensorial, and motor processes, perform diagnosis and assess the efficacy of therapeutic interventions.

Although EEG and MEG have their strength and weaknesses, they both allow recording the dynamic changes in the electromagnetic activity that occur within and across sensorimotor regions with exceptional time resolution. As such, they constitute precious tools for characterizing different types of brain processes as they unfold and unveil their role in sensorimotor functioning in healthy and pathological conditions. However, once recorded, the raw EEG/MEG signal appears as a continuous series of complex waves, thus requiring further processing to extract valuable information. Among the various information that can be extracted from EEG/MEG recordings at rest or during tasks, an important distinction consists in the differentiation between evoked and induced cortical oscillations [133, 134]. Evoked activity is reliably time-locked and phase-locked to an internal change (e.g., a thought) or external stimulus (e.g., light flash) and is typically measured across many trials through Event-Related Potentials, in the case of EEG, or Evoked-Magnetic Fields (EMFs), in the case of MEG measurements. Conversely, induced brain oscillatory activity is not phase-locked to a stimulus and displays trial-to-trial differences in its latency, thus it is considered to reflect the ongoing brain activity that is gradually modulated by cognitive, sensory, or motor processes [134]. Induced oscillations are characterized by different frequencies, and their amplitude and phase are modulated by the ongoing processing demands or brain state (e.g., alertness, deep sleep). Traditionally, evoked and induced brain activity are characterized either in the time domain, through Event-Related Potential (ERP) paradigms, or in the time-frequency domain with the study of brain rhythms (Fig. 6.4b).

## 6.10 Event-Related Potentials (ERP) and Event-Related Fields (ERF)

ERPs and ERFs, measured with EEG and MEG, respectively, represent small changes in the electric voltage of the brain in response to external stimuli (e.g., touch, light, sound) and are associated with sensory, motor, and cognitive processing [135, 136]. ERPs/ERFs are obtained by time-locking the ongoing electromagnetic signal to the stimuli and averaging several presentations of the same stimuli (i.e., trials) to obtain a visible waveform. This is necessary because, typically, ERPs/ERFs are 10–100 times smaller than the background EEG/MEG signal [137]. ERPs/ERFs are typically characterized according to their topography, direction (positive or negative), latency, and amplitude [138]. ERPs protocols have dominated the EEG/MEG research and clinical practice for over 50 years and are established tools for the evaluation of neurological and psychiatric disorders. Indeed, while the electromagnetic activity of the brain was first characterized in terms of ongoing oscillatory activity in the alpha and beta frequency ranges [139], the discovery of the evoked responses of the brain in the 60 s [140–146] opened the doors to a fruitful era of ERPs research, providing important advancements in our understanding of brain functioning. Although several ERP components have been identified over the years (Fig. 6.4b), we briefly mention the "Bereitschaftspotential" (BP) or readiness potential/field [141, 147] for its implication in motor functioning. The "Bereitschaftspotential" is a low frequency (0–5 Hz), negative-going potential consisting of two subcomponents; the early BP begins ~2 s before movement onset in the bilateral pre-Supplementary Motor Area (pre-SMA) and SMA regions and in the premotor cortex shortly after and is triggered by self-generated and cued voluntary movements. The late BP component originates ~400 ms before movement onset in the motor cortex contralateral to the moving limb, with a precise somatotopic representation [148]. The BP is thought to arise from the integration of sensory, cognitive, and motor signals in the brain, leading to the generation of an internal representation of the movement [148], with the early BP probably reflecting a general subconscious readiness signal and the late BP a conscious will to move. Importantly, the amplitude, slope, and/or latency of the BP is altered in some neurological disorders such as PD [149–152], in which the reduced or delayed BP has been interpreted as a failure in activating the pre-SMA region [153]. While ERPs provide important information on the timing and sequence of sensorimotor and cognitive processing on a macro level, they do not provide information about the underlying neural mechanisms that support sensorimotor and cognitive functioning. For this reason, in the last two decades, research has begun to gain interest in the rhythmic patterns of brain activity and their role in the orchestration of brain processes [131].

## 6.11 Brain Oscillations and Their Role in Sensorimotor Functioning

Neural oscillations, also known as brain waves, are rhythmic patterns of oscillatory activity that occur both spontaneously (resting oscillations) and when the brain is engaged in a task. These oscillations can range from lower frequencies, such as delta waves (0.5–4 Hz) and theta waves (4–8 Hz), to higher frequencies, such as alpha (8–13 Hz), beta (13–30 Hz), and gamma waves (>30 Hz) (Fig. 6.4b). While each of these brain rhythms has been associated with different states of brain functioning (e.g., sleep, arousal) and specific roles in cognition and sensorimotor functioning, oscillations in the mu (8–13 Hz), beta (14–25 Hz), and gamma (>30) frequency ranges consistently displayed their role in sensorimotor control [154–157]. Besides being recorded in resting conditions, these rhythms are also time-locked to movement onset and are characterized by a gradual decrease or increase in their amplitude during motor planning and execution, which are respectively known as event-related desynchronization (ERD) and event-related synchronization (ERS).

## 6.12 Mu Alpha and Beta Movement-Related Oscillatory Dynamics

Human and animal studies extensively demonstrated that self-initiated—internally or externally triggered—movements display an amplitude reduction (ERD) of both mu alpha (8–12 Hz) and beta (13–25 Hz) oscillations, beginning during the planning phase and reaching the maximum peak during movement execution [154, 158] (Fig. 6.5). Importantly, such ERD has been observed during motor imagery, passive movements, movement observation, and kinesthetic illusions, thus suggesting that amplitude reduction in mu and beta do not merely reflect muscular activation per se, and it has been interpreted as a general mechanism to release the sensorimotor network from its inhibited state [156, 159, 160]. Indeed, it is widely accepted that mu and beta ERD result from thalamocortical stimulation and reflect a state of increased cortical excitability in the sensorimotor network [161].

The Rolandic alpha rhythm (8–13 Hz), generally called mu rhythm to distinguish it from the occipital alpha rhythm, is the first frequency band described as related to motor control [162, 163]. Typically observed in resting conditions over the central sensorimotor regions, it desynchronizes (mu ERD) during the early stages of movement planning (~2 s before movement initiation) over the contralateral postcentral somatosensory area and spreads bilaterally during movement execution [158, 164]. Shortly after the beginning of mu ERD, the sensorimotor regions also display a gradual decrease in beta oscillatory activity that starts ~500 ms before movement

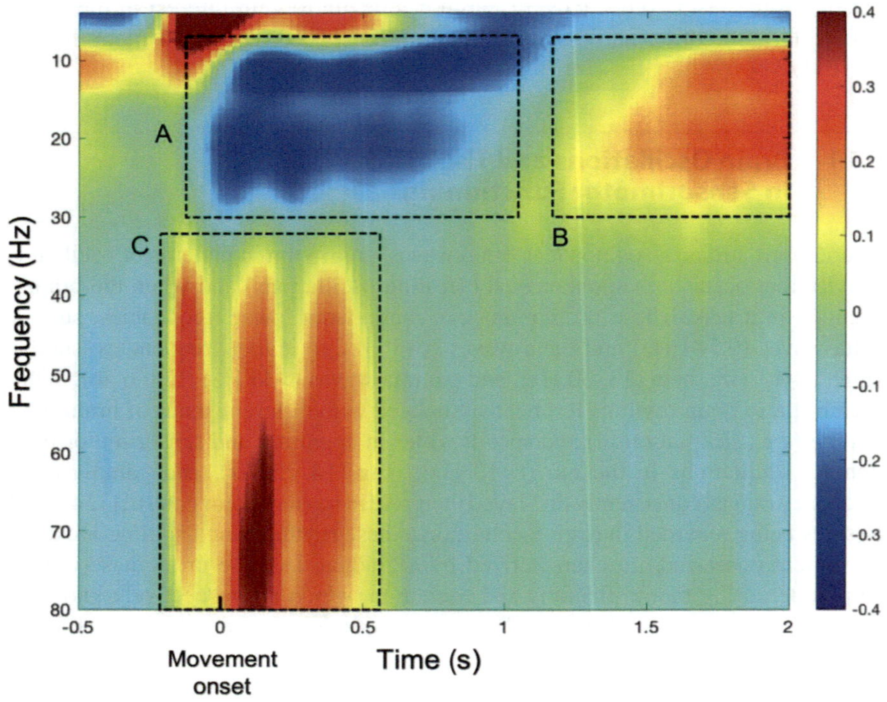


**Fig. 6.5** (**a**) Characteristic mu and beta event-related desynchronization (ERD) and event-related synchronization (ERS) in the beta (**b**) and gamma (**c**) frequency range in a healthy subject during a cued reaching movement with the right arm. Movement onset is at $t = 0$. The color bar represents the amplitude change compared to the average spectral power over the entire time window (dimensionless) (Adapted from Tatti et al. 2022) Creative Commons Attribution BY 4.0

initiation and has a contralateral dominance [165, 166]. As both mu and beta rhythms are predominant within the sensorimotor network at rest and during static hold posture, they have been generally interpreted as states of cortical "idling" or the maintenance of the *status quo* in the sensorimotor network [167]. As such, the mu/beta ERD during movement ideation and execution might reflect the release of the sensorimotor network from inhibition to allow movement [156, 159, 160]. Nonetheless, the fact that mu and beta display different movement-related dynamics and neural generators (mu over the postcentral somatosensory cortex and beta over the precentral motor cortex) indicates that they signal more than an idling state and serve distinct and more complex physiological functions. Although their functional role is still widely debated, there is an increasing agreement on mu ERD being an aggregation of different processes involved in transforming "perception into action" [168, 169] whereas beta ERD is more directly related to motor planning processes. Nonetheless, despite the beta ERD lasts as long as the effector is moving, research has shown no clear association between beta ERD magnitude and several parameters such as movement direction, length, force, and velocity speed [164, 170–175], type of praxis movement [176], and the number of activated muscles [177].

Furthermore, unlike mu, beta oscillations do not gradually return to baseline amplitude values once the movement is completed but rather display a sharp rebound (event-related synchronization, ERS) [158, 159] (Fig. 6.5b). This transient rebound, generally peaking ~300 ms after movement termination, is visible over the bilateral sensorimotor cortices with a stronger contralateral dominance [159, 177, 178]. Although the observed rebound is associated with a state of neuronal inhibition linked to the activity of the Gamma-Aminobutyric Acid (GABA) neurotransmitter [161, 179], compelling evidence in favor of a more comprehensive function of beta oscillatory dynamics is emerging. Set aside the view of beta ERD/ERS dynamic as the simple release (ERD) and subsequent return (ERS) to an idling state of the sensorimotor network [159, 180], beta oscillations may regulate the fine balance between promoting change and stability in neural networks. This wider functional role is supported by evidence of beta ERD/ERS modulation beyond motor tasks but also during working memory [181, 182] and attentional tasks [183, 184]. This suggests that the beta ERD/ERS dynamic might be a mechanism shared between motor and cognitive functions, where beta ERD would release cortical inhibition, thus enabling movement execution or cognitive flow, and the subsequent ERS might signal the need to maintain the current motor or cognitive set [167, 185, 186]. Indeed, research has shown that beta ERS magnitude is modulated by the degree of confidence in the activated forward model (i.e., the transformation of a set of motor commands into a prediction of the outcome of the action) [187–189]. Furthermore, movement accuracy is thought to be encoded by dopamine-mediated synaptic gain processes in the pyramidal neurons (i.e., plastic changes in the responsiveness of a neuron to an input) [190–192], thus beta oscillations might have a direct mechanistic effect on neural activity.

In line with this view, some authors recently proposed that the beta ERD/ERS dynamic might represent a biomarker of the capacity of a neural network to activate plasticity mechanisms [193, 194]. This suggestion comes from two sets of evidence based on non-invasive brain stimulation protocols (e.g., transcranial magnetic stimulation, TMS) and motor learning paradigms. First, intermittent Theta Burst Stimulation in humans, a TMS protocol that increases corticospinal excitability by triggering Long-Term Plasticity processes, results in increased beta ERS amplitude over the primary motor cortex [195]. Additionally, the entrainment of resting beta oscillations with transcranial alternating current stimulation (tACS) decreased the amplitude of motor-evoked potentials and enhanced cortical inhibition [196, 197]. Second, intensive motor learning protocols progressively increased beta ERD-ERS peak-to-peak amplitude (i.e., movement-related beta modulation) [174, 198–202] which also predicted the degree of skill retention on the following day [199]. These findings, together with the demonstrated link between beta power changes and dopamine, GABA, and lactate levels [203–214], as well as to measures of learning and plasticity [187, 195, 199], suggests that beta oscillations, in particular movement-related beta ERS, might represent a functional index of energy availability to engage use-dependent plasticity [193]. Although additional evidence is needed to confirm this hypothesis, evidence of abnormal beta oscillatory activity in neurological and psychiatric conditions characterized by sensorimotor impairment confirms the importance of mu and beta oscillations as reliable biomarkers of sensorimotor functioning.

## 6.13 Movement-Related Gamma ERS

Along the modulation of the mu alpha and beta-band activity, a prominent synchronization of gamma oscillations (>30 Hz) is also observed in cortical and subcortical regions in a wide range of movements, including simple [215–219], complex [220, 221], and imagined movements [222, 223]. Extensively investigated in vision, memory, and attention research, gamma oscillatory activity is generally interpreted as a rhythm that allows the integration of information required to accomplish a task and, as such, might be a mechanism that allows the organization and integration of motor processes. Differently from the movement-related mu and beta dynamics, gamma oscillations display a strong relationship with the motor output, such as with movement extension, velocity, applied force level, and duration [157, 224–231]. While these findings support the prokinetic role of gamma, experimental evidence has demonstrated that the role of gamma oscillations goes beyond the control of the motor output. Indeed, gamma oscillations have been suggested to have a role in modulating synaptic plasticity through GABAergic activity. Memory paradigms in human and animal research demonstrated that the magnitude of gamma activity predicts learning [157, 232]. Furthermore, transcranial alternating current stimulation in the gamma frequency range (75 Hz) was negatively associated with GABA activity [233] and, in turn, with motor learning [234]. In summary, while their role in synaptic plasticity is actively explored in novel investigations, gamma oscillations are thought to play an important role in coordinating the activity of the different motor regions involved in the planning and execution of movements, possibly by providing a mechanism for the precise temporal coordination of neuronal firing [235, 236].

## 6.14 EEG/MEG Oscillatory Abnormalities in Stroke

Regardless of the ongoing debate on the functional meaning of brain oscillations for motor control, literature extensively documented alterations in mu, beta, and gamma oscillations in a range of neuropsychiatric conditions, including PD, stroke, dystonia, major depression, and schizophrenia [237].

By investigating such alterations, research can not only foster our understanding of the relationship between brain oscillations and sensorimotor functioning but also gain insight into the underlying mechanisms of the disease and develop new rehabilitative interventions. Stroke is the leading cause of long-term disability worldwide and motor impairment occurs in 80% of cases, among which approximately

30% result in a permanent disability [238]. During the first weeks after a stroke, the brain undergoes a significant, plasticity-mediated, functional reorganization of the affected networks which is reflected in alterations of the brain oscillatory activity at rest and during a motor task [239, 240]. Among the various biomarkers of unilateral post-stroke functional reorganization, the degree of interhemispheric asymmetry is an established index of sensorimotor impairment. Indeed, in healthy subjects, the activation of the contralateral sensorimotor cortex is associated with an active interhemispheric inhibition of the ipsilateral one to avoid its mirror activation [241]. In stroke patients, the degree of interhemispheric asymmetry is typically lower, thus suggesting that the ipsilesional motor cortex is unable to effectively inhibit the contralesional hemisphere. At the same time, research has shown that the unaffected regions actively downregulate the cortical excitability of the lesioned hemisphere, thus reinforcing the hypoactivation of the affected region [242]. Importantly, the magnitude of such ipsilesional/contralesional unbalance has been positively correlated with the resulting motor impairment [243]. Such functional reorganization is visible in the magnitude of the alpha and beta ERD [53, 244–251]. Specifically, stroke patients display reduced ERD in the ipsilesional hemisphere, and such reduction is negatively correlated with the severity of the acquired motor impairment [247, 252]. Furthermore, a finger-tapping study on chronic stroke patients also reported markedly reduced beta ERS [249]. Altogether, these findings suggest that the reduced alpha- and beta-band ERD during movement planning and execution might reflect the difficulty in activating the affected sensorimotor network [247]. Besides the possible interpretations of such changes in mu and beta ERD, research demonstrated a positive relationship between ERD magnitude, residual motor functionality, and the electromyographic activity in the paretic limb, and a negative relationship with the total volume loss [253]. Given the positive relationship between beta oscillations and GABA concentration, the observed reduced ERD might result from the increased inhibition of the ipsilesional hemisphere. In parallel to the reduced ERD functionality of the lesioned sensorimotor area, a common finding in stroke patients is the increased activity in the contralesional hemisphere [254]. Thus, evidence of increased beta ERD over the ipsilateral sensorimotor cortex might be linked to the loss of interhemispheric suppression through the transcallosal fibers, thus resulting in the hyperexcitability of the contralesional hemisphere [247]. Indeed, the improvement of the functionality of the affected limb obtained with the reduction of such hyperactivity, support the negative impact of the contralesional area for motor recovery. Yet, another line of evidence supports the idea that the increased ERD in the contralesional cortex might result from an attempt to compensate for the suffered damage [255], as reflected by evidence of a positive relationship between contralesional activity and motor recovery [256]. As new investigations keep exploring the functional role of the contralesional area in motor functioning after a stroke and how to address it with rehabilitation, alpha and beta ERD remain established reliable biomarkers in stroke patients and can help assess motor function and rehabilitation progress.

## 6.15 EEG/MEG Oscillatory Abnormalities in Parkinson's Disease

Changes in sensorimotor oscillatory activity have been extensively investigated in PD. Since the first description of PD as "shaking palsy" (Paralysis Agitans), due to the combination of resting tremors with bradykinesia and muscle rigidity, researchers identified PD as an "oscillopathy" (i.e., a disorder characterized by alterations in the brain oscillatory activity). Indeed, PD is associated with increased beta oscillatory activity at rest throughout the basal ganglia-thalamocortical circuits [257–262] and such elevated beta activity positively correlates with bradykinesia severity [263–265]. Importantly, treatments with levodopa and Deep Brain Stimulation (DBS) reduce beta oscillatory activity to functional levels and improve motor functioning, thus confirming the relationship between beta and the active inhibition of the sensorimotor network. Alterations of the movement-related ERD-ERS oscillatory dynamics are also prevalent in PD patients; compared to healthy controls, beta ERD is markedly reduced and delayed in PD [198, 266–274]. Furthermore, the post-movement beta ERS, which is linked with motor skill retention in PD [269] and, possibly, to early plasticity processes [193] is also reduced [268–271, 275]. Despite PD patients displaying a progressive increase in beta ERS with motor practice as healthy individuals, the rate of ERS amplitude increase is prominently smaller in PD patients [174, 200, 201, 269, 275–277] a phenomenon that was interpreted in terms of a reduced energetic availability to induce experience-dependent plasticity [193].

Along with changes in mu and beta oscillatory activity, PD patients are also characterized by a reduction in movement-related gamma ERS in the sensorimotor network, which contributes to motor symptoms such as rigidity, bradykinesia, and tremor. As with beta oscillations, levodopa medication and DBS restore functional levels (i.e., increase) of gamma oscillatory activity in the sensorimotor network, improving motor functionality [278, 279]. Although the study of subcortical gamma oscillations and cortical movement-related gamma oscillations is still in its early days, they have been successfully implemented in non-invasive brain stimulation protocols to facilitate corticospinal excitability and improve motor functioning [280], and are actively explored as a biomarker of levodopa-induced dyskinesia [281].

## 6.16 Brain Oscillations and Their Use for Neurorehabilitation

After a stroke, a traumatic brain injury, a spinal cord injury, or a neurodegenerative disorder the brain undergoes significant changes in its function and structure, some because of the damage inflicted to the brain, and others as adaptive compensation mechanisms to reduce the resulting functional impairment. As research highlighted the existence and relevance of such neuroplastic changes for sensorimotor rehabilitation, the traditional approach of teaching compensatory skills to offset the acquired impairment has been progressively replaced by neuroscience-based interventions aimed at targeting neural plasticity. Indeed, a key goal of neurorehabilitation is to create learning protocols (e.g., motor imagery) to promote experience-dependent neural plasticity and manage maladaptive plastic changes that may hinder functional recovery [282]. Recent advancements in neuroimaging research have led to a greater understanding of the neurophysiological changes that characterize a neurological disorder, the development of novel, more efficacious, neurorehabilitation treatments [1, 283], and the opportunity of prognosticating the recovery and monitoring the functional and structural changes occurring in the brain during a rehabilitation treatment. Both EEG and MEG proved to be effective tools to support neurorehabilitation [284, 285]. A recent systematic review and meta-analysis on the prognostic utility of EEG for post-stroke outcome established that, compared to standard clinical assessment, EEG improved prognostic abilities in over 95% of the retrieved studies ($N = 75$). Importantly, such investigations examined the effects of different types of interventions, including standard inpatient rehabilitation therapy, robot-assisted training, BCIs, and virtual reality (VR) [285]. For example, Calabrò and colleagues [286] explored the efficacy of combining robotic-assisted gait training with virtual reality in 24 patients with chronic ischemic stroke. Besides the observed increased efficacy of adding VR to the rehabilitation intervention, EEG recordings highlighted greater movement-related beta and gamma oscillatory activity after the VR-augmented robot-assisted gait training, which was positively correlated with clinical improvement and, for gamma waves only, in muscle strength. Similarly, a recent study on stroke survivors investigated the use of EEG connectivity (see paragraph below) and movement-related ERD-ERS dynamics to assess the efficacy of 2 weeks of upper-extremity therapy coupled with random-frequency vibratory stimulation [287]. Results showed that the intervention resulted in increased connectivity in the sensorimotor regions and reduced alpha ERD and increased ERS which was maintained at follow-up. While the interpretation of these changes remains speculative, these findings support the use of electrophysiological techniques to understand and assess the effects of rehabilitation on the brain.

## 6.17 EEG/MEG Connectivity in Neurorehabilitation

Besides ERPs and spectral EEG activity, research is gaining interest in the use of connectivity measures to explore how different brain regions interact with each other and how functional brain networks are organized to support sensorimotor functioning. Although several methods have been recently developed to quantify functional connectivity with EEG/MEG [288], two major classes of estimates are used for connectivity in both the time and frequency domains; non-directed functional connectivity measures (i.e., correlation and mutual information) analyze any interdependence between the signals, without any inference on the direction (i.e., causation) of such relationship. Conversely, the goal of directed estimates (i.e., cross-correlation, Granger causality, and transfer entropy) is to establish a statistical cause-and-effect relationship (see [289] for a tutorial review on functional connectivity methods). Independently from the chosen metrics, connectivity analyses are based on the concept that brain functions arise from the activity of specialized areas that are functionally integrated into a network. Different from standard ERP/ERF and spectral/time-frequency analyses, often limited to specified regions of interest or sources, connectivity analyses offer the possibility of understanding how the brain functionally reorganizes within and across its networks after an injury and in response to rehabilitation. Emerging evidence demonstrated the prognostic utility of analyzing the integrity of the sensorimotor network through connectivity measures as they offer a window into the plasticity potential of the affected brain regions [290, 291]. As the number of studies on connectivity in neurological disorders such as stroke and PD rapidly increases [291–294], investigations on the effects of rehabilitation on functional connectivity measures are also emerging [295–298]. For example, in Kawano et al. [297], the contralesional intrahemispheric Phase Synchronization Index (PSI) in the alpha and theta bands at rest was significantly higher in patients; further, theta PSI correlated with the impairment reduction after standard physical and occupational therapy. Importantly, connectivity measures can support neurorehabilitation not only indirectly by identifying sensorimotor networks and monitoring brain activity during and after rehabilitation intervention but also directly as real-time feedback indices for neurofeedback and brain-computer interfaces (BCIs) systems.

## 6.18 Using Brain Oscillations for Neuroprosthetics

Despite the good efficacy of neurorehabilitation treatments in the management of sensorimotor deficits, a complete functional recovery of sensorimotor functions is not achievable in certain conditions, such as a neurodegenerative disorder (e.g., amyotrophic lateral sclerosis) or severe brain damage (e.g., spinal cord injury, brainstem stroke). For these patients, alternative methods to compensate for such deficits and reduce the resulting disability are fundamental, and BCIs are increasingly becoming reliable, fast, and effective tools to provide autonomy in everyday life. In neurorehabilitation, BCIs are implemented together with robot-aided and/or motor imagery therapy to train the patients to activate the brain regions implicated in a task, thus engaging adaptive plasticity processes. For example, in a motor imagery-based BCI system, the patient is asked to imagine performing an action (e.g., grabbing a spoon). The implemented neuroimaging technique records and decodes the brain features elicited by the imagined action to control a robotic arm that executes it. This type of intervention aims to help the brain restore the activity of the affected network, thus recovering the impaired behavioral function. Additionally, BCIs can be used as simple assistive tools to support the patient and caregiver in reducing the burden of the acquired disability in daily life activities, by allowing them to move, communicate, and interact with the external world [299]. Among the various non-invasive neuroimaging techniques that could provide a window on the motor intentions of a patient, at present, only EEG combines the cost-effectiveness, temporal precision, and portability to be feasible for such rehabilitative and assistive applications [300]. Indeed, thanks to (1) the knowledge increase on the relationship between brain activity and motor control, and (2) the rapid technological improvement in terms of inexpensive computer hardware and software to compute EEG analyses in real time, EEG is increasingly implemented for the creation of BCIs for rehabilitative and assistive purposes. An EEG-based BCI is a non-invasive system that automatically extracts evoked and/or induced electrophysiological activity in real time and uses it to control an external device. Two types of BCI can be implemented, dependent and independent. A dependent BCI relies on peripheral activity (i.e., muscles) to obtain the brain signal to be decoded. One example could be the use of gaze direction to obtain a visual-evoked potential that is going to be used to control the external device. Conversely, an independent BCI is designed to provide direct communication between brain activity and the output device, thus it does not require muscular activity to be implemented [300]. Recently, hybrid BCI systems, which combine evoked and induced brain signals or brain and muscular outputs have also been implemented [301]. Based on the evidence of similar movement-related ERD-ERS dynamics during motor imagery, BCIs that use movement-related sensorimotor oscillations are also ideal for those neurological conditions characterized by severe motor disability or complete paralysis (e.g., Amyotrophic Lateral Sclerosis, tetraplegia, spinal cord injury) [302–304]. Importantly, although two electrodes over the primary motor cortices (C3, C4) are deemed sufficient to obtain a high classification accuracy of the intended action [305–307], with the recent technological advancements in the online signal analysis, multichannel EEG data recordings are increasingly used to increase signal quality and better characterize the different processes occurring during motor imagery [301, 308–310]. Several studies investigated the use of sensorimotor mu and beta rhythm to control devices for assistive purposes, such as a 2D or 3D cursor [309, 311–318], wheelchair (Palumbo et al. [319] for a systematic review on BCI-controlled wheelchair studies), and exoskeletons [320].

Besides controlling external devices, BCI systems are implemented to support neural plasticity by using neurofeedback protocols. In general, neurofeedback can be used as part of a comprehensive rehabilitation treatment plan and consist of providing real-time feedback on brain activity to help patients regulate their brain function and improve their cognitive, emotional, and behavioral abilities. Neurofeedback-based protocols aimed at rehabilitating motor functions typically involve: (a) regulating sensorimotor rhythms, (b) functional electrical stimulation (FES), whereby the patients learn to use their brain activity to regulate FES applied to the affected muscles, and (c) combining neurofeedback with mirror therapy [321–324]. Despite the proven effectiveness in improving motor learning [325], motor and cognitive symptoms, and neuroplasticity [324, 326, 327], more systematic investigations are needed to refine and personalize these protocols, as well as to address some challenges that limit the efficacy of BCI interventions in general. One example is the inability of some patients (about 20% according to [328]) to learn how to modulate their brain activity to control the device or perform a task [329]. Finally, fatigue is another important factor that should be considered and addressed to improve the efficacy of neurorehabilitation interventions [330]. Indeed, fatigue is a prevalent symptom in neurological disorders and, besides being a disabling symptom per se, it may also affect BCI performance [331].

## 6.19 Conclusions

This chapter aimed at providing a simple but comprehensive overview of the impact of functional neuroimaging in the neurorehabilitation setting, with particular attention to stroke and PD. The current evidence suggests that stroke patients show rehabilitation-induced clinical improvement accompanied by brain reorganization aimed at restoring interhemispheric balance and ipsilesional hemisphere activity, as well as activation of functionally related contralateral areas. As new investigations keep exploring the functional role of the contralesional area in motor functioning after stroke and how to address it with rehabilitation, alpha and beta ERD remain established reliable biomarkers in stroke patients and can help assess motor function and rehabilitation progress.

We have also provided evidence that motor rehabilitation has clinical benefits for PD patients, which are typically associated with functional brain reorganization supported by neuroplastic mechanisms. Although various physical treatments have been employed for PD patients, the small number of studies prevents firm conclusions from being drawn on each specific rehabilitation approach. However, it is possible to identify a global tendency of rehabilitation to facilitate the restoration of the abnormal oscillatory activity of the cortico-striatal pathway and sensorimotor network, as well as the activation of compensatory systems such as the frontoparietal network.

Furthermore, we highlighted the role of functional connectivity measures in supporting neurorehabilitation not only indirectly by identifying sensorimotor networks and monitoring brain activity during and after rehabilitation intervention but also directly as real-time feedback indices for neurofeedback and BCIs systems. In this regard, most of the current literature on BCIs aiming at improving motor function relies on techniques that measure electromagnetic variations (i.e., EEG and MEG), and only a few studies have explored the potential of fMRI-based neurofeedback in motor rehabilitation [332].

Despite these advances, robust imaging markers are still lacking due to the use of different variables, such as rehabilitation settings, type of treatment, and inclusion criteria. Future clinical trials on the effects of neuromotor rehabilitation may benefit from more homogeneous populations and study designs. Currently, there is still a significant gap in translating findings from large-scale studies to individual patients, and future research is needed to properly design and validate neuroimaging markers to capture brain changes in response to rehabilitation, providing qualitative and quantitative information that can be used to track treatment efficacy in the setting of neurorehabilitation.

## References


1. Maier M, Ballester BR, Verschure PFMJ. Principles of neurorehabilitation after stroke based on motor learning and brain plasticity mechanisms. Front Syst Neurosci. 2019;13:74.
2. Nahum M, Lee H, Merzenich MM. Principles of neuroplasticity-based rehabilitation. Prog Brain Res. 2013;207:141–71.
3. Kelly C, Foxe JJ, Garavan H. Patterns of normal human brain plasticity after practice and their implications for neurorehabilitation. Arch Phys Med Rehabil. 2006;87:S20–9.
4. Takeuchi N, Izumi S-I. Maladaptive plasticity for motor recovery after stroke: mechanisms and approaches. Neural Plast. 2012;2012:359728.
5. Baglio F, Pirastru A, Bergsland N, Cazzoli M, Tavazzi E. Neuroplasticity mediated by motor rehabilitation in Parkinson's disease: a systematic review on structural and functional MRI markers. Rev Neurosci. 2022;33:213–26.
6. Tavazzi E, Bergsland N, Pirastru A, Cazzoli M, Blasi V, Baglio F. MRI markers of functional connectivity and tissue microstructure in stroke-related motor rehabilitation: a systematic review. NeuroImage Clin. 2022;33:102931.
7. Carey LM, Seitz RJ. Functional neuroimaging in stroke recovery and neurorehabilitation: conceptual issues and perspectives. Int J Stroke. 2007;2:245–64.
8. Basile GA, Bertino S, Nozais V, Bramanti A, Ciurleo R, Anastasi GP, et al. White matter substrates of functional connectivity dynamics in the human brain. NeuroImage. 2022;258:119391.
9. Nozais V, Forkel SJ, Foulon C, Petit L, Thiebaut de Schotten M. Functionnectome as a framework to analyse the contribution of brain circuits to fMRI. Commun Biol. 2021;4:1–12.
10. Kim S-G, Richter W, Ugurbil K. Limitations of temporal resolution in functional MRI. Magn Reson Med. 1997;37:631–6.
11. Gore JC. Principles and practice of functional MRI of the human brain. J Clin Invest. 2003;112:4–9.
12. Khanna N, Altmeyer W, Zhuo J, Steven A. Functional neuroimaging: fundamental principles and clinical applications. Neuroradiol J. 2015;28:87–96.
13. Tsiamalou A, Dardiotis E, Paterakis K, Fotakopoulos G, Liampas I, Sgantzos M, et al. EEG in neurorehabilitation: a bibliometric analysis and content review. Neurol Int. 2022;14:1046–61.
14. Dronkers NF. A new brain region for coordinating speech articulation. Nature. 1996;384:159–61.
15. Broca P. Perte de la parole, ramollissement chronique et destruction partielle du lobe anterieur gauche du cerveau. Bull Soc Anthrop Paris. 1861;2:235–8.
16. Kleist K. Gehirnpathologie. Leipzig: Barth; 1934.
17. Lashley KS. In search of the engram. In: Society for Experimental Biology, Physiological mechanisms in animal behavior. (Society's Symposium IV.). 1950, pp 454–482.
18. Meynert T. Clinical treatise on disease of the fore-brain based upon a study of its structure, functions, and nutrition. New York, NY: G.P. Putnam's Sons; 1885.
19. Geschwind N. Disconnexion syndromes in animals and man. I. Brain. 1965;88:237.
20. Geschwind N. Disconnexion syndromes in animals and man. II. Brain. 1965;88:585.
21. Catani M, Ffytche DH. The rises and falls of disconnection syndromes. Brain. 2005;128:2224–39.
22. Cicerone KD, Langenbahn DM, Braden C, Malec JF, Kalmar K, Fraas M, et al. Evidence-based cognitive rehabilitation: updated review of the literature from 2003 through 2008. Arch Phys Med Rehabil. 2011;92:519–30.
23. Nudo RJ. Neural bases of recovery after brain injury. J Commun Disord. 2011;44:515–20.
24. Huettel SA, Song AW, McCarthy G. Functional magnetic resonance imaging. Sunderland, MA: Sinauer Associates, Inc.; 2004.
25. Logothetis NK. What we can do and what we cannot do with fMRI. Nature. 2008;453:869–78.
26. Lindquist MA, Meng Loh J, Atlas LY, Wager TD. Modeling the hemodynamic response function in fMRI: efficiency, bias and mis-modeling. NeuroImage. 2009;45:S187–98.

27. Soares JM, Magalhães R, Moreira PS, Sousa A, Ganz E, Sampaio A, et al. A Hitchhiker's guide to functional magnetic resonance imaging. Front Neurosci. 2016;10:515. https://doi.org/10.3389/fnins.2016.00515.
28. Esteban O, Markiewicz CJ, Blair RW, Moodie CA, Isik AI, Erramuzpe A, et al. fMRIPrep: a robust preprocessing pipeline for functional MRI. Nat Methods. 2019;16:111–6.
29. Monti M. Statistical analysis of fMRI time-series: a critical review of the GLM approach. Front Hum Neurosci. 2011;5:28. https://doi.org/10.3389/fnhum.2011.00028.
30. Poline J-B, Brett M. The general linear model and fMRI: does love last forever? NeuroImage. 2012;62:871–80.
31. Poldrack RA. Region of interest analysis for fMRI. Soc Cogn Affect Neurosci. 2007;2:67–70.
32. McKeown MJ, Hansen LK, Sejnowski terrence J. Independent component analysis of functional MRI: what is signal and what is noise? Curr Opin Neurobiol. 2003;13:620–9.
33. Rogers BP, Morgan VL, Newton AT, Gore JC. Assessing functional connectivity in the human brain by fMRI. Magn Reson Imaging. 2007;25:1347–57.
34. Nudo RJ. Postinfarct cortical plasticity and behavioral recovery. Stroke. 2007;38:840–5.
35. Nudo RJ. Recovery after damage to motor cortical areas. Curr Opin Neurobiol. 1999;9:740–7.
36. Jaillard A, Martin CD, Garambois K, Lebas JF, Hommel M. Vicarious function within the human primary motor cortex? Brain. 2005;128:1122–38.
37. Schaechter JD, Moore CI, Connell BD, Rosen BR, Dijkhuizen RM. Structural and functional plasticity in the somatosensory cortex of chronic stroke patients. Brain. 2006;129:2722–33.
38. Carey LM, Abbott DF, Egan GF, Bernhardt J, Donnan GA. Motor impairment and recovery in the upper limb after stroke: behavioral and neuroanatomical correlates. Stroke. 2005;36:625–9.
39. Ward NS. The neural substrates of motor recovery after focal damage to the central nervous system. Arch Phys Med Rehabil. 2006;87:30–5.
40. Staines WR, Black SE, Graham SJ, McIlroy WE. Somatosensory gating and recovery from stroke involving the thalamus. Stroke. 2002;33:2642–51.
41. Saur D, Lange R, Baumgaertner A, Schraknepper V, Willmes K, Rijntjes M, et al. Dynamics of language reorganization after stroke. Brain. 2006;129:1371–84.
42. Corbetta M, Kincade MJ, Lewis C, Snyder AZ, Sapir A. Neural basis and recovery of spatial attention deficits in spatial neglect. Nat Neurosci. 2005;8:1603–10.
43. Luft AR, Waller S, Forrester L, Smith GV, Whitall J, Macko RF, et al. Lesion location alters brain activation in chronically impaired stroke survivors. NeuroImage. 2004;21:924–35.
44. Floel A, Cohen LG. Translational studies in neurorehabilitation: from bench to bedside. Cogn Behav Neurol. 2006;19:1–10.
45. Mesulam M. From sensation to cognition. Brain. 1998;121:1013–52.
46. Krakauer JW, Pine ZM, Ghilardi MF, Ghez C. Learning of visuomotor transformations for vectorial planning of reaching trajectories. J Neurosci. 2000;20:8916–24.
47. Johansen-Berg H, Matthews PM. Attention to movement modulates activity in sensori-motor areas, including primary motor cortex. Exp Brain Res. 2002;142:13–24.
48. Iacoboni M, Dapretto M. The mirror neuron system and the consequences of its dysfunction. Nat Rev Neurosci. 2006;7:942–51.
49. Calautti C, Naccarato M, Jones PS, Sharma N, Day DD, Carpenter AT, et al. The relationship between motor deficit and hemisphere activation balance after stroke: a 3T fMRI study. NeuroImage. 2007;34:322–31.
50. Nair DG, Hutchinson S, Fregni F, Alexander M, Pascual-Leone A, Schlaug G. Imaging correlates of motor recovery from cerebral infarction and their physiological significance in well-recovered patients. NeuroImage. 2007;34:253–63.
51. Seitz RJ, Azari NP, Knorr U, Binkofski F, Herzog H, Freund HJ. The role of diaschisis in stroke recovery. Stroke. 1999;30:1844–50.
52. Seitz RJ, Knorr U, Azari NP, Herzog H, Freund H-J. Visual network activation in recovery from sensorimotor stroke. Restor Neurol Neurosci. 1999;14:25–33.

53. Gerloff C, Bushara K, Sailer A, Wassermann EM, Chen R, Matsuoka T, et al. Multimodal imaging of brain reorganization in motor areas of the contralesional hemisphere of well recovered patients after capsular stroke. Brain. 2006;129:791–808.
54. Rehme AK, Eickhoff SB, Wang LE, Fink GR, Grefkes C. Dynamic causal modeling of cortical activity from the acute to the chronic stage after stroke. NeuroImage. 2011;55:1147–58.
55. Rehme AK, Fink GR, Von Cramon DY, Grefkes C. The role of the contralesional motor cortex for motor recovery in the early days after stroke assessed with longitudinal FMRI. Cereb Cortex. 2011;21:756–68.
56. Bütefisch CM, Netz J, Weßling M, Seitz RJ, Hömberg V. Remote changes in cortical excitability after stroke. Brain. 2003;126:470–81.
57. Liepert J, Restemeyer C, Kucinski T, Zittel S, Weiller C. Motor strokes: the lesion location determines motor excitability changes. Stroke. 2005;36:2648–53.
58. Ward NS. Restoring brain function after stroke—bridging the gap between animals and humans. Nat Rev Neurol. 2017;13:244–55.
59. Grefkes C, Fink GR. Reorganization of cerebral networks after stroke: new insights from neuroimaging with connectivity approaches. Brain. 2011;134:1264–76.
60. Grefkes C, Nowak DA, Eickhoff SB, Dafotakis M, Küst J, Karbe H, et al. Cortical connectivity after subcortical stroke assessed with functional magnetic resonance imaging. Ann Neurol. 2008;63:236–46.
61. Pool EM, Leimbach M, Binder E, Nettekoven C, Eickhoff SB, Fink GR, et al. Network dynamics engaged in the modulation of motor behavior in stroke patients. Hum Brain Mapp. 2018;39:1078–92.
62. Ward NS, Cohen LG. Mechanisms underlying recovery of motor function after stroke. Arch Neurol. 2004;61:1844–8.
63. Zangaladze A, Epstein CM, Grafton ST, Sathian K. Involvement of visual cortex in tactile discrimination of orientation. Nature. 1999;401:587–90.
64. Staines WR, Graham SJ, Black SE, McIlroy WE. Task-relevant modulation of contralateral and ipsilateral primary somatosensory cortex and the role of a prefrontal-cortical sensory gating system. NeuroImage. 2002;15:190–9.
65. Crafton KR, Mark AN, Cramer SC. Improved understanding of cortical injury by incorporating measures of functional anatomy. Brain. 2003;126:1650–9.
66. Marshall RS, Zarahn E, Alon L, Minzer B, Lazar RM, Krakauer JW. Early imaging correlates of subsequent motor recovery after stroke. Ann Neurol. 2009;65:596–602.
67. Zarahn E, Alon L, Ryan SL, Lazar RM, Vry MS, Weiller C, et al. Prediction of motor recovery using initial impairment and fMRI 48 h poststroke. Cereb Cortex. 2011;21:2712–21.
68. Swayne OBC, Rothwell JC, Ward NS, Greenwood RJ. Stages of motor output reorganization after hemispheric stroke suggested by longitudinal studies of cortical physiology. Cereb Cortex. 2008;18:1909–22.
69. Ameli M, Grefkes C, Kemper F, Riegg FP, Rehme AK, Karbe H, et al. Differential effects of high-frequency repetitive transcranial magnetic stimulation over ipsilesional primary motor cortex in cortical and subcortical middle cerebral artery stroke. Ann Neurol. 2009;66:298–309.
70. You SH, Jang SH, Kim YH, Hallett M, Ahn SH, Kwon YH, et al. Virtual reality-induced cortical reorganization and associated locomotor recovery in chronic stroke: an experimenter-blind randomized study. Stroke. 2005;36:1166–71.
71. Deng H, Durfee WK, Nuckley DJ, Rheude BS, Severson AE, Skluzacek KM, et al. Complex versus simple ankle movement training in stroke using telerehabilitation: a randomized controlled trial. Phys Ther. 2012;92:197–209.
72. Landsmann B, Pinter D, Pirker E, Pichler G, Schippinger W, Weiss EM, et al. An exploratory intervention study suggests clinical benefits of training in chronic stroke to be paralleled by changes in brain activity using repeated fMRI. Clin Interv Aging. 2016;11:97–103.
73. Enzinger C, Dawes H, Johansen-Berg H, Wade D, Bogdanovic M, Collett J, et al. Brain activity changes associated with treadmill training after stroke. Stroke. 2009;40:2460–7.

74. Luft AR, MacKo RF, Forrester LW, Villagra F, Ivey F, Sorkin JD, et al. Treadmill exercise activates subcortical neural networks and improves walking after stroke: a randomized controlled trial. Stroke. 2008;39:3341–50.
75. Whitall J, McCombe Waller S, Sorkin JD, Forrester LW, Macko RF, Hanley DF, et al. Bilateral and unilateral arm training improve motor function through differing neuroplastic mechanisms: a single-blinded randomized controlled trial. Neurorehabil Neural Repair. 2011;25:118–29.
76. Ramos-Murguialday A, Broetz D, Rea M, Läer L, Yilmaz Ö, Brasil FL, et al. Brain-machine interface in chronic stroke rehabilitation: a controlled study. Ann Neurol. 2013;74:100–8.
77. Takahashi CD, Der-Yeghiaian L, Le V, Motiwala RR, Cramer SC. Robot-based hand motor therapy after stroke. Brain. 2008;131:425–37.
78. Johansen-Berg H, Rushworth MFS, Bogdanovic MD, Kischka U, Wimalaratna S, Matthews PM. The role of ipsilateral premotor cortex in hand movement after stroke. Proc Natl Acad Sci. 2002;99:14518–23.
79. Saleh S, Fluet G, Qiu Q, Merians A, Adamovich SV, Tunik E. Neural patterns of reorganization after intensive robot-assisted virtual reality therapy and repetitive task practice in patients with chronic stroke. Front Neurol. 2017;8:452. https://doi.org/10.3389/FNEUR.2017.00452.
80. Luft AR, McCombe-Waller S, Whitall J, Forrester LW, Macko R, Sorkin JD, et al. Repetitive bilateral arm training and motor cortex activation in chronic stroke: a randomized controlled trial. JAMA. 2004;292:1853–61.
81. Lin KC, Chung HY, Wu CY, Liu HL, Hsieh YW, Chen IH, et al. Constraint-induced therapy versus control intervention in patients with stroke: a functional magnetic resonance imaging study. Am J Phys Med Rehabil. 2010;89:177–85.
82. Schaechter JD, Kraft E, Hilliard TS, Dijkhuizen RM, Benner T, Finklestein SP, et al. Motor recovery and cortical reorganization after constraint-induced movement therapy in stroke patients: a preliminary study. Neurorehabil Neural Repair. 2002;16:326–38.
83. Page SJ, Harnish SM, Lamy M, Eliassen JC, Szaflarski JP. Affected arm use and cortical change in stroke patients exhibiting minimal hand movement. Neurorehabil Neural Repair. 2010;24:195–203.
84. Fan YT, Wu CY, Liu HL, Lin KC, Wai YY, Chen YL. Neuroplastic changes in resting-state functional connectivity after stroke rehabilitation. Front Hum Neurosci. 2015;9:546. https://doi.org/10.3389/FNHUM.2015.00546.
85. Sun L, Yin D, Zhu Y, Fan M, Zang L, Wu Y, et al. Cortical reorganization after motor imagery training in chronic stroke patients with severe motor impairment: a longitudinal fMRI study. Neuroradiology. 2013;55:913–25.
86. Zheng X, Sun L, Yin D, Jia J, Zhao Z, Jiang Y, et al. The plasticity of intrinsic functional connectivity patterns associated with rehabilitation intervention in chronic stroke patients. Neuroradiology. 2016;58:417–27.
87. Koganemaru S, Sawamoto N, Aso T, Sagara A, Ikkaku T, Shimada K, et al. Task-specific brain reorganization in motor recovery induced by a hybrid-rehabilitation combining training with brain stimulation after stroke. Neurosci Res. 2015;92:29–38.
88. Ward NS, Brown MM, Thompson AJ, Frackowiak RSJ. Neural correlates of motor recovery after stroke: a longitudinal fMRI study. Brain. 2003;126:2476–96.
89. Wu Q, Yue Z, Ge Y, Ma D, Yin H, Zhao H, et al. Brain functional networks study of subacute stroke patients with upper limb dysfunction after comprehensive rehabilitation including BCI training. Front Neurol. 2020;10:1419. https://doi.org/10.3389/FNEUR.2019.01419.
90. Gauthier LV, Taub E, Perkins C, Ortmann M, Mark VW, Uswatte G. Remodeling the brain: plastic structural brain changes produced by different motor therapies after stroke. Stroke. 2008;39:1520–5.
91. Bajaj S, Butler AJ, Drake D, Dhamala M. Brain effective connectivity during motor-imagery and execution following stroke and rehabilitation. NeuroImage Clin. 2015;8:572–82.
92. Liu H, Song LP, Zhang T. Changes in brain activation in stroke patients after mental practice and physical exercise: a functional MRI study. Neural Regen Res. 2014;9:1474–84.

93. Silva S, Borges LRDM, Santiago L, Lucena L, Lindquist AR, Ribeiro T. Motor imagery for gait rehabilitation after stroke. Cochrane Database Syst Rev. 2020;9(9):CD013019. https://doi.org/10.1002/14651858.CD013019.PUB2.
94. Ramos-Murguialday A, Curado MR, Broetz D, Yilmaz Ö, Brasil FL, Liberati G, et al. Brain-machine Interface in chronic stroke: randomized trial Long-term follow-up. Neurorehabil Neural Repair. 2019;33(3):188–98. https://doi.org/10.1177/1545968319827573.
95. Van den Heuvel MP, Hulshoff Pol HE. Exploring the brain network: a review on resting-state fMRI functional connectivity. Eur Neuropsychopharmacol. 2010;20(8):519–34. https://doi.org/10.1016/j.euroneuro.2010.03.008.
96. Preti MG, Bolton TA, Van De Ville D. The dynamic functional connectome: state-of-the-art and perspectives. NeuroImage. 2017;160:41–54.
97. Karahanoğlu FI, Van De Ville D. Dynamics of large-scale fMRI networks: deconstruct brain activity to build better models of brain function. Curr Opin Biomed Eng. 2017;3:28–36.
98. Bonkhoff AK, Espinoza FA, Gazula H, Vergara VM, Hensel L, Michely J, et al. Acute ischaemic stroke alters the brain's preference for distinct dynamic connectivity states. Brain. 2020;143:1525–40.
99. Bonkhoff AK, Schirmer MD, Bretzner M, Etherton M, Donahue K, Tuozzo C, et al. Abnormal dynamic functional connectivity is linked to recovery after acute ischemic stroke. Hum Brain Mapp. 2021;42:2278–91.
100. Tarun A, Wainstein-Andriano D, Sterpenich V, Bayer L, Perogamvros L, Solms M, et al. NREM sleep stages specifically alter dynamical integration of large-scale brain networks. iScience. 2020;24(1):101923. https://doi.org/10.1016/J.ISCI.2020.101923.
101. Calamante F, Smith RE, Liang X, Zalesky A, Connelly A. Track-weighted dynamic functional connectivity (TW-dFC): a new method to study time-resolved functional connectivity. Brain Struct Funct. 2017;222:3761–74.
102. Domingos J, Keus SHJ, Dean J, De Vries NM, Ferreira JJ, Bloem BR. The European physiotherapy guideline for Parkinson's disease: implications for neurologists. J Parkinsons Dis. 2018;8:499–502.
103. Park A, Stacy M. Non-motor symptoms in Parkinson's disease. J Neurol. 2009;256 https://doi.org/10.1007/S00415-009-5240-1.
104. Abbruzzese G, Marchese R, Avanzino L, Pelosin E. Rehabilitation for Parkinson's disease: current outlook and future challenges. Parkinsonism Relat Disord. 2016;22(Suppl 1):S60–4.
105. Goodwin VA, Richards SH, Taylor RS, Taylor AH, Campbell JL. The effectiveness of exercise interventions for people with Parkinson's disease: a systematic review and meta-analysis. Mov Disord. 2008;23:631–40.
106. Petzinger GM, Fisher BE, McEwen S, Beeler JA, Walsh JP, Jakowec MW. Exercise-enhanced neuroplasticity targeting motor and cognitive circuitry in Parkinson's disease. Lancet Neurol. 2013;12:716–26.
107. Zigmond MJ, Smeyne RJ. Exercise: is it a neuroprotective and if so, how does it work? Parkinsonism Relat Disord. 2014;20(Suppl):1. https://doi.org/10.1016/S1353-8020(13)70030-0.
108. Ferrazzoli D, Ortelli P, Cucca A, Bakdounes L, Canesi M, Volpe D. Motor-cognitive approach and aerobic training: a synergism for rehabilitative intervention in Parkinson's disease. Neurodegener Dis Manag. 2020;10:41–55.
109. Petzinger GM, Fisher BE, Van Leeuwen JE, Vukovic M, Akopian G, Meshul CK, et al. Enhancing neuroplasticity in the basal ganglia: the role of exercise in Parkinson's disease. Mov Disord. 2010;25(0 1):S141–5. https://doi.org/10.1002/mds.22782.
110. Nackaerts E, D'Cruz N, Dijkstra BW, Gilat M, Kramer T, Nieuwboer A. Towards understanding neural network signatures of motor skill learning in Parkinson's disease and healthy aging. Br J Radiol. 2019;92(1101):20190071. https://doi.org/10.1259/BJR.20190071.
111. Milardi D, Quartarone A, Bramanti A, Anastasi G, Bertino S, Basile GA, et al. The cortico-basal ganglia-cerebellar network: past, present and future perspectives. Front Syst Neurosci. 2019;13:61.

112. Quartarone A, Cacciola A, Milardi D, Ghilardi MF, Calamuneri A, Chillemi G, et al. New insights into cortico-basal-cerebellar connectome: clinical and physiological considerations. Brain. 2020;143:396–406.
113. Pyatigorskaya N, Gallea C, Garcia-Lorenzo D, Vidailhet M, Lehericy S. A review of the use of magnetic resonance imaging in Parkinson's disease. Ther Adv Neurol Disord. 2014;7:206–20.
114. Tessitore A, Cirillo M, De Micco R. Functional connectivity signatures of Parkinson's disease. J Parkinsons Dis. 2019;9:637–52.
115. Wu T, Wang L, Hallett M, Chen Y, Li K, Chan P. Effective connectivity of brain networks during self-initiated movement in Parkinson's disease. NeuroImage. 2011;55:204–15.
116. Simioni AC, Dagher A, Fellows LK. Effects of levodopa on corticostriatal circuits supporting working memory in Parkinson's disease. Cortex. 2017;93:193–205.
117. Herz DM, Eickhoff SB, Løkkegaard A, Siebner HR. Functional neuroimaging of motor control in Parkinson's disease: a meta-analysis. Hum Brain Mapp. 2014;35:3227–37.
118. Filippi M, Elisabetta S, Piramide N, Agosta F. Functional MRI in idiopathic Parkinson's disease. 1st ed. Elsevier Inc.; 2018. https://doi.org/10.1016/bs.irn.2018.08.005.
119. Maidan I, Rosenberg-Katz K, Jacob Y, Giladi N, Deutsch JE, Hausdorff JM, et al. Altered brain activation in complex walking conditions in patients with Parkinson's disease. Parkinsonism Relat Disord. 2016;25:91–6.
120. Tessa C, Lucetti C, Diciotti S, Paoli L, Cecchi P, Giannelli M, et al. Hypoactivation of the primary sensorimotor cortex in de novo Parkinson's disease : a motor fMRI study under controlled conditions. Neuroradiology. 2012;54:261–8.
121. Messa LV, Ginanneschi F, Momi D, Monti L, Battisti C, Cioncoloni D, et al. Functional and brain activation changes following specialized upper-limb exercise in Parkinson's disease. Front Hum Neurosci. 2019;13:350. https://doi.org/10.3389/FNHUM.2019.00350.
122. Sacheli MA, Neva JL, Lakhani B, Murray DK, Vafai N, Shahinfard E, et al. Exercise increases caudate dopamine release and ventral striatal activation in Parkinson's disease. Mov Disord. 2019;34:1891–900.
123. Boldrini M, Fulmore CA, Tartt AN, Simeon LR, Pavlova I, Poposka V, et al. Human hippocampal neurogenesis persists throughout aging. Cell Stem Cell. 2018;22:589–599.e5.
124. Silva-Batista C, Corcos DM, Horak FB, Ugrinowitsch C. Reply to: 'letter to the editor on "a randomized controlled trial of exercise for parkinsonian individuals with freezing of gait". Mov Disord. 2020;35:2123–4.
125. Agosta F, Gatti R, Sarasso E, Volonté MA, Canu E, Meani A, et al. Brain plasticity in Parkinson's disease with freezing of gait induced by action observation training. J Neurol. 2017;264:88–101.
126. Maidan I, Rosenberg-Katz K, Jacob Y, Giladi N, Hausdorff JM, Mirelman A. Disparate effects of training on brain activation in Parkinson disease. Neurology. 2017;89:1804–10.
127. Bunzeck N, Singh-Curry V, Eckart C, Weiskopf N, Perry RJ, Bain PG, et al. Motor phenotype and magnetic resonance measures of basal ganglia iron levels in Parkinson's disease. Parkinsonism Relat Disord. 2013;19:1136–42.
128. Zhang JR, Feng T, Hou YN, Chan P, Wu T. Functional connectivity of vim nucleus in tremor- and akinetic–/rigid-dominant Parkinson's disease. CNS Neurosci Ther. 2016;22:378–86.
129. Wang M, Jiang S, Yuan Y, Zhang L, Ding J, Wang J, et al. Alterations of functional and structural connectivity of freezing of gait in Parkinson's disease. J Neurol. 2016;263:1583–92.
130. Botvinik-Nezer R, Holzmeister F, Camerer CF, Dreber A, Huber J, Johannesson M, et al. Variability in the analysis of a single neuroimaging dataset by many teams. Nature. 2020;582:84–8.
131. Thut G, Miniussi C, Gross J. The functional importance of rhythmic activity in the brain. Curr Biol. 2012;22:R658–63.
132. Singh SP. Magnetoencephalography: Basic principles. Ann Indian Acad Neurol. 2014;17(Suppl 1):S107–12. https://doi.org/10.4103/0972-2327.128676.
133. Tallon-Baudry C, Bertrand O. Oscillatory gamma activity in humans and its role in object representation. Trends Cogn Sci. 1999;3:151–62.

134. David O, Kilner JM, Friston KJ. Mechanisms of evoked and induced responses in MEG/EEG. NeuroImage. 2006;31:1580–91.
135. Blackwood DH, Muir WJ. Cognitive brain potentials and their application. Br J Psychiatry Suppl. 1990;(9):96–101.
136. Sur S, Sinha VK. Event-related potential: an overview. Ind Psychiatry J. 2009;18:70–3.
137. Helfrich RF, Knight RT. Cognitive neurophysiology: event-related potentials. 1st ed. Elsevier B.V; 2019. https://doi.org/10.1016/B978-0-444-64032-1.00036-9.
138. Luck SJ. An introduction to the event-related potential technique. MIT press; 2014.
139. Berger H. Über das elektroenkephalogramm des menschen. Arch Psychiatr Nervenkr. 1929;87(1):527–570.
140. Walter WG, Cooper R, Aldridge VJ, McCallum WC, Winter AL. Contingent negative variation : an electric sign of Sensori-motor association and expectancy in the human brain. Nature. 1964;203:380–4.
141. Kornhuber HH, Deecke L. Hirnpotentialänderungen bei Willkürbewegungen und passiven Bewegungen des Menschen: Bereitschaftspotential und reafferente Potentiale. Pflüger's Arch Gesamte Physiol Menschen Tiere. 1965;284:1–17.
142. Cooper R, Winter AL, Crow HJ, Walter WG. Comparison of subcortical, cortical and scalp activity using chronically indwelling electrodes in man. Electroencephalogr Clin Neurophysiol. 1965;18:217–28.
143. Sutton S, Braren M, Zubin J, John ER. Evoked-Potential Correlates of Stimulus Uncertainty. Science (80-). 1965;150:1187–8.
144. Naatanen R, Gaillard AW, Mantysalo S. Early selective-attention effect on evoked potential reinterpreted. Acta Psychol. 1978;42:313–29.
145. Kutas M, Hillyard SA. Reading senseless sentences: brain potentials reflect semantic incongruity. Science (80-). 1980;207:203–5.
146. Luck SJ, Hillyard SA. Electrophysiological correlates of feature analysis during visual search. Psychophysiology. 1994;31:291–308.
147. Barrett G, Shibasaki H, Neshige R. Cortical potentials preceding voluntary movement: evidence for three periods of preparation in man. Electroencephalogr Clin Neurophysiol. 1986;63:327–39.
148. Shibasaki H, Hallett M. What is the Bereitschaftspotential? Clin Neurophysiol. 2006;117(11):2341–56. https://doi.org/10.1016/j.clinph.2006.04.025.
149. Dick JPR, Rothwell JC, Day BL, Cantello R, Buruma O, Gioux M, et al. The bereitschaftspotential is abnormal in parkinson's disease. Brain. 1989;112(Pt 1):233–44. https://doi.org/10.1093/brain/112.1.233.
150. Jahanshahi M, Jenkins IH, Brown RG, Marsden CD, Passingham RE, Brooks DJ. Self-initiated versus externally triggered movements. I. An investigation using measurement of regional cerebral blood flow with PET and movement-related potentials in normal and Parkinson's disease subjects. Brain. 1995;118(Pt 4):913–33.
151. Praamstra P, Cools AR, Stegeman DF, Horstink MW. Movement-related potential measures of different modes of movement selection in Parkinson's disease. J Neurol Sci. 1996;140:67–74.
152. Cunnington R, Iansek R, Bradshaw JL, Phillips JG. Movement-related potentials in Parkinson's disease. Presence and predictability of temporal and spatial cues. Brain. 1995;118(Pt 4):935–50.
153. Colebatch JG. Bereitschaftspotential and movement-related potentials: origin, significance, and application in disorders of human movement. Mov Disord. 2007;22:601–10.
154. Pfurtscheller G. Central beta rhythm during sensorimotor activities in man. Electroencephalogr Clin Neurophysiol. 1981;51:253–64.
155. Feige B, Kristeva-Feige R, Rossi S, Pizzella V, Rossini P-M. Neuromagnetic study of movement-related changes in rhythmic brain activity. Brain Res. 1996;734:252–60.
156. Kilavik BE, Zaepffel M, Brovelli A, MacKay WA, Riehle A. The ups and downs of beta oscillations in sensorimotor cortex. Exp Neurol. 2013;245:15–26.

157. Nowak M, Zich C, Stagg CJ. Motor cortical gamma oscillations: what have we learnt and where are we headed? Curr Behav Neurosci Rep. 2018;5(2):136–42. https://doi.org/10.1007/s40473-018-0151-z.
158. Neuper C, Wortz M, Pfurtscheller G. ERD/ERS patterns reflecting sensorimotor activation and deactivation. Prog Brain Res. 2006;159:211–22.
159. Pfurtscheller G, Lopes Da Silva FH. Event-related EEG/MEG synchronization and desynchronization: basic principles. Clin Neurophysiol. 1999;110(11):1842–57. https://doi.org/10.1016/S1388-2457(99)00141-8.
160. Toma K, Mima T, Matsuoka T, Gerloff C, Ohnishi T, Koshy B, et al. Movement rate effect on activation and functional coupling of motor cortical areas. J Neurophysiol. 2002;88(6):3377–85. https://doi.org/10.1152/jn.00281.2002.
161. Chen R, Yaseen Z, Cohen LG, Hallett M. Time course of corticospinal excitability in reaction time and self-paced movements. Ann Neurol. 1998;44:317–25.
162. Gastaut HJ, Bert J. EEG changes during cinematographic presentation (moving picture activation of the EEG). Electroencephalogr Clin Neurophysiol. 1954;6:433–44.
163. Gastaut HJ. Electrocorticographic study of the reactivity of rolandic rhythm. Rev Neurol (Paris). 1952;87:176–82.
164. Stancák A, Pfurtscheller G. Event-related desynchronisation of central beta-rhythms during brisk and slow self-paced finger movements of dominant and nondominant hand. Cogn Brain Res. 1996;4(3):171–83. https://doi.org/10.1016/S0926-6410(96)00031-6.
165. Nagamine T, Kajola M, Salmelin R, Shibasaki H, Hari R. Movement-related slow cortical magnetic fields and changes of spontaneous MEG- and EEG-brain rhythms. Electroencephalogr Clin Neurophysiol. 1996;99:274–86.
166. Hari R, Salmelin R, Hari R. Human cortical oscillations : a neuromagnetic view through the skull. Trends Neurosci. 1997;20:44–9.
167. Engel AK, Fries P. Beta-band oscillations-signalling the status quo? Curr Opin Neurobiol. 2010;20:156–65.
168. Pineda JA. The functional significance of mu rhythms: translating 'seeing' and 'hearing' into 'doing'. Brain Res Rev. 2005;50:57–68.
169. Hari R. Action–perception connection and the cortical mu rhythm. In: Neuper C, Klimesch WBT-P, editors. in BR Event-related dynamics of brain oscillations. Elsevier; 2006. p. 253–60.
170. Stančák A, Riml A, Pfurtscheller G. The effects of external load on movement-related changes of the senorimotor EEG rhythms. Electroencephalogr Clin Neurophysiol. 1997;102:495–504.
171. Stancák A, Pfurtscheller G. Desynchronization and recovery of β rhythms during brisk and slow self-paced finger movements in man. Neurosci Lett. 1995;196(1-2):21–4. https://doi.org/10.1016/0304-3940(95)11827-J.
172. Pistohl T, Schulze-Bonhage A, Aertsen A, Mehring C, Ball T. Decoding natural grasp types from human ECoG. NeuroImage. 2012;59(1):248–60. https://doi.org/10.1016/j.neuroimage.2011.06.084.
173. Tombini M, Zappasodi F, Zollo L, Pellegrino G, Cavallo G, Tecchio F, et al. Brain activity preceding a 2D manual catching task. NeuroImage. 2009;47:1735–46.
174. Tatti E, Ricci S, Mehraram R, Lin N, George S, Nelson AB, et al. Beta modulation depth is not linked to movement features. Front Behav Neurosci. 2019;13:49. https://doi.org/10.3389/fnbeh.2019.00049.
175. Mima T, Sadato N, Yazawa S, Hanakawa T, Fukuyama H, Yonekura Y, et al. Brain structures related to active and passive finger movements in man. Brain. 1999;122:1989–97.
176. Wheaton L, Fridman E, Bohlhalter S, Vorbach S, Hallett M. Left parietal activation related to planning, executing and suppressing praxis hand movements. Clin Neurophysiol. 2009;120:980–6.
177. Salmelin R, Forss N, Knuutila J, Hari R. Bilateral activation of the human somatomotor cortex by distal hand movements. Electroencephalogr Clin Neurophysiol. 1995;95(6):444–52. https://doi.org/10.1016/0013-4694(95)00193-X.

178. Pfurtscheller G, Stancák A, Neuper C. Post-movement beta synchronization. A correlate of an idling motor area? Electroencephalogr Clin Neurophysiol. 1996;98:281–93.
179. Gaetz W, Edgar JC, Wang DJ, Roberts TPL. Relating MEG measured motor cortical oscillations to resting gamma-aminobutyric acid (GABA) concentration. NeuroImage. 2011;55:616–21.
180. Solis-Escalante T, Müller-Putz GR, Pfurtscheller G, Neuper C. Cue-induced beta rebound during withholding of overt and covert foot movement. Clin Neurophysiol. 2012;123:1182–90.
181. Miller EK, Lundqvist M, Bastos AM. Working memory 2.0. Neuron. 2018;100:463–75.
182. Lundqvist M, Rose J, Herman P, Brincat SL, Buschman TJ, Miller EK. Gamma and Beta bursts underlie working memory. Neuron. 2016;90:152–64.
183. Hanslmayr S, Matuschek J, Fellner M-C. Entrainment of prefrontal beta oscillations induces an endogenous echo and impairs memory formation. Curr Biol. 2014;24:904–9.
184. Zavala B, Damera S, Dong JW, Lungu C, Brown P, Zaghloul KA. Human subthalamic nucleus theta and beta oscillations entrain neuronal firing during sensorimotor conflict. Cereb Cortex. 2017;27:496–508.
185. Zaepffel M, Trachel R, Kilavik BE, Brochier T. Modulations of EEG Beta power during planning and execution of grasping movements. PLoS One. 2013;8:e60060.
186. Spitzer B, Haegens S. Beyond the status quo: a role for beta oscillations in endogenous content (RE)activation. eNeuro. 2017;4(4):ENEURO.0170-17.2017. https://doi.org/10.1523/ENEURO.0170-17.2017.
187. Tan H, Wade C, Brown P. Post-movement beta activity in sensorimotor cortex indexes confidence in the estimations from internal models. J Neurosci. 2016;36:1516–28.
188. Tan H, Zavala B, Pogosyan A, Ashkan K, Zrinzo L, Foltynie T, et al. Human subthalamic nucleus in movement error detection and its evaluation during visuomotor adaptation. J Neurosci. 2014;34:16744–54.
189. Palmer CE, Auksztulewicz R, Ondobaka S, Kilner JM. Sensorimotor beta power reflects the precision-weighting afforded to sensory prediction errors. NeuroImage. 2019;200:59–71.
190. Friston K. A theory of cortical responses. Philos Trans R Soc London Ser B Biol Sci. 2005;360:815–36.
191. Burrone J, Murthy VN. Synaptic gain control and homeostasis. Curr Opin Neurobiol. 2003;13:560–7.
192. Friston K, Kiebel S. Predictive coding under the free-energy principle. Philos Trans R Soc London Ser B Biol Sci. 2009;364:1211–21.
193. Ghilardi MF, Tatti E, Quartarone A. Beta power and movement-related beta modulation as hallmarks of energy for plasticity induction: implications for Parkinson's disease. Park Relat Disord. 2021;88:136–9.
194. Démas J, Bourguignon M, Périvier M, De Tiège X, Dinomais M, Van Bogaert P. Mu rhythm: state of the art with special focus on cerebral palsy. Ann Phys Rehabil Med. 2020;63:439–46.
195. Hsu Y-F, Liao K-K, Lee P-L, Tsai Y-A, Yeh C-L, Lai K-L, et al. Intermittent theta burst stimulation over primary motor cortex enhances movement-related beta synchronisation. Clin Neurophysiol. 2011;122:2260–7.
196. Suzuki M, Tanaka S, Gomez-Tames J, Okabe T, Cho K, Iso N, et al. Nonequivalent after-effects of alternating current stimulation on motor cortex oscillation and inhibition: simulation and experimental study. Brain Sci. 2022;12(2):195. https://doi.org/10.3390/brainsci12020195.
197. Leunissen I, Van Steenkiste M, Heise K-F, Monteiro TS, Dunovan K, Mantini D, et al. Effects of beta-band and gamma-band rhythmic stimulation on motor inhibition. iScience. 2022;25:104338.
198. Moisello C, Blanco D, Fontanesi C, Lin J, Biagioni M, Kumar P, et al. TMS enhances retention of a motor skill in Parkinson's disease. Brain Stimul. 2015;8:224–30.
199. Nelson AB, Moisello C, Lin J, Panday P, Ricci S, Canessa A, et al. Beta oscillatory changes and retention of motor skills during practice in healthy subjects and in patients with Parkinson's disease. Front Hum Neurosci. 2017;11:104. https://doi.org/10.3389/fnhum.2017.00104.

200. Tatti E, Ferraioli F, Peter J, Alalade T, Nelson AB, Ricci S, et al. Frontal increase of beta modulation during the practice of a motor task is enhanced by visuomotor learning. Sci Rep. 2021;11(1):17441. https://doi.org/10.1038/S41598-021-97004-0.
201. Tatti E, Ricci S, Nelson AB, Mathew D, Chen H, Quartarone A, et al. Prior practice affects movement-related beta modulation and quiet wake restores it to baseline. Front Syst Neurosci. 2020;14:61.
202. Ricci S, Tatti E, Mattia D, Cincotti F, Sanguineti V, Morasso P et al. Beta oscillations during adaptation to inertial and velocity dependent perturbations. In: 2020 8th IEEE RAS/EMBS International Conference for Biomedical Robotics and Biomechatronics (BioRob). 2020, pp 1210–1215.
203. Iskhakova L, Rappel P, Deffains M, Fonar G, Marmor O, Paz R, et al. Modulation of dopamine tone induces frequency shifts in cortico-basal ganglia beta oscillations. Nat Commun. 2021;12:7026.
204. Schwerdt HN, Amemori K, Gibson DJ, Stanwicks LL, Yoshida T, Bichot NP, et al. Dopamine and beta-band oscillations differentially link to striatal value and motor control. Sci Adv. 2020;6(39):eabb9226. https://doi.org/10.1126/sciadv.abb9226.
205. Jenkinson N, Brown P. New insights into the relationship between dopamine, beta oscillations and motor function. Trends Neurosci. 2011;34(12):611–8. https://doi.org/10.1016/j.tins.2011.09.003.
206. Jensen O, Goel P, Kopell N, Pohja M, Hari R, Ermentrout B. On the human sensorimotor-cortex beta rhythm: sources and modeling. NeuroImage. 2005;26:347–55.
207. Roopun AK, Middleton SJ, Cunningham MO, LeBeau FEN, Bibbig A, Whittington MA, et al. A beta2-frequency (20-30 Hz) oscillation in nonsynaptic networks of somatosensory cortex. Proc Natl Acad Sci USA. 2006;103:15646–50.
208. Muthukumaraswamy SD, Myers JFM, Wilson SJ, Nutt DJ, Lingford-Hughes A, Singh KD, et al. The effects of elevated endogenous GABA levels on movement-related network oscillations. NeuroImage. 2013;66:36–41.
209. Rossiter HE, Davis EM, Clark EV, Boudrias M-H, Ward NS. Beta oscillations reflect changes in motor cortex inhibition in healthy ageing. NeuroImage. 2014;91:360–5.
210. Legon W, Adams S, Bansal P, Patel PD, Hobbs L, Ai L, et al. A retrospective qualitative report of symptoms and safety from transcranial focused ultrasound for neuromodulation in humans. Sci Rep. 2020;10:5573.
211. Yamawaki N, Stanford IM, Hall SD, Woodhall GL. Pharmacologically induced and stimulus evoked rhythmic neuronal oscillatory activity in the primary motor cortex in vitro. Neuroscience. 2008;151:386–95.
212. Hall SD, Prokic EJ, McAllister CJ, Ronnqvist KC, Williams AC, Yamawaki N, et al. GABA-mediated changes in inter-hemispheric beta frequency activity in early-stage Parkinson's disease. Neuroscience. 2014;281:68–76.
213. Hall SD, Stanford IM, Yamawaki N, McAllister CJ, Rönnqvist KC, Woodhall GL, et al. The role of GABAergic modulation in motor function related neuronal network activity. NeuroImage. 2011;56:1506–10.
214. Hall SD, Barnes GR, Furlong PL, Seri S, Hillebrand A. Neuronal network pharmacodynamics of GABAergic modulation in the human cortex determined using pharmaco-magnetoencephalography. Hum Brain Mapp. 2010;31:581–94.
215. Crone NE, Miglioretti DL, Gordon B, Lesser RP. Functional mapping of human sensorimotor cortex with electrocorticographic spectral analysis. II. Event-related synchronization in the gamma band. Brain. 1998;121(Pt 1):2301–15.
216. Pfurtscheller G, Graimann B, Huggins JE, Levine SP, Schuh LA. Spatiotemporal patterns of beta desynchronization and gamma synchronization in corticographic data during self-paced movement. Clin Neurophysiol. 2003;114:1226–36.
217. Miller KJ, Leuthardt EC, Schalk G, Rao RPN, Anderson NR, Moran DW, et al. Spectral changes in cortical surface potentials during motor movement. J Neurosci. 2007;27:2424 LP–2432.

218. Cheyne D, Bells S, Ferrari P, Gaetz W, Bostan AC. Self-paced movements induce high-frequency gamma oscillations in primary motor cortex. NeuroImage. 2008;42:332–42.
219. Pfurtscheller G, Neuper C. Simultaneous EEG 10 Hz desynchronization and 40 Hz synchronization during finger movements. Neuroreport. 1992;3:1057–60.
220. Gwin JT, Gramann K, Makeig S, Ferris DP. Electrocortical activity is coupled to gait cycle phase during treadmill walking. NeuroImage. 2011;54:1289–96.
221. Seeber M, Scherer R, Müller-Putz GR. EEG oscillations are modulated in different behavior-related networks during rhythmic finger movements. J Neurosci. 2016;36:11671–81.
222. Grosse-Wentrup M, Schölkopf B, Hill J. Causal influence of gamma oscillations on the sensorimotor rhythm. NeuroImage. 2011;56:837–42.
223. Miller KJ, Schalk G, Fetz EE, den Nijs M, Ojemann JG, Rao RPN. Cortical activity during motor execution, motor imagery, and imagery-based online feedback. Proc Natl Acad Sci USA. 2010;107:4430–5.
224. Ball T, Demandt E, Mutschler I, Neitzel E, Mehring C, Vogt K, et al. Movement related activity in the high gamma range of the human EEG. NeuroImage. 2008;41(2):302–10. https://doi.org/10.1016/j.neuroimage.2008.02.032.
225. Brücke C, Huebl J, Schönecker T, Neumann WJ, Yarrow K, Kupsch A, et al. Scaling of movement is related to pallidal γ oscillations in patients with dystonia. J Neurosci. 2012;32(3):1008–19. https://doi.org/10.1523/JNEUROSCI.3860-11.2012.
226. Gunduz A, Brunner P, Sharma M, Leuthardt EC, Ritaccio AL, Pesaran B, et al. Differential roles of high gamma and local motor potentials for movement preparation and execution. Brain-computer Interfaces. 2016;3(2):88–102. https://doi.org/10.1080/2326263X.2016.1179087.
227. Joundi R, Jenkinson N, Brittain J-S, Aziz TZ, Brown P. Driving oscillatory activity in the human cortex enhances motor performance. Curr Biol. 2012;22:403–7.
228. Lofredi R, Neumann WJ, Bock A, Horn A, Huebl J, Siegert S, et al. Dopamine-dependent scaling of subthalamic gamma bursts with movement velocity in patients with Parkinson's disease. elife. 2018;7:e31895. https://doi.org/10.7554/eLife.31895.
229. Muthukumaraswamy SD. Functional properties of human primary motor cortex gamma oscillations. J Neurophysiol. 2010;104(5):2873–85. https://doi.org/10.1152/jn.00607.2010.
230. Tatti E, Ferraioli F, Cacciola A, Chan C, Quartarone A, Ghilardi MF. Modulation of gamma spectral amplitude and connectivity during reaching predicts peak velocity and movement duration. Front Neurosci. 2022;16:836703.
231. Wang P, McCrimmon C, King C, Shaw S, Millett D, Gong H, et al. Characterization of electrocorticogram high-gamma signal in response to varying upper extremity movement velocity. Brain Struct Funct. 2017;222(8):3705–48. https://doi.org/10.1007/s00429-017-1429-8.
232. Perfetti B, Moisello C, Landsness EC, Kvint S, Lanzafame S, Onofrj M, et al. Modulation of gamma and theta spectral amplitude and phase synchronization is associated with the development of visuo-motor learning. J Neurosci. 2011;31:14810–9.
233. Wyss C, Tse DHY, Kometer M, Dammers J, Achermann R, Shah NJ, et al. GABA metabolism and its role in gamma-band oscillatory activity during auditory processing: an MRS and EEG study. Hum Brain Mapp. 2017;38:3975–87.
234. Nowak M, Hinson E, Van Ede F, Pogosyan A, Guerra A, Quinn A, et al. Driving human motor cortical oscillations leads to behaviorally relevant changes in local GABAA inhibition: a tACS-TMS study. J Neurosci. 2017;37:4481–92.
235. Buzsáki G, Wang X-J. Mechanisms of gamma oscillations. Annu Rev Neurosci. 2012;35:203–25.
236. Ray S, Maunsell JHR. Do gamma oscillations play a role in cerebral cortex? Trends Cogn Sci. 2015;19:78–85.
237. Peter J, Ferraioli F, Mathew D, George S, Chan C, Alalade T, et al. Movement-related beta ERD and ERS abnormalities in neuropsychiatric disorders. Front Neurosci. 2022;16:1–19.

238. Ingwersen T, Wolf S, Birke G, Schlemm E, Bartling C, Bender G, et al. Long-term recovery of upper limb motor function and self-reported health: results from a multicenter observational study 1 year after discharge from rehabilitation. Neurol Res Pract. 2021;3:66.
239. Cramer SC. Repairing the human brain after stroke: I. Mechanisms of spontaneous recovery. Ann Neurol. 2008;63:272–87.
240. Hallett M. Plasticity of the human motor cortex and recovery from stroke. Brain Res Brain Res Rev. 2001;36:169–74.
241. Ferbert A, Priori A, Rothwell JC, Day BL, Colebatch JG, Marsden CD. Interhemispheric inhibition of the human motor cortex. J Physiol. 1992;453:525–46.
242. Marshall RS. Neuroimaging in stroke recovery: a position paper from the first international workshop on neuroimaging and stroke recovery. Cerebrovasc Dis. 2004;18:260–7.
243. Murase N, Duque J, Mazzocchio R, Cohen LG. Influence of interhemispheric interactions on motor function in chronic stroke. Ann Neurol. 2004;55:400–9.
244. Daly JJ, Fang Y, Perepezko EM, Siemionow V, Yue GH. Prolonged cognitive planning time, elevated cognitive effort, and relationship to coordination and motor control following stroke. IEEE Trans Neural Syst Rehabil Eng. 2006;14:168–71.
245. Ibáñez J, del Castillo MD, Serrano JI, Rueda FM, Pereira EM, Montero FMR, et al. In: Roa Romero LM, editor. Single-trial detection of the event-related desynchronization to locate with temporal precision the onset of voluntary movements in stroke patients BT–XIII Mediterranean conference on medical and biological engineering and computing 2013. Cham: Springer International Publishing; 2014. p. 1651–4.
246. Platz T, Kim IH, Pintschovius H, Winter T, Kieselbach A, Villringer K, et al. Multimodal EEG analysis in man suggests impairment-specific changes in movement-related electric brain activity after stroke. Brain. 2000;123(Pt 12):2475–90.
247. Rossiter HE, Boudrias M-H, Ward NS. Do movement-related beta oscillations change after stroke? J Neurophysiol. 2014;112:2053–8.
248. Serrien DJ, Brown P. The functional role of interhemispheric synchronization in the control of bimanual timing tasks. Exp Brain Res. 2002;147:268–72.
249. Shiner CT, Tang H, Johnson BW, McNulty PA. Cortical beta oscillations and motor thresholds differ across the spectrum of post-stroke motor impairment, a preliminary MEG and TMS study. Brain Res. 2015;1629:26–37.
250. Stępień M, Conradi J, Waterstraat G, Hohlefeld FU, Curio G, Nikulin VV. Event-related desynchronization of sensorimotor EEG rhythms in hemiparetic patients with acute stroke. Neurosci Lett. 2011;488:17–21.
251. Wiese H, Stude P, Sarge R, Nebel K, Diener H-C, Keidel M. Reorganization of motor execution rather than preparation in poststroke hemiparesis. Stroke. 2005;36:1474–9.
252. Espenhahn S, Rossiter HE, van Wijk BCM, Redman N, Rondina JM, Diedrichsen J, et al. Sensorimotor cortex beta oscillations reflect motor skill learning ability after stroke. Brain Commun. 2020;2(2):fcaa161. https://doi.org/10.1093/braincomms/fcaa161.
253. Bartur G, Pratt H, Soroker N. Changes in mu and beta amplitude of the EEG during upper limb movement correlate with motor impairment and structural damage in subacute stroke. Clin Neurophysiol. 2019;130:1644–51.
254. Berenguer-Rocha M, Baltar A, Rocha S, Shirahige L, Brito R, Monte-Silva K. Interhemispheric asymmetry of the motor cortex excitability in stroke: relationship with sensory-motor impairment and injury chronicity. Neurol Sci. 2020;41:2591–8.
255. Dodd KC, Nair VA, Prabhakaran V. Role of the contralesional vs. Ipsilesional hemisphere in stroke recovery. Front Hum Neurosci. 2017;11:1–9.
256. Weiller C, Ramsay SC, Wise RJ, Friston KJ, Frackowiak RS. Individual patterns of functional reorganization in the human cerebral cortex after capsular infarction. Ann Neurol. 1993;33:181–9.

257. Chen CC, Litvak V, Gilbertson T, Kühn A, Lu CS, Lee ST, et al. Excessive synchronization of basal ganglia neurons at 20 Hz slows movement in Parkinson's disease. Exp Neurol. 2007;205:214–21.
258. Devergnas A, Pittard D, Bliwise D, Wichmann T. Relationship between oscillatory activity in the cortico-basal ganglia network and parkinsonism in MPTP-treated monkeys. Neurobiol Dis. 2014;68:156–66.
259. Hammond C, Bergman H, Brown P. Pathological synchronization in Parkinson's disease: networks, models and treatments. Trends Neurosci. 2007;30:357–64.
260. Kühn AA, Tsui A, Aziz T, Ray N, Brücke C, Kupsch A, et al. Pathological synchronisation in the subthalamic nucleus of patients with Parkinson's disease relates to both bradykinesia and rigidity. Exp Neurol. 2009;215:380–7.
261. Nini A, Feingold A, Slovin H, Bergman H. Neurons in the globus pallidus do not show correlated activity in the normal monkey, but phase-locked oscillations appear in the MPTP model of parkinsonism. J Neurophysiol. 1995;74:1800–5.
262. Wichmann T, Bergman H, DeLong MR. The primate subthalamic nucleus. III. Changes in motor behavior and neuronal activity in the internal pallidum induced by subthalamic inactivation in the MPTP model of parkinsonism. J Neurophysiol. 1994;72:521–30.
263. Little S, Brown P. The functional role of beta oscillations in Parkinson's disease. Parkinsonism Relat Disord. 2014;20:S44–8.
264. Lofredi R, Tan H, Neumann W-J, Yeh C-H, Schneider G-H, Kühn AA, et al. Beta bursts during continuous movements accompany the velocity decrement in Parkinson's disease patients. Neurobiol Dis. 2019;127:462–71.
265. Prokic EJ, Stanford IM, Woodhall GL, Williams AC, Hall SD. Bradykinesia is driven by cumulative Beta power during continuous movement and alleviated by gabaergic modulation in Parkinson's disease. Front Neurol. 2019;10:1298.
266. Pfurtscheller G, Pichler-Zalaudek K, Ortmayr B, Diez J, Reisecker. Postmovement beta synchronization in patients with Parkinson's disease. J Clin Neurophysiol. 1998;15(3):243–50.
267. Praamstra P, Pope P. Slow brain potential and oscillatory EEG manifestations of impaired temporal preparation in Parkinson's disease. J Neurophysiol. 2007;98:2848–57.
268. Heinrichs-Graham E, Wilson TW, Santamaria PM, Heithoff SK, Torres-Russotto D, Hutter-Saunders JAL, et al. Neuromagnetic evidence of abnormal movement-related beta desynchronization in Parkinson's disease. Cereb Cortex. 2014;24(10):2669–78. https://doi.org/10.1093/cercor/bht121.
269. Nelson AB, Moisello C, Lin J, Panday P, Ricci S, Canessa A, et al. Beta oscillatory changes and retention of motor skills during practice in healthy subjects and in patients with Parkinson's disease. Front Hum Neurosci. 2017;11:1–12.
270. Vinding MC, Tsitsi P, Piitulainen H, Waldthaler J, Jousmäki V, Ingvar M, et al. Attenuated beta rebound to proprioceptive afferent feedback in Parkinson's disease. Sci Rep. 2019;9:2604.
271. Wu H-M, Hsiao F-J, Chen R-S, Shan D-E, Hsu W-Y, Chiang M-C, et al. Attenuated NoGo-related beta desynchronisation and synchronisation in Parkinson's disease revealed by magnetoencephalographic recording. Sci Rep. 2019;9:7235.
272. Weersink JB, Gefferie SR, van Laar T, Maurits NM, de Jong BM. Pre-movement cortico-muscular dynamics underlying improved Parkinson gait initiation after instructed arm swing. J Parkinsons Dis. 2020;10:1675–93.
273. McLinden J, Deligani RJ, Abtahi MR, Akbar U, Mankodiya K, Shahriari Y. Disruptions of cortico-kinematic interactions in Parkinson's disease. Behav Brain Res. 2021;404:113153.
274. Canessa A, Pozzi NG, Arnulfo G, Brumberg J, Reich MM, Pezzoli G, et al. Striatal dopaminergic innervation regulates subthalamic beta-oscillations and cortical-subcortical coupling during movements: preliminary evidence in subjects with Parkinson's disease. Front Hum Neurosci. 2016;10:1–12.
275. Moisello C, Blanco D, Lin J, Panday P, Kelly SP, Quartarone A, et al. Practice changes beta power at rest and its modulation during movement in healthy subjects but not in patients with Parkinson's disease. Brain Behav. 2015;5(10):e00374. https://doi.org/10.1002/brb3.374.

276. Marchesi G, Albanese GA, Ferrazzoli D, George S, Ricci S, Tatti E et al. Effects of rTMS and intensive rehabilitation in Parkinson's disease on learning and retention. In: IEEE International Conference on Rehabilitation Robotics 2019. https://doi.org/10.1109/ICORR.2019.8779471.
277. Ricci S, Mehraram R, Tatti E, Nelson AB, Bossini-Baroggi M, Panday P, et al. Aging does not affect beta modulation during reaching movements. Neural Plast. 2019;2019:1619290. https://doi.org/10.1155/2019/1619290.
278. Wiest C, Torrecillos F, Tinkhauser G, Pogosyan A, Morgante F, Pereira EA, et al. Finely-tuned gamma oscillations: spectral characteristics and links to dyskinesia. Exp Neurol. 2022;351:113999. https://doi.org/10.1016/j.expneurol.2022.113999.
279. Androulidakis AG, Kühn AA, Chu Chen C, Blomstedt P, Kempf F, Kupsch A, et al. Dopaminergic therapy promotes lateralized motor activity in the subthalamic area in Parkinson's disease. Brain. 2007;130:457–68.
280. Guerra A, Asci F, D'Onofrio V, Sveva V, Bologna M, Fabbrini G, et al. Enhancing gamma oscillations restores primary motor cortex plasticity in Parkinson's disease. J Neurosci. 2020;40:4788–96.
281. Swann NC, de Hemptinne C, Miocinovic S, Qasim S, Wang SS, Ziman N, et al. Gamma oscillations in the hyperkinetic state detected with chronic human brain recordings in Parkinson's disease. J Neurosci. 2016;36:6445–58.
282. Kleim JA, Jones TA. Principles of experience-dependent neural plasticity: implications for rehabilitation after brain damage. J Speech Lang Hear Res. 2008;51:S225–39.
283. Berthier ML, Pulvermüller F. Neuroscience insights improve neurorehabilitation of post-stroke aphasia. Nat Rev Neurol. 2011;7:86–97.
284. Paggiaro A, Birbaumer N, Cavinato M, Turco C, Formaggio E, Del Felice A, et al. Magnetoencephalography in stroke recovery and rehabilitation. Front Neurol. 2016;7:1–10.
285. Vatinno AA, Simpson A, Ramakrishnan V, Bonilha HS, Bonilha L, Seo NJ. The prognostic utility of electroencephalography in stroke recovery: a systematic review and meta-analysis. Neurorehabil Neural Repair. 2022;36:255–68.
286. Calabrò RS, Naro A, Russo M, Leo A, De Luca R, Balletta T, et al. The role of virtual reality in improving motor performance as revealed by EEG: a randomized clinical trial. J Neuroeng Rehabil. 2017;14:53.
287. Schranz C, Vatinno A, Ramakrishnan V, Seo NJ. Neuroplasticity after upper-extremity rehabilitation therapy with sensory stimulation in chronic stroke survivors. Brain Commun. 2022;4:1–10.
288. Wang HE, Bénar CG, Quilichini PP, Friston KJ, Jirsa VK, Bernard C. A systematic framework for functional connectivity measures. Front Neurosci. 2014;8:405.
289. Bastos AM, Schoffelen JM. A tutorial review of functional connectivity analysis methods and their interpretational pitfalls. Front Syst Neurosci. 2016;9:1–23.
290. Hosp JA, Luft AR. Cortical plasticity during motor learning and recovery after ischemic stroke. Neural Plast. 2011;2011:871296.
291. Cassidy JM, Mark JI, Cramer SC. Functional connectivity drives stroke recovery: shifting the paradigm from correlation to causation. Brain. 2022;145:1211–28.
292. Conti M, Bovenzi R, Garasto E, Schirinzi T, Placidi F, Mercuri NB, et al. Brain functional connectivity in de novo Parkinson's disease patients based on clinical EEG. Front Neurol. 2022;13:844745.
293. Bočková M, Rektor I. Impairment of brain functions in Parkinson's disease reflected by alterations in neural connectivity in EEG studies: a viewpoint. Clin Neurophysiol. 2019;130:239–47.
294. Bistriceanu CE, Danciu FA, Cuciureanu DI. Cortical connectivity in stroke using signals from resting-state EEG: a review of current literature. Acta Neurol Belg. 2023;123(2):351–7. https://doi.org/10.1007/s13760-022-02102-z.
295. Vatinno AA, Schranz C, Simpson AN, Ramakrishnan V, Bonilha L, Seo NJ. Predicting upper extremity motor improvement following therapy using EEG-based connectivity in chronic stroke. NeuroRehabilitation. 2022;50:105–13.

296. Naro A, Pignolo L, Calabrò RS. Brain network organization following Post-stroke neurorehabilitation. Int J Neural Syst. 2022;32(4):2250009. https://doi.org/10.1142/S0129065722500095.
297. Kawano T, Hattori N, Uno Y, Hatakenaka M, Yagura H, Fujimoto H, et al. Electroencephalographic phase synchrony index as a biomarker of poststroke motor impairment and recovery. Neurorehabil Neural Repair. 2020;34:711–22.
298. Vecchio F, Tomino C, Miraglia F, Iodice F, Erra C, Di Iorio R, et al. Cortical connectivity from EEG data in acute stroke: a study via graph theory as a potential biomarker for functional recovery. Int J Psychophysiol. 2019;146:133–8.
299. Jamil N, Belkacem AN, Ouhbi S, Lakas A. Noninvasive electroencephalography equipment for assistive, adaptive, and rehabilitative brain–computer interfaces: a systematic literature review. Sensors. 2021;21(14):4754. https://doi.org/10.3390/s21144754.
300. Wolpaw JR, Millán JD, Ramsey NF. Chapter 2–Brain-computer interfaces: definitions and principles. In: Ramsey NF, Millán JD, editors. RBT-H of CN Brain-computer interfaces. Elsevier; 2020. p. 15–23.
301. Ma Y, Gong A, Nan W, Ding P, Wang F, Fu Y. Personalized brain–computer Interface and its applications. J Pers Med. 2023;13(1):46. https://doi.org/10.3390/jpm13010046.
302. Yuan H, He B. Brain-computer interfaces using sensorimotor rhythms: current state and future perspectives. IEEE Trans Biomed Eng. 2014;61:1425–35.
303. He B, Baxter B, Edelman BJ, Cline CC, Ye W. Noninvasive brain-computer interfaces based on sensorimotor rhythms. Proc IEEE Inst Electr Electron Eng. 2015;103:907–25.
304. Abiri R, Borhani S, Sellers EW, Jiang Y, Zhao X. A comprehensive review of EEG-based brain-computer interface paradigms. J Neural Eng. 2019;16(1):011001. https://doi.org/10.1088/1741-2552/aaf12e.
305. Pfurtscheller G, Neuper C, Flotzinger D, Pregenzer M. EEG-based discrimination between imagination of right and left hand movement. Electroencephalogr Clin Neurophysiol. 1997;103:642–51.
306. Pfurtscheller G, Guger C, Müller G, Krausz G, Neuper C. Brain oscillations control hand orthosis in a tetraplegic. Neurosci Lett. 2000;292:211–4.
307. Guger C, Edlinger G, Harkam W, Niedermayer I, Pfurtscheller G. How many people are able to operate an EEG-based brain-computer interface (BCI)? IEEE Trans neural Syst Rehabil Eng. 2003;11:145–7.
308. Guger C, Ramoser H, Pfurtscheller G. Real-time EEG analysis with subject-specific spatial patterns for a brain-computer interface (BCI). IEEE Trans Rehabil Eng. 2000;8:447–56.
309. Wolpaw JR, McFarland DJ. Multichannel EEG-based brain-computer communication. Electroencephalogr Clin Neurophysiol. 1994;90:444–9.
310. Wu SL, De LL, Liou CH, Chen SA, Ko LW, Chen BW, et al. Design of the multi-channel electroencephalography-based brain-computer interface with novel dry sensors. Annu Int Conf IEEE Eng Med Biol Soc IEEE Eng Med Biol Soc Annu Int Conf. 2012;2012:1793–7.
311. Wolpaw JR, McFarland DJ, Neat GW, Forneris CA. An EEG-based brain-computer interface for cursor control. Electroencephalogr Clin Neurophysiol. 1991;78:252–9.
312. Wolpaw JR, McFarland DJ. Control of a two-dimensional movement signal by a noninvasive brain-computer interface in humans. Proc Natl Acad Sci USA. 2004;101:17849–54.
313. Müller-Putz GR, Scherer R, Pfurtscheller G, Rupp R. EEG-based neuroprosthesis control: a step towards clinical practice. Neurosci Lett. 2005;382:169–74.
314. Baxter BS, Decker A, He B. Noninvasive control of a robotic arm in multiple dimensions using scalp electroencephalogram. In: 2013 6th International IEEE/EMBS Conference on Neural Engineering (NER). 2013, pp 45–47.
315. Sarac M, Koyas E, Erdogan A, Cetin M, Patoglu V. Brain computer Interface based robotic rehabilitation with online modification of task speed. IEEE Int Conf Rehabil Robot. 2013;2013:6650423.
316. Hayta Ü, Irimia DC, Guger C, Erkutlu İ, Güzelbey İH. Optimizing motor imagery parameters for robotic arm control by brain-computer interface. Brain Sci. 2022;12(7):833. https://doi.org/10.3390/brainsci12070833.

317. Li Y, Long J, Yu T, Yu Z, Wang C, Zhang H, et al. An EEG-based BCI system for 2-D cursor control by combining Mu/beta rhythm and P300 potential. IEEE Trans Biomed Eng. 2010;57:2495–505.
318. Huang D, Lin P, Fei D-Y, Chen X, Bai O. EEG-based online two-dimensional cursor control. Annu Int Conf IEEE Eng Med Biol Soc IEEE Eng Med Biol Soc Annu Int Conf. 2009;2009:4547–50.
319. Palumbo A, Gramigna V, Calabrese B, Ielpo N. Motor-imagery EEG-based BCIs in wheelchair movement and control: a systematic literature review. Sensors (Basel). 2021;21(18):6285. https://doi.org/10.3390/s21186285.
320. Colucci A, Vermehren M, Cavallo A, Angerhöfer C, Peekhaus N, Zollo L, et al. Brain-computer interface-controlled exoskeletons in clinical neurorehabilitation: ready or not? Neurorehabil Neural Repair. 2022;36:747–56.
321. Le Franc S, Herrera Altamira G, Guillen M, Butet S, Fleck S, Lécuyer A, et al. Toward an adapted neurofeedback for post-stroke motor rehabilitation: state of the art and perspectives. Front Hum Neurosci. 2022;16:917909. https://doi.org/10.3389/fnhum.2022.917909.
322. Onagawa R, Muraoka Y, Hagura N, Takemi M. Neurofeedback training for improving motor performance in healthy adults: ah systematic review and meta-analysis. bioRxiv. 2022:2022.04.26.487963.
323. Behboodi A, Lee WA, Hinchberger VS, Damiano DL. Determining optimal mobile neurofeedback methods for motor neurorehabilitation in children and adults with non-progressive neurological disorders: a scoping review. J Neuroeng Rehabil. 2022;19:104.
324. Mane R, Chouhan T, Guan C. BCI for stroke rehabilitation: motor and beyond. J Neural Eng. 2020;17(4):041001. https://doi.org/10.1088/1741-2552/aba162.
325. Ros T, Baars BJ, Lanius RA, Vuilleumier P. Tuning pathological brain oscillations with neurofeedback: a systems neuroscience framework. Front Hum Neurosci. 2014;8:1–22.
326. Shindo K, Kawashima K, Ushiba J, Ota N, Ito M, Ota T, et al. Effects of neurofeedback training with an electroencephalogram-based brain-computer interface for hand paralysis in patients with chronic stroke: a preliminary case series study. J Rehabil Med. 2011;43:951–7.
327. Vilou I, Varka A, Parisis D, Afrantou T, Ioannidis P. EEG-neurofeedback as a potential therapeutic approach for cognitive deficits in patients with dementia, multiple sclerosis, stroke and traumatic brain injury. Life. 2023;13(2):365. https://doi.org/10.3390/life13020365.
328. Shu X, Chen S, Yao L, Sheng X, Zhang D, Jiang N, et al. Fast recognition of BCI-inefficient users using physiological features from EEG signals: a screening study of stroke patients. Front Neurosci. 2018;12:93.
329. Orban M, Elsamanty M, Guo K, Zhang S, Yang H. A review of brain activity and EEG-based brain–computer interfaces for rehabilitation application. Bioengineering. 2022;9(12):768. https://doi.org/10.3390/bioengineering9120768.
330. Welch K, Mead G. The impact of fatigue on neurorehabilitation. Oxford Textbook of NeuroRehabilitation; 2015. p. 359.
331. Myrden A, Chau T. Effects of user mental state on EEG-BCI performance. Front Hum Neurosci. 2015;9:308.
332. Linden DEJ, Turner DL. Real-time functional magnetic resonance imaging neurofeedback in motor neurorehabilitation. Curr Opin Neurol. 2016;29:412–8.

# Motion Analysis in Neurological Rehabilitation: From the Lab to the Clinic

7

Renato Baptista, Bernard Liew, Serena Pizzocaro, Xiaojun Zhai, Svonko Galasso, David Rügamer, Tamsin Waterkeyn, Issam Boukhennoufa, Xuqi Zhu, and Alessandro Marco De Nunzio

R. Baptista (✉) · A. M. De Nunzio
Research and Development Department, LUNEX, Differdange, Luxembourg

Luxembourg Health and Sport Sciences Research Institute A.s.b.l, Differdange, Luxembourg
e-mail: rbaptista@lunex.lu; adenunzio@lunex.lu

B. Liew · I. Boukhennoufa
University of Essex, Colchester, UK
e-mail: bl19622@essex.ac.uk; ib20472@essex.ac.uk

S. Pizzocaro
Research and Development Department, LUNEX, Differdange, Luxembourg

Luxembourg Health and Sport Sciences Research Institute A.s.b.l, Differdange, Luxembourg

Department of Electrical, Computer, and Biomedical Engineering, University of Pavia, Pavia, Italy
e-mail: spizzocaro@lunex.lu

X. Zhai · X. Zhu
Embedded Intelligent Systems (EIS) Laboratory, University of Essex, Colchester, UK
e-mail: xzhai@essex.ac.uk; xz18173@essex.ac.uk

S. Galasso
Research and Development Department, LUNEX, Differdange, Luxembourg

Luxembourg Health and Sport Sciences Research Institute A.s.b.l, Differdange, Luxembourg

Department of Electrical and Information Engineering, University of Cassino and Southern Lazio, Cassino, Italy
e-mail: sgalasso@lunex.lu

D. Rügamer
LMU Munich, Munich, Germany

Munich Center for Machine Learning, Munich, Germany
e-mail: david.ruegamer@stat.uni-muenchen.de

T. Waterkeyn
University of Technology and Applied Science, Al Musannah, Sultanate of Oman
e-mail: tamsin@act.edu.com


R. S. Calabrò (ed.), *Translational Neurorehabilitation*,
https://doi.org/10.1007/978-3-031-63604-2_7

## 7.1 Motor Impairments and Assessment in Neurological Rehabilitation

Neurological disorders, like Stroke and Parkinson's Disease (PD), are heterogeneous groups of conditions that lead to significant disability [1]. Many of these disorders result in complex multi-system impairments that negatively impact activities of daily living (ADLs), exercise, and sports. In most neurological disorders, walking is commonly impacted, where performance, quantity, and quality are affected [2, 3]. For example, the walking speed of post-stroke patients ranges from 0.18 to 1.03 m/s, compared with that of healthy age-matched adults who walked with an average speed of 1.4 m/s [4].

Motion analysis is commonly used in laboratories to quantify motor and functional impairments in patients with neurological disorders that specifically affect movement performance. An advantage of laboratory-based analysis is the presence of sophisticated equipment that can quantify mechanistic variables, such as three-dimensional (3D) joint kinematics, muscle activation patterns, muscle forces, and coordination patterns. Such variables provide a non-invasive means of understanding the complex human physiology and motor control of human behaviour and how the neurological disorder alters it. For example, post-stroke patients have reduced leg propulsion capacity on their paretic limb, contributing to a reduced walking speed [5]. However, post-stroke patients also walk using energetically more inefficient strategies compared to healthy individuals. For example, the former relied less on their ankle extensors for propulsion and more on their hip muscles [6, 7]. Individuals with neurological disorders, such as stroke and cerebral palsy (CP), experience problems swinging their legs [8, 9], which not only impedes forward progression but also increases the risk of tripping and falling. Specifically, patients suffering from neurological conditions have modelling studies which reduced peak knee flexion angle during swing (i.e. stiff knee gait) [8, 9]. The biomechanical cause of stiff knee gait is multifactorial, including reduced ankle plantarflexor push-off power during pre-swing, reduced knee flexion velocity during pre-swing [10], and elevated quadriceps activity during pre-swing [8], to name just a few.

In contrast to disorders like stroke, neurodegenerative disorders like PD, exhibit a unique set of motor impairments, such as freezing, tremors, and rigidity, which ultimately impair balance [11]. The severity of these motor impairments is typically assessed under supervised conditions by a clinician using semi-quantitative methods like the Unified Parkinson's Disease Rating Scale (UPDRS) [12]. Although the UPDRS has acceptable psychometric properties [13], it is limited by the need for an external assessor to evaluate it, it also takes a long time [14], and it is limited by the partially subjective nature of the evaluation. The static nature of using questionnaire-based assessment tools means that treatments for such neurological conditions cannot be rapidly adjusted based on the day-to-day changes in symptoms and signs. For example, freezing occurs more frequently during the medication off-state than on-state. Real-time detection of the onset of freezing, its frequency, and duration throughout the day is the potential for rapid adjustment of pharmacological management of the disorder [15].

Neurological disorders do not only affect the quality of movement, but also it negatively impact the quantity and type of physical activity participation [16]. The lack of adequate physical activity can have a major influence on the patient's mental health [17], social participation [18], and health-related quality of life [19]. The World Health Organization (WHO) recommends that adults living with some form of disability undertake at least 150–300 min of moderate-intensity physical activity per week [20]. However, neurological patients such as stroke survivors routinely do not meet the recommended physical activity guidelines. For example, a meta-analysis reported that patients living with stroke achieve approximately 60% of the recommended daily step counts for people with disability [21]. Individuals with early PD have been reported to walk also about 56% of the steps undertaken per day by healthy adults [22].

## 7.2 Markerless Motion Capture Assessment

### 7.2.1 Challenges and Solutions for Monocular Camera Gait Analysis System

Traditionally, optoelectronic motion capture systems have stood as the most prominent and effective measurement systems employed in clinical gait analysis [23]. Despite their proven capabilities, however, these systems encounter significant limitations when deployed in real-world environments. These systems are bulky, costly, and require professional installation, thus making them impractical for widespread clinical use. Another solution to clinical gait analysis involves a multi-camera or camera network [24] to analyse kinematics and reconstruct 3D human postures. Nevertheless, these methods also have their limitations, including time-consuming installation, complex calibration procedures, and largely restricting subjects to collect gait data under controlled laboratory conditions.

In contrast to optoelectronic and multi-camera systems, a single-camera-based *markerless* system represents an attractive alternative solution. However, monocular cameras only capture a two-dimensional view of the subject, which makes it difficult to accurately measure three-dimensional joint angles and other parameters related to gait analysis, such as step width. Recent advances in computer vision and *Deep Learning* (DL) provide powerful deep neural network models to reconstruct 3D human posture by estimating the joint locations in 2D perspectives [25].

Human pose estimation models provide predicted joint locations that can be used to fill the lost information during limb overlapping, but the uncertainty of the prediction leads to undesired *noise* in the joint angle signal. To address this issue, the Kalman Filter (KF) and Frequency Domain Filter (FDF) can be used [26]. The KF treats the joint angle changes as a classical kinematic model (linear function) during a short period of time and obtains the posterior joint angle by combining noised measurement and predicted value under the guidance of a Gaussian distribution model. If a pose estimation model returns a low confidence prediction (i.e. ≤40%), the KF will replace uncertain measurement by the latest posterior joint angle.

However, the KF algorithm has a limited effect on removing extreme values and waveform distortion. To mitigate against extreme values, the FDF can be used to de-noise and smooth the joint angle signals. The FDF is implemented by Discrete Fourier Transform (DFT), inverse-DFT, and filter strategy, where the filter strategy plays an important role to reduce noise. The most common filter strategy is to pick the components with the highest energy in the frequency domain to recover the denoising signal, because of the fact that effective components usually have most of the energy.

## 7.3 Wearable Sensors

### 7.3.1 Activity Recognition

The use of sensors, algorithms, and machine learning to automatically recognise and classify physical activities a human does is called activity recognition (see Fig. 7.1). This technology can be used in neurorehabilitation to monitor and study the movements and behaviours of individuals suffering from neurological illnesses. These systems aim to identify and differentiate specific movements performed by patients during rehabilitation for record-keeping and monitoring. In this category, researchers have focused on monitoring Activities of Daily Living (ADL), with a particular emphasis on detecting general activities such as standing, sitting, lying down, standing up, and sitting down [27]. Activity recognition technology can offer

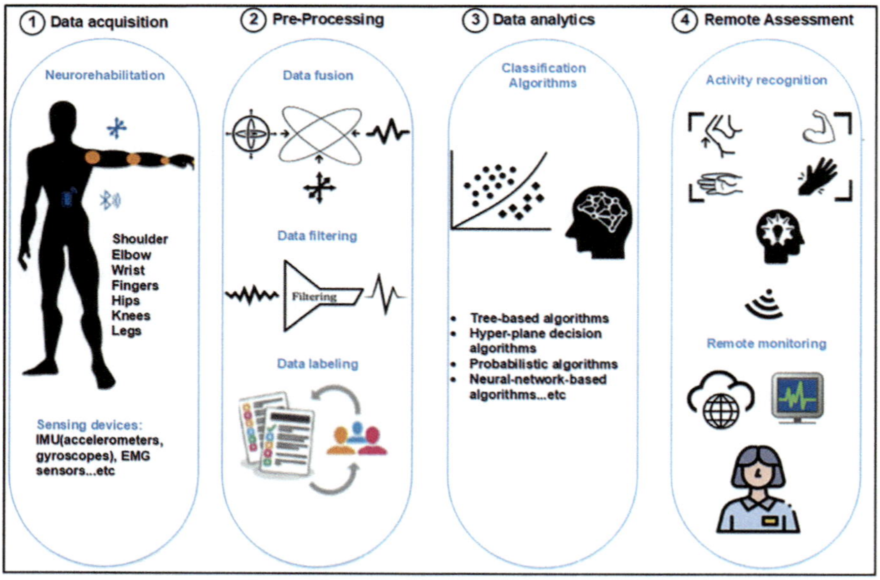


**Fig. 7.1** Remote activity recognition

patients and physicians real-time feedback on the quality and number of movements done during rehabilitation activities, hence boosting rehabilitation efficacy. It can also give objective and quantitative indicators of patients' progress during rehabilitation, allowing physicians to tailor treatment strategies to match the requirements of individual patients. Moreover, this technology allows patients to undertake rehabilitation activities outside of clinical settings, increasing the frequency of rehabilitation.

Researchers have primarily focused on monitoring ADL, including detecting general activities such as standing, sitting, lying, standing up, and sitting down, as well as performing kitchen tasks like making a drink and chopping food, routine activities like making the bed, reading, tying shoe laces, sweeping, and brushing teeth [27]. Additionally, some researchers have studied specific body parts, such as recognising different hand gestures and arm gestures, as well as exercises to strengthen shoulders and arms [28].

The placement of sensing technology on the human body has demonstrated a diverse range of positions, which is influenced by the specific type of technology employed and the intended purpose of the monitoring system [27]. In the context of upper-limb rehabilitation, sensing technology is typically placed on the wrists, arms, forearms, fingers, and hands to monitor activities involving hand usage [29]. Conversely, monitoring systems designed for lower-limb activities, such as walking, often utilise placement on the chest, shank, thighs, feet, hip, waist, or lower back. The selection of these specific locations is primarily determined by the monitoring system's intended objective [27].

### 7.3.2 Motor Deficits Monitoring

Research has demonstrated the effectiveness of wearable devices in evaluating PD symptoms, with results comparable to those obtained using standard clinical evaluation scales. Wearable devices overcome the limitations associated with clinical evaluation scales, such as subjectivity, lack of accuracy, and poor sensitivity [30]. Assessment scales are limited by the clinician's subjective judgement and professional knowledge, as well as by patients' emotional state, level of compliance, and recall bias. On the other hand, wearable technology improves the accuracy of evaluation by objectively quantifying PD symptoms. Wearable devices are highly sensitive and capable of detecting subtle motor abnormalities, facilitating early diagnosis or differential diagnosis, and monitoring changes in the motion state of patients over time [31].

Numerous attempts exist to quantify bradykinesia in PD, with many of them utilising accelerometry, sometimes in conjunction with gyrostat or gait analysis [32]. Among these, stride length has been found to be the most effective indicator of bradykinesia [33]. Current approaches involve utilising wearable inertial sensors (accelerometer, with or without gyroscopes) integrated into shoes, watch-like devices, or sensors placed on different body parts to measure movements, including gait patterns [34]. In most cases, the device is used in conjunction with the investigation of specific motor tasks. For example, using the standard "Timed-Up-and-Go" test in combination with a single accelerometer fixed to the lower back, it was reported that just three features combined had the best accuracy for discriminating between healthy and mild PD patients [35]. Recent studies have shown that various gait parameters measured by wearable sensor-based devices strongly correlate not only with the disease state but also with bradykinesia scores [36]. Thus, the data provided so far using wearable sensors and specific motor tasks suggest that these methods can characterise PD motor impairment and may be useful for prospective follow-up and monitoring of disease progression.

The evaluation of tremor can be done through clinical methods such as the UPDRS [13] or by quantifying drawn spirals using scales such as the Bain and Findley scale or the Fahn-Tolosa-Marin tremor rating scale [37]. Automated tremor analysis using accelerometry and electromyography can also assess tremor frequency, amplitude, and central oscillator, but it is still challenging to differentiate between different types of tremors [38]. Digitising tablets can also be used to detect and quantify tremor characteristics. Inertial sensors such as a 3-axis accelerometer and a 3-axis gyroscope have been developed to assess tremors automatically. Still, the differentiation between PD tremor and tremor due to essential tremor cannot yet be consistently made [39]. Several accelerometer-equipped watches or mobile medical apps are available on the market to record tremors in home settings, but caution must be exercised when using these apps in professional patient care or clinical studies as they may not provide accurate and diagnostically relevant information about various types of tremors. While the portability and ease of use of these devices may help translate them into routine clinical use, their suitability to improve patient care, including health-related quality of life, must be demonstrated in prospective clinical studies [40].

### 7.3.3 Kinematics-Based Approach for Kinetic Assessment—Joint Moments

Kinetic measures, such as joint moments, powers, and contact forces, are commonly calculated variables in clinical biomechanics that provide indirect measures of muscular effort [7], tissue loading [41], and motor control strategies [42]. However, the traditional method of calculating kinetic biomechanical measures relies on specialised equipment that cannot be replicated clinically [43]. Quantifying kinetic variables outside the biomechanics laboratory is increasingly possible using machine learning (ML), which is coupled with variables that can more reasonably be collected clinically or in the field. Briefly, ML creates statistical models that map the relationship between a set of predictors—which are typically easier to collect than kinetic variables, and the outcome—kinetic variables [43, 44]. Predictors that have been used in ML models include optical camera-based kinematics [43–46], inertial measurement units (IMUs) based kinematics [47–50], kinematics from markerless motion capture [51], or using a combination of wearable sensors like IMUs, electromyography (EMG), and pressure insoles [52, 53]. Neural networks are the most common ML method used in contemporary biomechanics research for predicting kinetic variables [44, 46, 49–51, 53, 54], although methods suggesting gradient boosting have been used [50, 52].

Many ML studies have focused on predicting knee moments of force [44, 49, 50, 53, 55], where predictive performance has ranged from 0.06 Nm/kg (root mean squared error [RMSE]) [53] to 0.15 Nm/kg (mean absolute error [MAE]) [51]. A recent study achieved an average performance of 0.06 Nm/kg for joint moment prediction across all joints and varying walking situations (e.g. stairs, treadmill, ramp) [52]. The type of variables collected and used for prediction influences prediction accuracy. For example, kinematics from markerless motion cameras may suffer from greater inaccuracy [51], compared to kinematics from other sources like IMUs [52]. This is because markerless motion capture methods themselves rely on ML models for generating kinematic features from skin landmarks.

The performance of ML models not only depends on the type of input predictors and ML learning algorithms but also the methods of validation and performance metrics used [56]. Two distinct validation methods have been used, including splitting the training and testing data set by subjects ($\text{Split}_{\text{subject}}$) or splitting the dataset by trials ($\text{Split}_{\text{trials}}$). For example, early work by our group reported that knee flexion-extension joint moment had an average RMSE of 0.069 Nm/kg using $\text{Split}_{\text{trials}}$ and 0.208 Nm/kg when using $\text{Split}_{\text{subject}}$. In contrast, when the ML model performance is already very good, for example, the ankle inversion-eversion moment, the difference between $\text{Split}_{\text{trials}}$ and $\text{Split}_{\text{subject}}$ may not be clinically relevant, at 0.033 Nm/kg and 0.081 Nm/kg, respectively. Whether to train and validate ML models using a $\text{Split}_{\text{trials}}$ or $\text{Split}_{\text{subject}}$ approach depends on the intended use case of the model. If the model is to be used in a clinical environment where predictions are required on new patients, a $\text{Split}_{\text{subject}}$ is required. If the model is to be used in a scenario in that a participant comes for a baseline biomechanics assessment to develop a personalised model for the prediction, then a $\text{Split}_{\text{trials}}$ may be appropriate [44].

## 7.4 Explainable Artificial Intelligence

Explainable AI (XAI) methods are designed to help humans better understand the decision-making process of complex AI models. There are several approaches to XAI, such as post-hoc interpretation and intrinsic interpretable models [57]. Post-hoc interpretation methods involve analysing the output of an AI model after it has made a decision, such as by creating visualisations of the important features used in the decision or using local surrogate models to approximate the behaviour of the original model [58]. Intrinsic interpretable models, such as linear models, are designed to be transparent from the start, with features that are easily understandable by humans. The basic principle behind all these XAI methods is to make AI models more understandable, interpretable, and trustworthy so that they can be more effectively integrated into decision-making processes.

For more complex data types, such as time series signals from motion sensors, one interpretable approach successfully used in biomechanical studies is *functional data boosting* (FDboost) [59–61]. FDboost is based on functional regression models. These models consider repeated time series observations as so-called functional data and extend classic statistical models by allowing for functional covariates and/or functional responses in the model. Functional regression models thereby provide statements on how different signals can influence another signal of interest (the outcome) for every possible time point combination. Functional regression models can also be used to determine a scalar outcome, such as a classification of the patient's pain category [60, 61] by integrating over all time points of the input signals. FDboost combines the idea of functional regression models with a fitting routine called gradient boosting from machine learning to inherit the interpretability of functional regression models while making use of the prediction performance and scalability of gradient boosting. Using a special type of gradient boosting, FDboost also allows for feature (signal) selection and thereby provides insights into the importance of signals and further simplifies the model's interpretation.

As it is not possible to break down all complex relationships, methods combining interpretable models with more complex ones have emerged in the past. One example is semi-structured regression models, combining statistical regression models and neural networks [62]. These methods allow us to partly interpret relationships yet provide enough flexibility to model more complex interactions beyond comprehensible limits. By combining the deep neural network and the statistical regression predictor additively (in one unifying neural network) and ensuring orthogonality between the two parts, semi-structured regression models also allow for decomposition of variance of the predictions. This gives insights into how much of the network's prediction and accuracy can be related to the interpretable model part and how much is due to more complex interactions between features.

As discussed previously, another direction to increase the intelligibility of models is by inducing sparsity. This facilitates better interpretation and gives insights into the importance of predictors. Obtaining sparse models in deep learning is not straightforward as networks naturally come with less structure. Their optimisation is usually done with methods that do not work well with non-smooth optimisation problems, such as those induced by sparsity assumptions. One possible approach to enforce sparsity among certain neuron groups in neural networks or for network inputs is via a so-called Hadamard product parameterization [62]. This generic approach can be used to transform the optimisation problem and thereby allows standard deep learning toolkits to provide sparsity at various levels of the network. One particularly interesting case of usage is the sparsity in the input signals for multivariate time series networks. Using HPP, this can be implemented independent of the network architecture and results in a sparse, better interpretable deep network by only using some of the multiple input signals.

## 7.5 Overview of Machine Learning Models

- *Multilayer Perceptron* (MLP) is a feed-forward artificial neural network widely used for solving supervised learning problems, characterised by several layers of input nodes connected as a directed graph between the input and output layers. The number of hidden layers between the input and output layer depends on the problem and remains one of the main challenges while tuning MLP [63].
- *Logistic Regression* is a machine learning classification technique that uses a logistic function to predict a binary outcome. It is an extension of the linear regression model that cannot be used for classification problems [64].
- *Support Vector Machine* (SVM) algorithm is a supervised learning classifier that constructs a transformed space of the input feature space dividing each class by constructing a boundary based on a specific kernel function (linear, polynomial, gaussian, sigmoid, etc.) [65].
- *The K-Nearest Neighbors* (KNN) algorithm assigns each new instance of the test set to one of the classes in the training set by evaluating its distance from the nearest point in the feature space. The Euclidean distance was the metric used to evaluate the distance between points [66].
- *Adaptive Boosting* (AdaBoost) is an ensemble learning method that combines a set of weak learners (decision stumps) into strong learners to minimise training errors. At each iteration, the weights of AdaBoost instances are re-assigned based on the misclassification error of the previous iteration. Misclassified instances are assigned higher weights [67].
- *Random Forest* is a tree-based machine learning algorithm that leverages the power of multiple decision trees for making decisions. The random forest algorithm combines the output of multiple, randomly created decision trees (ensemble approach) to generate the final output [68].
- *Random Tree* combines the principles of single-model trees with the ideas of random forest. It selects a K number of attributes to calculate the entropy at each node without performing pruning [69].
- *Reduced Error Pruning Tree* (REPTree) is fast decision tree learning which uses regression tree logic to create multiple trees and selects the best among all generated trees. The mean square error on the predictions made by the tree is used to prune the tree [70].

## 7.6 Machine Learning: Limitations

Machine learning (ML) based systems, such as those used in neuro telerehabilitation, require rigorous computational models to achieve the desired results and accurately estimate the needed parameters. However, constructing an efficient model requires a massive amount of data, and the most sophisticated algorithms (e.g. deep learning) require at least ten times the number of samples as parameters in the network. While these algorithms thrive in domains with large amounts of data (e.g. computer vision), the number of patients in neurorehabilitation is limited, and they may not always be willing to participate in research projects due to the extra burden it places on them. Potential solutions to overcome the lack of big data in neurorehabilitation are using deep learning methods or biomechanics solutions to artificially augment data [71, 72].

Despite Artificial Intelligence (AI) being widely used in many fields, however, due to interpretability and privacy issues, many ethical and safety issues within healthcare remain. In addition, due to the complexity of AI, it is challenging to implement it in low-cost portable devices that are constrained by energy and computing efficiency [73]. Reducing AI complexity by retaining the most important features can make AI more transparent and increase portability, but this may make AI more vulnerable to privacy attacks and less protective [74]. Furthermore, AI is a data-driven solution that often needs multiple participants to train local models cooperatively, which imposes other privacy challenges of disclosing sensitive data to a central cloud server. These challenges cannot be addressed excepted through an interdisciplinary approach, and so this project draws together proven experts in computer science and electronic engineering, machine learning and networking, sport science and clinicians, and physiotherapists to foster new understanding and derive concrete guidelines for successful trust, efficient, and protective edge AI solution in healthcare.

Currently, there is little definite structural information for considering AI performance, privacy, reliability, and portability for their complex and independent deployment environments, which has been identified in important human–computer interaction pathways [75]. This is because of the technical problems inherent in data-driven AI optimisation techniques. For example, a trained AI model often contains much redundant information, which includes sensitive and non-sensitive information, to deploy the model on low-cost portable devices. The most common solution is removing redundant information, which leaves sensitive information more exposable to privacy breaches [76]. Interacting with the model with end users makes it hard to handle. It becomes an optimisation problem requiring multi-disciplines scientists to work together to produce a unified and dynamic solution from low-level hardware/software to the end users.

Although deep learning-based AI applications show attractive potential capabilities in edge computing, the considerable runtime computational resources cost is a significant challenge for employing such AI applications on resource-limited edge equipment. Pruning and quantizing are mostly adopted optimisation methods to allow the DNNs to be implemented on resource-constrained embedded systems with reasonable energy costs [77]. These types of methods modify the network structure to remove redundant weight and network connections, which will reduce the size of the model as well as the accuracy of the model, but with an improvement in its running efficiency [78]. Neutral Architecture Searching (NAS) is widely used to find the optimised neural networks for the given computing hardware. Together with optimised computing hardware for the given network architecture, it is possible to achieve better neural network running efficiency with improved speed and energy efficiency.

## 7.7 Markerless Motion Capture Assessment of Kinematics

### 7.7.1 Introduction: Contextualising Commercial Solutions for Motion Analysis

As mentioned previously, in Sect. 7.1, human motion analysis and biomechanics are fundamental for clinical assessments, and together they provide relevant information towards diagnosing numerous neurodegenerative conditions, not only strokes [79] but also Parkinson's Disease (PD), Alzheimer's Disease (AD), multiple sclerosis, Huntington's Disease (HD), and Amyotrophic Lateral Sclerosis (ALS). Human motion analysis aims at understanding the cause of altered motion patterns, mainly assisting with the prevention, identification, and rehabilitation of a comprehensive list of neurodegenerative pathologies [80]. Traditionally, motion analysis assessment relies on the patient's self-report and the practitioner's visually assessed observations [80]. Therefore, such assessments are often subjective and susceptible to human-induced error [81, 82]. In contrast, sophisticated devices can provide practitioners with quantitative accuracy by equipping practitioners with precise, reliable, and objective measurements [79, 80] to simultaneously monitor an extensive set of parameters for gait analysis. These devices are usually based on floor-mounted measurement systems, body-mounted devices, or image-processing techniques.

### 7.7.2 Comparison: Marker-Based and Markerless Motion Capture

However, analysing gait data with such sophisticated devices requires significant technical expertise and is expensive for most clinics and hospitals, especially in rural areas [79]. The amount of information these systems bring to the practitioner includes many variables which may represent a real challenge for the practitioners to interpret in clinical settings [79, 83]. For the above reasons, an easy-to-use, objective, portable, and low-cost solution that quantifies gait analysis can allow practitioners to focus their resources on prescribing effective and patient-tailored treatments [79, 84]. Hence, video-based motion capture systems represent a viable solution for recording and processing images to efficiently identify limb location and orientation in three dimensions (3D) [80]. Video-based motion capture systems are divided into two categories: *marker-based* and *markerless* motion capture.

- *Marker-based motion capture systems* precisely track human motion in 3D space [85] and are considered the gold standard in terms of accurately tracking human gait analysis [85, 86]. *Marker-based motion capture systems* are usually configured with predefined cameras and lightning conditions to observe and track reflective markers attached to the human body in order to infer the configuration of a skeleton model that best fits the marker positions [86].

- *Markerless* systems do not rely on the markers attached to the human to infer the skeleton model. Instead, they rely on significantly challenging processes of gathering information from the images [86]. *Markerless motion capture systems* have the potential to alleviate the required technical and practical issues by replacing physical palpation and skin surface marker tracking of body landmarks [87].

### 7.7.3 Markerless Motion Capture

In general, *markerless systems* consist of two stages: **Stage one** is the offline stage, where the model is designed and trained using deep learning-based algorithms considering manually annotated data**. Stage two:** the visual information is input into the trained model to estimate the human body pose (so-called human pose estimation) [86]. *Markerless* motion capture systems can be categorised into two primary families of camera systems, with and without depth-sensing cameras [80, 86]. Systems with depth-sensing cameras record a standard video and simultaneously record the distance between each pixel and the camera. Depth-sensing cameras have a few drawbacks, such as capture rate, limited working range, and controlled lighting conditions [80, 88, 89]. Figure 7.2 below shows a general *markerless* motion capture system overview:

*Markerless motion* captures comprising single- or multi-camera systems that use standard video cameras are relatively new and quickly emerging as a viable

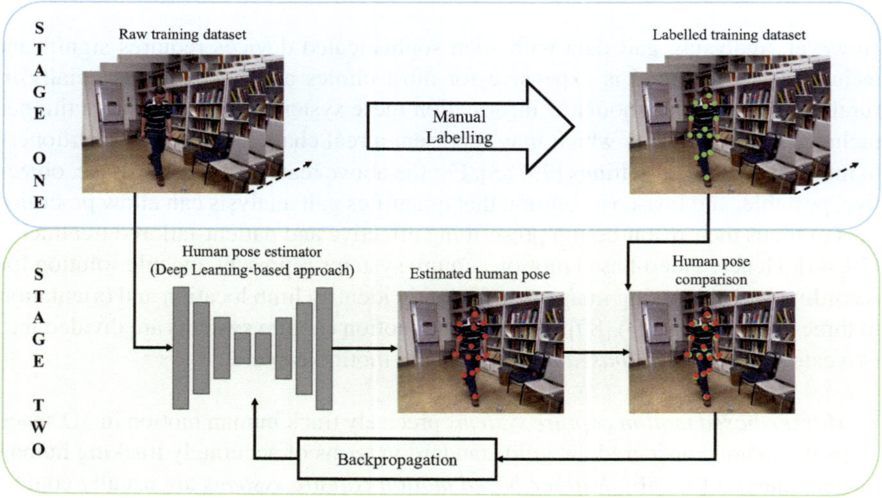


**Fig. 7.2** Schematic drawings of a general *markerless* motion capture system overview (Stage 1) manual labelling and (Stage 2) backpropagation method. (Modified from "Applications and limitations of current markerless motion capture methods for clinical gait biomechanics," by Wade, L., 2022 2007, Developmental Psychology, 43, p. 1515 [80] Creative Commons Attribution BY 4.0)

technology for clinical rehabilitation and injury prevention applications [80]. Single-camera markerless motion capture involves obtaining joint centre locations from a single image or video using two-dimensional (2D) human pose estimation algorithms [80, 87, 90–92]. Single-camera markerless systems have benefits; for example, they are cheaper, limit errors in incorrectly placed markers, and occupy less environmental space; however, self-occlusion errors are inevitable [80]. Self-occlusion errors may happen when a joint centre location is missing for one or more video frames and when the opposite limb is obstructed or incorrectly detected [80, 93, 94]. Furthermore, to obtain biomechanically relevant 2D joint angles, an assumption must be taken into account. Such an assumption is based on the fact that the camera is perfectly aligned with frontal or sagittal plane movements [80, 94]. Obtaining 3D joint centre location from single-camera systems aims to estimate the corresponding joint location in 3D, considering a 2D video as input information [95]. Since the direction of the movement can vary (plane), entire limbs may be occluded for more extended periods. Multi-camera motion capture systems are a progression of single-camera methods that overcome self-occlusion errors by adopting a multi-camera setup. These systems combine 2D pose estimation with an additional multi-camera reconstruction step to estimate 3D joint centre locations [80, 96–98]. There are three open-source deep learning approaches: OpenPose [90], DeepLabCut [87], and AlphaPose [91].

These open-source deep learning approaches aim to perform human pose estimation from visual information. Recently, several studies have studied the performance of human pose estimation from *multi-camera markerless systems* and compared this with *marker-based systems* [96, 98, 99]. Such performance measures are reported in joint centre location and temporospatial differences by comparing slower and faster movements. In line with [100], the results are now approaching the error rates of marker-based motion capture systems. Interestingly, a significant systematic difference appears across the previously mentioned open-source methods, and it occurs at the hip joint due to the poorly labelled open-access datasets [25, 86, 101]. The hip joint is difficult to identify without physical palpation, and such inaccuracies can lead to unreliable joint centre locations.

### 7.7.4 Markerless Motion Capture for Clinical-Related Application

While *markerless motion capture systems* are a new and quickly emerging technology, several studies have demonstrated the corresponding potential for clinical-related applications [102–108]. Eichler et al. [104] developed a multi-camera system used in medical and telemedicine applications [104]. Their system focused on automatically assessing the medical condition of stroke patients via the stroke-specific performance-based impairment index, Fugl-Meyer Assessment (FMA) [109]. Their system consists of two Microsoft Kinect cameras ensuring each side of the patient's body is appropriately visible (positioned at a 45-degree angle to the patient's frontal view and a distance of around 1.5–2 m). Eichler et al. [104]

concluded that their results are consistent with the ones obtained using expensive *marker-based motion capture systems*. Additionally, their system detected trunk flexion (compensation) in patients attempting to move their hand to a target position compared to healthy participants [104]. Similar results were found in Michaelsen et al. [110], where infrared light-emitting diodes were placed on body landmarks of the hand, arm, and trunk [110]. Moro et al. [107] used the DeepLabCut approach to extract walking sagittal 2D joint angles in stroke patients, showing significant differences between the affected and unaffected sides. Additionally, they report no significant differences between marker-based gait analysis and their *markerless* solution, which consists of a simple RGB camera and the open-source human pose estimator [107]. As a concluding note, Moro et al. [107] claim that *markerless* solutions are a step forward in terms of performing gait analysis in a cheaper, less obtrusive, and simpler way, with a potential for a wide range of applications in clinical practice [107]. Similarly, Martinez et al. [106] combined a *markerless* system with OpenPose to assess walking cadence and automate the calculation of an abnormality score for Parkinson's disease patients, providing the practitioners with an unbiased and objective general overview of the patient's disease progression [106]. Instead, Shin et al. [108] analysed temporospatial outcome measures, such as step length, walking velocity, and turning time, from monocular frontal videos of Parkinson's disease patients. They validated the *markerless* solution with a pressure-sensitive walkway, and their findings demonstrated excellent reliability (ICC > 0.9). In addition, their system could detect minor gait disturbances unnoticed by the practitioner's subjective analysis, confirming its validity and importance for clinicians [108].

The before-mentioned solutions mainly analyse the 2D joint centre locations from single- or multi-camera system solutions. However, 3D human pose estimation is also being used to perform gait analysis, mainly because it adds an extra dimension to the 2D joint centre locations, the depth. Considering this, a few *markerless* systems can be identified and detailed:

The *iPi Motion Capture* [111] is a scalable *markerless* motion capture software that supports 1–4 depth sensors or a multi-camera setup (3–16 ordinary cameras) to track 3D human body motions, transfer motion to 3D characters and produce 3D animations [111]. Kotsifaki et al. [112] presented a study that aimed to examine the validity of the *iPi Motion Capture* system using 2 Microsoft Kinect v2 cameras during functional movements. The authors reported that a configuration with 2 Microsoft Kinect v2 is helpful for assessing sagittal plane knee range and peak angles during squat and jumping tests. Additionally, during the single-leg squat test, the agreement level between the thigh and foot adduction/abduction motion systems was excellent [112].

*The Captury* [113] system is a *markerless* and optical 3D motion capture system based on ordinary commercial video cameras. The system is based on sequential computer vision approaches to estimate the silhouette of the subject being captured [114]. Harsted et al. [114] conducted measurements using The Captury system, and they reported that the obtained results could not be considered interchangeable with marker-based measures (e.g. Vicon [115]). Nevertheless, their results suggest that The Captury can produce estimates with acceptable levels of agreement and reliability of jump length, jump height, knee flexion, knee-to-hip separation distance ratio, and others [114]. In a more recent study, Harsted et al. [116] used The Captury solution to explore the software-driven random error associated with *markerless* measurements in young children and the clinical relevance of these measurements. The authors reported that the jump length performance and four kinematic variables demonstrated acceptable instrumental variability (ICC > 0.76). Furthermore, the week-to-week reliability was excellent for jump length performance (ICC = 0.90) but poor to moderate (ICC < 0.55) for the kinematic variables [116].

The *SIMI Shape 3D* software enables motion capture without using markers attached to the human body [117]. The SIMI Shape 3D tracking pipeline can be divided into three steps: (1) segmentation, (2) model initialisation, and (3) tracking. In order to improve the tracking outcomes, a morph suit is recommended (e.g. orange colour) to provide a good contrast to the background environment [118]. Currently, no studies using the SIMI Shape 3D *markerless* motion capture have been found.

The *Theia3D* [119] is another *markerless* motion capture solution which uses deep convolutional neural networks to obtain 3D human pose estimation. Theia3D has been developed, tested for reliability, and validated against marker-based motion capture solutions in a controlled laboratory setting [120]. Compared to marker-based motion capture solutions and pressure-sensitive gait mats, Theia3D demonstrated good to excellent agreement for temporospatial parameters [121–123]. Riazati et al. [124] used the Theia3D *markerless* motion capture solution to investigate the test-retest measurement error of lower extremity kinematics and the temporospatial parameters of gait. Their data was collected in community and clinic-adjacent settings to address the feasibility of *markerless* motion capture systems for gait analysis as a clinical outcome in people with neuropathological conditions. The authors reported that the measurement error of the 3D kinematics was low, providing a basis for interpreting both clinical status and monitoring rehabilitation outcomes [124].

The usability of any of the three emerging technologies above, in terms of gait assessment, mobility, and balance in neurological conditions, lies in their capabilities to provide information regarding the detection, progression, and clinical outcome of the related pathology [120]. Specifically in neurorehabilitation, it is essential to acquire information regarding the underlying mechanisms contributing to altered gait and balance parameters [120]. From this review, the conclusion is that *Markerless motion capture systems* do have the potential to perform gait analysis and, additionally, reduce the time for data collection and data processing when compared to *marker-based motion capture systems*. Current pose estimation algorithms are approaching similar error rates regarding joint centre locations compared to marker-based motion capture. However, based on this evidence, a gap exists: a gold standard solution is still missing (e.g. bi-planar videoradiography) [80]. Hence, the main aspect to investigate further is the true accuracy of *markerless* systems, which is still unknown. It is also reasonable to conclude that because open-source human pose estimation approaches were never intended for biomechanical applications, this results in inconsistently and inaccurately labelled data [80]. Therefore, improvements in the labelling of publicly accessible data would be a crucial step forward for future developments in *markerless motion capture solutions* specifically for neurorehabilitation [80].

## 7.8 Wearable Sensors

### 7.8.1 Wearable Sensors: Contextualising Wearables Meant to Assess Muscular Activity

Surface electromyography (sEMG) is a non-invasive acquisition methodology that provides valuable information about the condition of the muscle and its role in movement and exercise [125]. sEMG can be used in rehabilitation medicine, prosthetic control, and gesture recognition of sign language [126]. Some examples of applications in clinics are as follows:

- Diagnosis and prevention of diseases (Huntington's, myopathies, or muscular dystrophies). The EMG signal helps detect neuromuscular disorders that affect the regular activity of motor units and identify the origin of these disorders [127–130]
- Control of prosthetics and motorised wheelchairs [131]
- Biofeedback therapy [132]
- Provides the instrument for a quantitative evaluation of progress in rehabilitation programmes [133, 134]
- Provides a quantitative approach in the field of sports medicine [135]
- Provides a quantitative approach in occupational medicine for the prevention of work-related neuromuscular disorders [136]

The sEMG signal represents the characteristics of muscle functions and provides information about their activities, which have been used in various fields, especially during rehabilitation [137]. The analysis of such signals can give the practitioners diagnostic information, which could be used to help the decision-making process in an appropriate treatment plan for muscle dysfunction [138].

In order to assess the gait function in orthopaedic patients, it is necessary to include specific kinematic and kinetic parameters. However, by combining the kinematic and kinetic parameters with *superficial EMG amplifiers* (sEMG) assessment, the overall evaluation may give a complete understanding of the patient's gait and muscle balances [139]. Moreover, the sEMG signal lets practitioners know when voluntary muscle activation is detectable. Together with kinematic and kinetic measurements from other wearable sensors, it may better characterise patient movements regarding gait function. Therefore, it could identify which muscles present weakness and are not playing an essential role during the gait analysis [137]. The beforementioned findings and clinical evaluation may present proper evidence to follow the patients in time and to plan tailored rehabilitation treatments after surgery, avoiding unnecessary intervention and obtaining the best possible gait performance [137, 140, 141].

### 7.8.2 Electromyographic Signal

Muscular contractions happen because electrical stimuli excite the muscle fibres. This electrical activity from the nerves propagates in muscular tissue due to its conductive ability and is measurable through the **electromyography** (EMG) technique [142]. sEMG is challenging to analyse as non-reproducible and hard to treat in a standardised way [125]. Despite these challenges, sEMG has a lot of applications [142, 143]. Electromyographic signals are studied to extract helpful information, such as the activation or non-activation of a muscle, the temporal relationships between the activation of different muscles, the muscles' activation percentage, muscle coordination, or the presence of fatigue [143].

Depending on the collection modality, there are two main types of EMG: surface EMG (sEMG), which uses skin surface electrodes (Fig. 7.3) and needle EMG which uses intramuscular electrodes [125].

A representative example of an sEMG raw signal collected by attaching electrodes to the vastus lateralis muscle during leg extension (knee at 90° angle) is shown in Fig. 7.4. The signal was recorded during a strong isometric contraction. The muscle contracts when the amplitude value is more prominent, and the signal is dense (between 1 and 7 s).

An example of the soleus's muscular activity during walking is shown in Fig. 7.5. Since walking has a cyclical pattern, the contractions rhythmically alternate with relaxation periods. The soleus contractions repeat in every gait cycle, but every contraction shape is different since EMG is a stochastic signal and never repeats exactly [125].

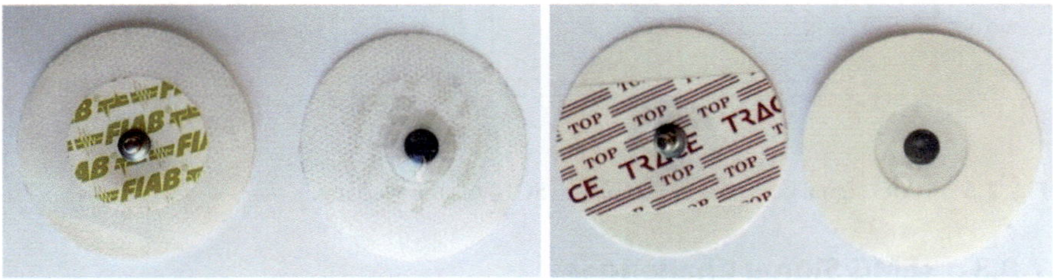


**Fig. 7.3** Examples of skin electrodes used for sEMG

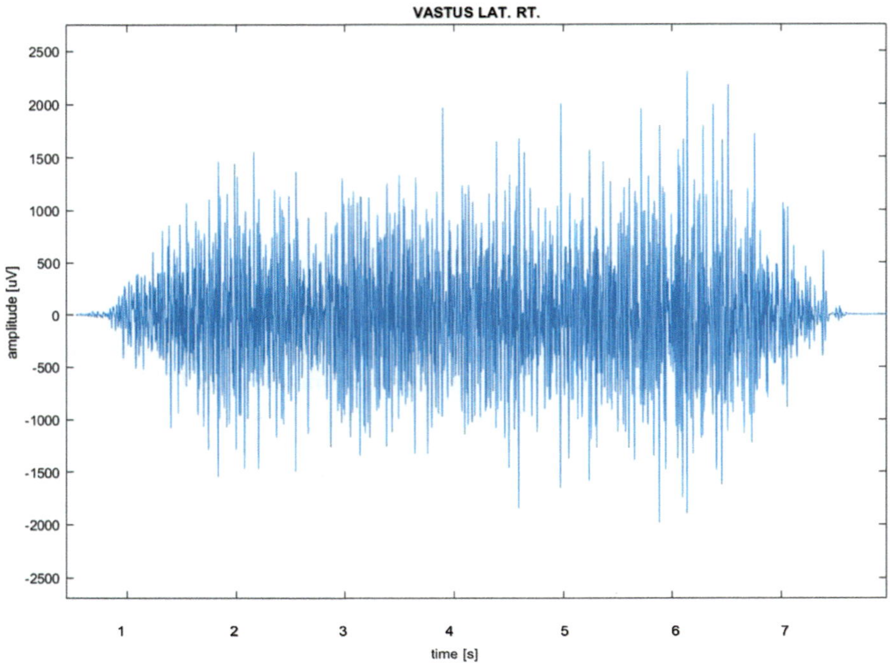


**Fig. 7.4** sEMG signal collected from vastus lateralis (VL) during a strong and continuous muscular contraction

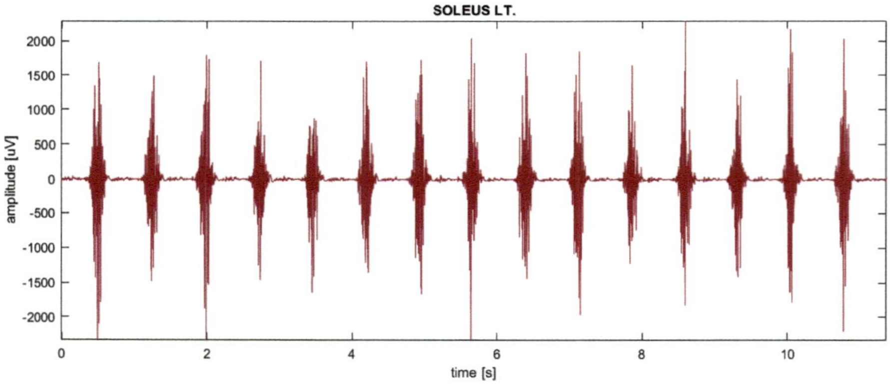


**Fig. 7.5** sEMG signal collected from soleus (SOL) during walking

### 7.8.3 sEMG Signal Challenges

The sEMG signal has to cross different subcutaneous tissue layers between the muscle and the electrodes. While travelling, various environmental factors can alter the signal in shape and characteristics [125]. Some examples are presented below.

One example is how the signal in shape and characteristics are altered by biological tissues that attenuate the signal during its propagation to the surface of the skin, such as other muscles, tendons, blood vessels, fat, and skin layers. For this reason, it is essential to prepare the skin area where electrodes will be applied. A solution to skin resistance is by using special abrasive conductive paste that helps removing dead skin cells and cleans the skin from dirt and sweat [125].

Another example of altering signals is how the joint and limb movements can change the relative position of the electrodes, and the muscle can migrate below the electrode site during joint movement. Such a relative motion could be evident in dynamic motion studies (e.g. running). These factors lead to a change in the value of the amplitude of the sEMG signal [125].

Other biological or electromagnetic signals are present in the environment (e.g. power line, Bluetooth, Wi-Fi) that can be collected by electrodes and EMG signal.

Finally, the movement between the electrode and tissue or the movement of cables connected to the collecting electronics can cause interferences called *moving artefacts*. The *power line* is always another source of noise due to alternating mode.

Ideally, all these unwanted signal components should be removed; otherwise, they can hide the informative content of EMG [142]. Denoising methods are widely used, both at hardware and software levels. At the hardware level, improving the performance of the acquisition instrument lowers the noise content of the signal [126], whereas, at the software, denoising includes filters [126]. These filters will be discussed subsequently. Moreover, power line interferences can be reduced thanks to accurate skin preparation, a correct amplifier, and moving subject and equipment far from power cables [144].

Special precautions should be applied when positioning the electrodes. They have to be applied in a direction parallel to the MFs and do not have to be over the ends of the muscle where the tendon is located or directly over the neuromuscular junction or innervation zone (Fig. 7.6).

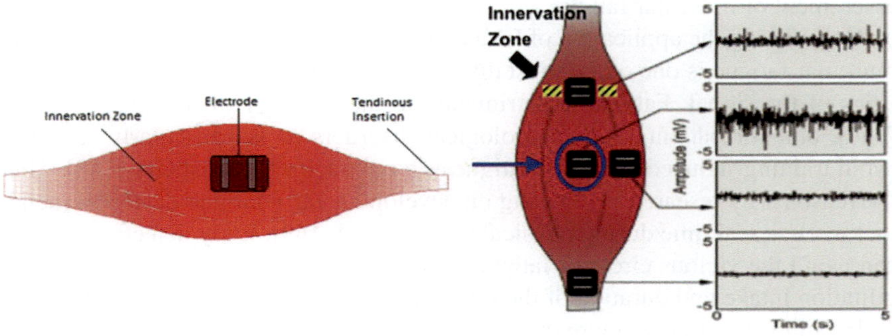


**Fig. 7.6** Best positioning of electrodes over a muscle avoiding the innervation zone, tendinous insertion, and following the muscular fibres (MFs) direction. The signal recorded from the electrode in the best position is the strongest. (Adapted from [145, 146])

### 7.8.4 Wearables to Assess Muscular Activity

The usefulness of sEMG in assessing novel and promising therapies for neurorehabilitation is evident [147]. sEMG provides data about the effectiveness of a rehabilitation intervention, treatment, or training [125]. Continuous research about these kinds of sensors brought to the development of progressively smaller and lighter wireless devices. These characteristics are important in motion analysis where capturing the subject's natural motion patterns is fundamental. Having sensors that do not interfere with the subject's movement allowing full range of motion is necessary for achieving this goal.

sEMG sensors market is various with multiple solutions available ranging from around 1000–25,000 euros. The simpler and relatively cheaper devices can be purchased for a bit more than a 1000 euros (e.g. Myo Gesture Control Armband Thalmic Labs—eBay.com). These devices are developed for users without any technical knowledge and in recent years are often associated with an app. More complex devices are composed of a set of probes for the analysis of several muscles at the same time. They are extremely sensitive and precise, they are developed for advanced users that need anatomical, physiological, and signal analysis knowledge to be able to use them. They are combined with powerful analysis software that allows the user to visualise signal waves and access raw data or even pre-process data, correct errors, and print reports. Some sEMG devices are certified medical devices which means that they have been validated for applications in clinical environment and can be used to assess patients.

Some examples of commercial sEMG solutions are listed in Appendix 3 with specifications.

## 7.9 sEMG Approaches to Assess Muscular Fatigue

The EMG signals collected from muscles are processed to extract features of interest (e.g. the activation instants, amplitude feature, spectral features, etc.) and for the assessment of muscular fatigue.

Considering the application of EMG analysis in rehabilitation, the assessment of muscular fatigue is one of the most difficult tasks due to the nonlinearity characteristics of the signal. Fatigue is detrimental to the recovery of impaired conditions [147], and for patients with neurological disorders, it is particularly relevant to avoid training-induced muscular fatigue as a risk factor for spasticity [148]. That is the reason why research is focusing on developing early fatigue detection methods that work in real time during physical activity [133, 149]. Early detection of fatigue can avoid the vicious circle of fatigue and spasticity [150] allowing to adapt rehabilitation intake and duration of the rehabilitation sessions to each patient's needs.

In fatigue conditions, more than one physiological process that enables muscular fibres to generate a force become impaired and, consequently, the muscle's electrical and mechanical characteristics change. The electrical factors include changes in the firing rate of action potentials and a different number of recruited motor units.

The mechanical variations include the reduction of the generated force from slow- and fast-twitching MFs [151].

When fatigue occurs, the metabolism inside the muscle changes, consequently, MFs properties modify, reducing their ability to conduct electricity [125]. Such modifications reflect in the sEMG signal. Piper [152], and after him, many other investigators [153, 154] observed that EMG frequencies shift when the muscle becomes fatigued [152]. Observations noted that this shift was directly derived from reducing the power in the high-frequency signal components. Moreover, it was demonstrated that as muscle fibres become fatigued during sustained activity, additional motor units are recruited to maintain the required level of force. The sEMG represents the summation of active MUs, so the more MUs that are recruited, the larger the EMG amplitude is observed [155].

The traditional approach to quantifying fatigue, which derives directly from the given definition of fatigue, measures changes in the force produced by the muscle undergoing Maximum Voluntary Isometric Contraction (MVIC). The test is repeated at two different time points, usually before and after performing the fatiguing task. Fatigue is indicated by a drop in the maximal voluntary contraction output. During the execution of an MVIC, before and after a fatiguing task, a series of features are extracted from the sEMG signal, both in the time and frequency domain, to study signs of local muscle fatigue [156]. The most used features are presented below in Table 7.1.

### 7.9.1 Time-Domain Features

The sEMG signal is band-pass filtered (typically 20–450 Hz) to eliminate the noise, rectified and the signal envelope extracted via a low-pass filter (cut-off frequency from 1 to 10 Hz) [157]. The average amplitude value of the envelope (*Averaged Rectified Value*, ARV) (Fig. 7.7 left) and the integral of the envelope curve (*Area Under the Envelope,* AUE) (Fig. 7.7 right) are studied at different time points to detect a possible increment due to fatigue.

**Table 7.1** Most used features to study sEMG signals during isometric constant-force contractions and how they change when muscular fatigue occurs

| Feature name | Feature description | Effects of fatigue |
|---|---|---|
| Averaged rectified value (ARV) | Average amplitude value of the envelope of the sEMG signal | Increase |
| Area under the envelope (AUE) | Envelope integral over an epoch | Increase |
| Median spectral frequency (MDF) | The frequency that divides PS into two regions of equal power | Decrease |
| Mean spectral frequency (MNF) | The frequency that corresponds to the barycentre of PS | Decrease |

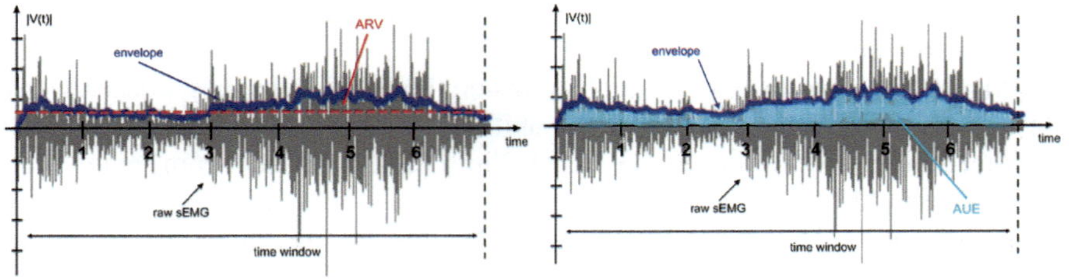


**Fig. 7.7** Average rectified value (left) and area under the envelope (right). (Adapted from [125])

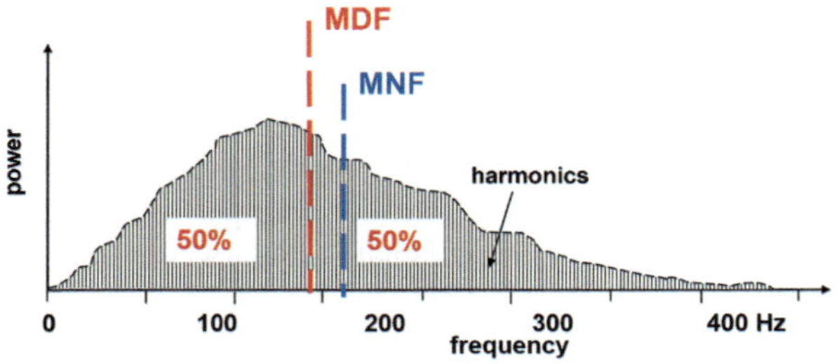


**Fig. 7.8** Median frequency (MDF) is a frequency at which the sEMG power spectrum is divided into two regions with equal area. Mean frequency (MNF) is the frequency that corresponds to the barycentre of the power spectrum (PS). (Adapted from [125])

### 7.9.2 Frequency Domain Features

EMG frequency spectrum shifts to lower frequencies when MFs conduction velocity declines (fatiguing conditions) [154]. Due to the decline of MFs conduction velocity, the collected signal has a higher power in the low frequency's harmonics [154]. This is the reason why fatigue can be detected by studying EMG frequency content.

The signal is band-pass filtered to eliminate the noise (typically 20–450 Hz), and the power spectrum (PS) is computed by Fast Fourier Transform (FFT). From the PS, *median* and *mean frequencies* are the most used features to determine fatigue presence [156, 158]. The Median frequency (MDF) divides the sEMG power spectrum (PS) into two regions with equal areas. Mean frequency (MNF) is the frequency that corresponds to the barycentre or centroid of PS (Fig. 7.8). MDF and MNF are monitored to detect a possible decrease in their values recorded at a non-fatigued stage.

These techniques rely on the assumption that the signals are stationary and linear [159]. While they remain important fatigue assessment techniques, each situation should be analysed individually to see if they are the best choice.

In clinical practice, there is a need to acquire sEMG signals under dynamic conditions [156, 160]. During dynamic contractions, several factors increase the non-stationarity of the sEMG signal, such as the change in the number of active MU, or changes in force/power through the range of motion, and also, changes in fibre and muscle length [158, 160].

Therefore, this implies that commonly used features (e.g. MDF and MNF) do not accurately reflect muscle fatigue in dynamic measuring conditions [160]. In these circumstances, to capture the signal frequency content, it is helpful to analyse it for short, perhaps overlapping, time windows. This is what time-frequency analysis does.

### 7.9.3 Time-Frequency Domain Features

It has been observed that sEMG features extracted from the time-frequency domain are more reliable than the frequency domain when studying dynamic contractions [161, 162]. Firstly, according to Kamaruddin et al. [158], time-frequency representation methods provide accurate representation for non-stationary signals such as sEMG [158]. Time-frequency methodologies are used to study signals with a relevant frequency content variability and change in time, which is similar to the situation seen in muscle fatigue [156].

Secondly, Cifrek et al. [154], used the median frequency calculated from the spectrogram to indicate muscle fatigue during cyclic dynamic contractions of the quadriceps muscle [154]. This time-frequency analysis method is measuring the power spectrum estimation using Short-Time Fourier Transform (STFT) or spectrogram estimation. This system uses a sliding window moves across the signal, and at each position, the PS of the signal inside the window is computed. The STFT generates a whole group of power spectra, one for each time position of the window. The window's width determines the frequency resolution, so a wider window corresponds to a higher frequency resolution. As an indicator of muscle fatigue, a change in power spectrum median frequency from the spectrogram can be used.

However, other features have been developed to address the need to assess fatigue development during fast and short contractions recorded during motion tasks.

For example, Georgakis et al. [163], compared Averaged Instantaneous Frequency (AIF) to MDF and MNF [163]. This study defined instantaneous frequency as the first derivative of a signal's phase and computed its average over consecutive, non-overlapping windows. The results demonstrated that AIF performed significantly better in terms of robustness to the length of the analysis window.

Instead, Cifrek et al. [153] applied the Hilbert–Huang transform (HHT), comprising the empirical mode decomposition (EMD) and Hilbert spectral analysis (HSA), to the sEMG signal recorded over quadriceps muscle during cyclic dynamic contractions [153]. As defined in the study, HSA is a procedure that enables frequency parameters estimation, while EMD is an iterative procedure introduced by Huang [164], to obtain a set of primary oscillatory functions that can be Hilbert transformed [164]. It was found that the HHT procedure provides the most consistent results compared to Fourier and wavelet-based procedures. Therefore, it was concluded that HHT is a reliable procedure for dealing with the non-stationary and non-linear properties of biomedical signals.

Other studies found that nonlinear analysis methods provide a better description of changes in the muscle characteristics than spectral methods [165]. Makaram et al. [151], utilised a transition network-based analysis to characterise dynamic nonlinear variations in the sEMG signal to develop a reliable fatigue index [151]. A transition network is a technique, in which the signals are symbolised, and the temporal variations of these symbols are analysed [151, 166–168]. Features extracted from the symbolic sequence and transition network (e.g. symbolic entropy, network entropy, uniformity, and minimum and maximum effective degrees) reflect both the temporal and spatial complexity of the signal equally. Moreover, these features indicate a lower complexity in fatigue conditions [151]. This have been attributed to the reduction in the conduction velocity of muscle fibres and synchronisation of motor units during fatigue [143]. The pros of the symbolisation procedure are a reduction of noise influence, intersubject variability, and computational cost [166, 167, 169].

### 7.9.4 Fatigue Assessment Methods in Clinics

Surface electromyography (sEMG) is commonly used as an electrophysiological outcome measure in clinical trials in neurorehabilitation of individuals living with SCI [170], constituting an informative complement to current clinical testing [171]. sEMG signal is currently used to understand the long-term effect of neurorehabilitation on neuromuscular activation and control [147]. The potential application of sEMG analysis in spotting muscular fatigue throughout a rehabilitation session should also be considered. Research is active in this field with the aim of developing advanced techniques that can monitor early fatigue onset in real time [133, 149, 151]. The early detection of muscle fatigue using sEMG-based electrical measurements is challenging due to the nonlinear variations of the signal [151], but it is an important goal since avoiding fatigue of the subject means a more effective rehabilitation.

Wearable sEMG devices such as REMO (Morecognition s.r.l., Turin, Italy) and Myo Armband, (Thalmic Labs, Canada) are available on the market as accessible and relatively cheap solutions for muscular activity monitoring. In particular, REMO is a Class 1 certified medical device that measures muscle activity as well as movement [172] and can be used in clinics to improve the rehabilitation process of patients with reduced motor functions. Both devices contain an IMU sensor (composed of accelerometer, gyroscope, and magnetometer). This inertial sensor completes the device allowing to conduct a comprehensive motion analysis with a single device. Moreover, the IMU is useful to standardise the armband position unifying the sEMG acquisition in all the subjects [173].

Recently, some research studies have analysed the efficacy and effectiveness of these simple wearable sEMG devices collecting preliminary results that can guide understanding of their potential use in rehabilitation application. Pregnolato et al. [173] tested this device for the control of a rehabilitative PC with the final goal of defining the most important clinical features to use for hand rehabilitation training of stroke survivors [173]. Castiblanco et al. [174] implemented an experimental protocol to test three different subject groups: (1) non-stroke without hand impairments, (2) stroke without hand impairments, and (3) stroke with hand impairments. Acquiring data from the participant's forearm with Myo Armband they were able to identify hand movements (patter recognition) from sEMG signals proposing a promising approach for robotic-based rehabilitation assistance [174]. Sathiyanarayanan and Rajan [175] analyse via two questionnaires the Myo Armband user experience, both for doctors and patients. The aim of the study was to define guidelines for the use of the armband and the related software (Myo Diagnostics) and to support the development of interactive physiotherapy analysis for patient self-examination at home [175]. Research about the application of devices to fatigue monitoring is still relatively limited but some promising results were achieved. Geryes et al. [176] recognised that Myo Armband could be considered as biomedical equipment with possible medical applications [176]. It recorded muscular activity with Myo Armband during an assigned simple task and analysing the sEMG signal energy and RMS envelope (Root Mean Square), it was noticeable that a faster decrease of energies and RMS, corresponded to a faster fatigue onset in a subject with an injured limb [176]. A summary of these studies is schematised in the appendices for more information.

## 7.10 Limitations/Gaps of the Technology and Methods in Neurorehabilitation and Future Research Activities

Despite the exponential amount of research on the topic of this new technology, dissemination and clinical use of sEMG technology are still noticeably limited [177]. The insufficient usage of sEMG in clinics and the gap between available technology and direct clinical applications is discussed in these works: "Surface Electromyography: Barriers limiting widespread use of sEMG in clinical assessment and neurorehabilitation" [178]. Here, the authors identified the barriers as mainly educational, technical, and administrative obstacles to overcome.

Because sEMG is a complex signal because of the large amount of information contained in it, the visual analysis and interpretation are quite complex, in contrast to other biological signals, such as ECG. Performing a proper analysis and fully understanding the invaluable information that an sEMG signal contains, requires a comprehensive biomechanical and signal processing knowledge, together with the necessary experience in application [179, 180].

Unfortunately, the first barrier to implementation is in the educational sector. Clinical operator training does not include such technical competencies [177]. The curricula of most rehabilitation clinicians does not have sufficient background in mathematics and physics [147]. Rarely do the schools training physiotherapists or neurophysiology technicians teach sEMG technology or delve into its clinical application. This results in fewer clinical publications and limited grant requests in the field of sEMG for large clinical studies. From an academic perspective, it seems arguable that the experience of sEMG remains low and there are insufficient numbers of qualified instructors that could transfer the competence in the use of sEMG in educational settings [178].

In addressing this barrier in the educational sector, a possible solution is the introduce elements to update the academic curriculum of physiotherapists, neurophysiology technicians, and clinical technologists [177], however, this process in higher education would take a long time and the generated impact on society could only be evaluated long term [177].

Some immediate steps in this direction in the training sector is evident with two examples of translating the sEMG available knowledge into clinical activities with some translational initiatives. The first example is the Consensus for Experimental Design in Electromyography (CEDE) project [181] that has generated the most up-to-date consensus papers available on sEMG relevant topics. The aim is to guide decision-making in recording, analysis, and interpretation of EMG across all applications. The papers contain not only recommendations, but also provide detailed justification for them ensuring that the user is fully informed. The second example is a series of tutorials aimed at readers without an engineering background from the International Society for Electromyography and Kinesiology (ISEK [182]) [178].

The latest technology is trying to adapt and address the gap between technology and clinical application, by incorporating intelligent error correction, automatic elimination of power line interference, movement artefacts, or other errors. Parallelly users' competence still needs to be increased as previously emphasised.

In addressing the technical barriers, much work still needs to be done to get to the point where EMG is used as a routine practice in support of neurorehabilitation. The main issue is a lack of time that stop the diffusion of sEMG in clinical settings [183, 184]. It has been noted that the application of electrodes is time-consuming and requires knowledge about proper electrode positioning. Clinical trials are limited because these would have to be conducted within environments, such as healthcare institutions, where engineers are rarely present and clinical operators lack the time to conduct them as well as enough technical expertise and external support [185]. Recently, more significant trials have not relied on sEMG, for example, when studying diagnosis or tracking outcomes of SCI patients' rehabilitation [147]. Therefore, a challenge is presented where only small or medium sample sizes in completed studies demonstrate its application, which are not enough if the goal is to support the role of sEMG in diagnostics and tracking neuro-recovery [147].

Finally, in addressing the administrative issues, what limits sEMG-based investigations in routine practice is the reimbursement system. All clinical activities are coded and reimbursed according to such codes. While gait analysis is coded (in a few countries), most of the sEMG-based assessment procedures are not. To attract investors and influence policy level commitment to prioritising such technology, further analysis should prove that these innovative sEMG-based procedures help reduce the weight of ineffective therapies, as well as address health insurance interests, and reduce costs for national health systems [178].

# Appendices

## Appendix 1

| Solution | Year | Camera setup | Clinical assessment | Efficiency | Experience |
|---|---|---|---|---|---|
| Eichler et al. [104] | 2018 | Multi-camera (two Microsoft Kinect cameras positioned at a 45-degree angle to the patient's frontal view and around 1.5–2 m distance) | Automated assessment of the medical condition of stroke patients via the stroke-specific performance-based impairment index, Fugl-Meyer Assessment (FMA) | | The system is able to detect trunk flexion in patients attempting to move their hand to a target position compared to healthy participants |
| Moro et al. [107] | 2020 | Single-camera with DeepLabCut open-source human pose estimator | | | They report no significant differences between marker-based gait analysis and their markerless solution |
| Martinez et al. [106] | 2018 | Single-camera with OpenPose open-source human pose estimator | Assessment of walking cadence and automated calculation of an abnormality score for Parkinson's disease patients | | The system showed a markerless 3D system correlated with UPDRS-III scores [186]. A significant negative correlation was found between gait and stride length in the Unified Parkinson's Disease Rating Scale part III (here) and body motion evaluation ($r= -0.833$, $p = 0.039$) as well as right step length ($r = -0.926$, $p = 0.008$) in the off-state |

| Solution | Year | Camera setup | Clinical assessment | Efficiency | Experience |
|---|---|---|---|---|---|
| Shin et al. [108] | 2021 | Single-camera with OpenPose open-source human pose estimator | Assessment of temporospatial outcome measures while performing timed up and go test of Parkinson's disease patients | The solution was validated with a pressure-sensitive walkway, demonstrating excellent reliability (ICC > 0.9) | The system could detect minor gait disturbances unnoticed by the practitioner's subjective analysis |
| The Captury [116] | 2022 | Multi-camera setup with eight tripod-mounted Blackfly digital cameras | System used to explore the software-driven random error associated with markerless measurements in young children and the clinical relevance of these measurements (here) | Jump length performance and four kinematic variables demonstrated acceptable instrumental variability (ICC > 0.76) | Week-to-week reliability was excellent for jump length performance (ICC = 0.90) but poor to moderate (ICC < 0.55) for the kinematic variables |

| Solution | Year | Camera setup | Clinical assessment | Efficiency | Experience |
|---|---|---|---|---|---|
| SIMI Shape 3D | 2015 | 8 Cameras recording both marker (SIMI Motion 3D) and markerless (SIMI Shape) | | SIMI marked based system (37 markers) was used to assess the accuracy of the markerless approach<br>Hybrid setting uses 19 markers instead of the 37 used for full body marker-based setting | Correlations have been defined to be very good when ≥0.9. Markerless tracking shows problems when:<br>• Silhouette appearance barely changes during segment rotation<br>• Elbow is performing a hyperextension instead of a flexion if the arm is strongly rotated in the shoulder joint<br>• The pelvis segment is difficult to track because of its nearly rotationally symmetric shape and by that hip angles are affected<br>• Foot movements (eversion/inversion and abduction/adduction) cannot be tracked accurately<br>To achieve very good results in all joint angles, hybrid tracking with few additional markers/features to support silhouette tracking is necessary |

| Solution | Year | Camera setup | Clinical assessment | Efficiency | Experience |
|---|---|---|---|---|---|
| Theia3D (McGuirk et al. [187]) | | Technology uses deep convolutional neural networks to perform 3D human pose estimation from a multi-camera setup | | Compared to marker-based motion capture solutions and pressure-sensitive gait mats, Theia3D demonstrated good to excellent agreement for spatiotemporal parameters | Data was collected in community and clinic-adjacent settings<br>The measurement error of the 3D kinematics was low, providing a basis for interpreting both clinical status and monitoring rehabilitation outcomes |
| iPi Soft | 2018 | Two Kinect v2 cameras placed one in front and one to the left side of the capture area at an angle of 70° between them | The study's objective was to determine whether a dual-camera markerless motion capture system can be used for lower limb kinematic evaluation in athletes in a preseason screening setting | The solution was validated with the marker-based BTS-SMART 1000 system showing no systematic error for most of the measures examined. However, hip flexion angle appeared to have a systematic error of approximately 10° when comparing the markerless and marker-based systems | Excellent agreement between systems for the flexion/extension range of motion of the shin during all tests and for the thigh abduction/adduction during single-leg squat were seen. For peak angles, results showed excellent agreement for knee flexion. Poor correlation was seen for the rotation movements<br>The study supports the use of dual Kinect v2 configuration with the iPi software as a valid tool for assessment of sagittal and frontal plane hip and knee kinematic parameters but not axial rotation in athletes |

## Appendix 2

| References | Year | Technology | Goals |
|---|---|---|---|
| Pregnolato et al. [173] | 2023 | REMO, Morecognition s.r.l., Turin, Italy In collaboration with the Italian Institute of Technology (Genoa, Italy) and the San Camillo IRCCS hospital (Venice, Italy) | Testing an innovative wearable (REMO) that detected the residual surface electromyography of forearm muscles. Using the total muscle activity with no need to detect the electromyography signal of a specific single muscle to control a rehabilitative PC interface. Defining the clinical features of stroke survivors able to perform ten, five, or no hand movements for rehabilitation training |
| Castiblanco et al. [174] | 2020 | Myo Armband, Thalmic Labs | Finding the best pattern recognition techniques to identify the wrist/hand motions from sEMG signals recorded in the forearm of three subject groups: (1) non-stroke without hand impairments, (2) stroke without hand impairments, and (3) stroke with hand impairments |
| Geryes et al. [176] | 2017 | Myo Armband, Thalmic Labs | Focusing on the Myo as a biomedical equipment that could be used for medical purposes. Evaluating energy and rms derived from STFT of sEMG |
| Sathiyanarayanan and Rajan [175] | 2016 | Myo Armband, Thalmic Labs | Defining guidelines and solutions for using the Myo Armband (physical device) and the Myo Diagnostics (software application). Designing two types of questionnaires for the 24 participants. Study the MyoArmband user experience, both for doctors and patients |

## Appendix 3

| Device | Company | Specifications |
|---|---|---|
| Ultium EMG | NORAXON, USA | Variable geometry electrodes, up to 16 probes<br>Sampling rate up to 2 kHz<br>Weight of 14 grams<br>(https://www.noraxon.com/our-products/ultium-emg/#1541097779421-89a192e6-7d8d) |

| Device | Company | Specifications |
|---|---|---|
| Core EMG | NORAXON, USA | Announced in April 2023<br>New direct connect sensor design, eliminating lead connections and the need for additional disposables<br>Sampling rate up to 2 kHz<br>(https://www.noraxon.com/our-products/semg/core-emg/#1541097779421-89a192e6-7d8d) |
| FREEEMG | BTS Bioengineering, Italy | Variable geometry electrodes with snap connectors, up to 20 probes—16-bit resolution<br>Sampling rate 1 kHz<br>Weight of 13 g<br>Class IIa Medical Device<br>(https://www.btsbioengineering.com/it/products/freeemg/) |
| FREEEMG *H2O version* | BTS Bioengineering, Italy | As above + waterproof up to a depth of 3 m |
| REMO | Morecognition s.r.l., Turin, Italy | 8 bipolar sEMG electrodes embedded in an armband<br>Transmission to a host device (PC, tablet) via Bluetooth<br>Can be configured to detect four types of signals in three different options: RAW, RAW-IMU and RMS value<br>Class I Medical Device<br>(https://morecognition.com/, [173]) |
| Myo Armband | Thalmic Labs, Canada | 8 bipolar sEMG electrodes embedded in an armband (frequency rate of 200 Hz). Bluetooth communication<br>Weight of about 250 g<br>(https://www.sciencedirect.com/science/article/pii/S1746809419303180) |

## References

1. Feigin VL, Vos T, Nichols E, Owolabi MO, Carroll WM, Dichgans M, Deuschl G, Parmar P, Brainin M, Murray C. The global burden of neurological disorders: translating evidence into policy. Lancet Neurol. 2020;19(3):255–65. https://doi.org/10.1016/S1474-4422(19)30411-9.
2. Wonsetler EC, Bowden MG. A systematic review of mechanisms of gait speed change post-stroke. Part 1: spatiotemporal parameters and asymmetry ratios. Top Stroke Rehabil. 2017a;24(6):435–46. https://doi.org/10.1080/10749357.2017.1285746.
3. Wonsetler EC, Bowden MG. A systematic review of mechanisms of gait speed change post-stroke. Part 2: exercise capacity, muscle activation, kinetics, and kinematics. Top Stroke Rehabil. 2017b;24(5):394–403. https://doi.org/10.1080/10749357.2017.1282413.
4. Hsu AL, Tang PF, Jan MH. Analysis of impairments influencing gait velocity and asymmetry of hemiplegic patients after mild to moderate stroke. Arch Phys Med Rehabil. 2003;84(8):1185–93. https://doi.org/10.1016/s0003-9993(03)00030-3.
5. Hsiao H, Awad LN, Palmer JA, Higginson JS, Binder-Macleod SA. Contribution of paretic and nonparetic limb peak propulsive forces to changes in walking speed in individuals poststroke. Neurorehabil Neural Repair. 2016;30(8):743–52. https://doi.org/10.1177/1545968315624780.

6. Awad LN, Lewek MD, Kesar TM, Franz JR, Bowden MG. These legs were made for propulsion: advancing the diagnosis and treatment of post-stroke propulsion deficits. J Neuroeng Rehabil. 2020;17(1):139. https://doi.org/10.1186/s12984-020-00747-6.
7. Farris DJ, Hampton A, Lewek MD, Sawicki GS. Revisiting the mechanics and energetics of walking in individuals with chronic hemiparesis following stroke: from individual limbs to lower limb joints. J Neuroeng Rehabil. 2015;12:24. https://doi.org/10.1186/s12984-015-0012-x.
8. Brough LG, Kautz SA, Neptune RR. Muscle contributions to pre-swing biomechanical tasks influence swing leg mechanics in individuals post-stroke during walking. J Neuroeng Rehabil. 2022;19(1):55. https://doi.org/10.1186/s12984-022-01029-z.
9. Hall AL, Peterson CL, Kautz SA, Neptune RR. Relationships between muscle contributions to walking subtasks and functional walking status in persons with post-stroke hemiparesis. Clin Biomech (Bristol, Avon). 2011;26(5):509–15. https://doi.org/10.1016/j.clinbiomech.2010.12.010.
10. Neptune RR, Zajac FE, Kautz SA. Muscle force redistributes segmental power for body progression during walking. Gait Posture. 2004;19(2):194–205. https://doi.org/10.1016/s0966-6362(03)00062-6.
11. Moustafa AA, Chakravarthy S, Phillips JR, Gupta A, Keri S, Polner B, Frank MJ, Jahanshahi M. Motor symptoms in Parkinson's disease: a unified framework. Neurosci Biobehav Rev. 2016;68:727–40. https://doi.org/10.1016/j.neubiorev.2016.07.010.
12. Mitchell SL, Harper DW, Lau A, Bhalla R. Patterns of outcome measurement in Parkinson's disease clinical trials. Neuroepidemiology. 2000;19(2):100–8. https://doi.org/10.1159/000026244.
13. Ramaker C, Marinus J, Stiggelbout AM, Van Hilten BJ. Systematic evaluation of rating scales for impairment and disability in Parkinson's disease. Mov Disord. 2002;17(5):867–76. https://doi.org/10.1002/mds.10248.
14. Martínez-Martín P, Gil-Nagel A, Gracia LM, Gómez JB, Martínez-Sarriés J, Bermejo F. Unified Parkinson's disease rating scale characteristics and structure. The Cooperative Multicentric Group. Mov Disord. 1994;9(1):76–83. https://doi.org/10.1002/mds.870090112.
15. Cui CK, Lewis SJG. Future therapeutic strategies for freezing of gait in Parkinson's disease. Front Hum Neurosci. 2021;15:741918. https://doi.org/10.3389/fnhum.2021.741918.
16. Gebruers N, Vanroy C, Truijen S, Engelborghs S, De Deyn PP. Monitoring of physical activity after stroke: a systematic review of accelerometry-based measures. Arch Phys Med Rehabil. 2010;91(2):288–97. https://doi.org/10.1016/j.apmr.2009.10.025.
17. Aaron SE, Gregory CM, Simpson AN. Lower odds of poststroke symptoms of depression when physical activity guidelines met: National Health and Nutrition Examination Survey 2011-2012. J Phys Act Health. 2016;13(8):903–9. https://doi.org/10.1123/jpah.2015-0446.
18. Obembe AO, Eng JJ. Rehabilitation interventions for improving social participation after stroke: a systematic review and meta-analysis. Neurorehabil Neural Repair. 2016;30(4):384–92. https://doi.org/10.1177/1545968315597072.
19. Rand D, Eng JJ, Tang PF, Hung C, Jeng JS. Daily physical activity and its contribution to the health-related quality of life of ambulatory individuals with chronic stroke. Health Qual Life Outcomes. 2010;8:80. https://doi.org/10.1186/1477-7525-8-80.
20. Fini NA, Bernhardt J, Said CM, Billinger SA. How to address physical activity participation after stroke in research and clinical practice. Stroke. 2021;52(6):e274–7. https://doi.org/10.1161/STROKEAHA.121.034557.
21. Fini NA, Holland AE, Keating J, Simek J, Bernhardt J. How physically active are people following stroke? Systematic review and quantitative synthesis. Phys Ther. 2017;97(7):707–17. https://doi.org/10.1093/ptj/pzx038.
22. Pradhan S, Kelly VE. Quantifying physical activity in early Parkinson disease using a commercial activity monitor. Parkinsonism Relat Disord. 2019;66:171–5. https://doi.org/10.1016/j.parkreldis.2019.08.001.
23. Topley M, Richards JG. A comparison of currently available optoelectronic motion capture systems. J Biomech. 2020;106:109820. https://doi.org/10.1016/j.jbiomech.2020.109820.

24. Pagnon D, Domalain M, Reveret L. Pose2Sim: an end-to-end workflow for 3D Markerless sports kinematics—part 2: accuracy. Sensors. 2022;22(7):2712. https://www.mdpi.com/1424-8220/22/7/2712.
25. Needham L, Evans M, Cosker DP, Wade L, McGuigan PM, Bilzon JL, Colyer SL. The accuracy of several pose estimation methods for 3D joint centre localisation. Sci Rep. 2021b;11(1):20673. https://doi.org/10.1038/s41598-021-00212-x.
26. Zhu X, Boukhennoufa I, Liew XWB, McDonald-Maier K, Zhai X. A Kalman filter based approach for markerless pose tracking and assessment. In: IEEE 27th international conference on automation and computing (ICAC); 2022.
27. Boukhennoufa I, Zhai X, Utti V, Jackson J, McDonald-Maier KD. Wearable sensors and machine learning in post-stroke rehabilitation assessment: a systematic review. Biomed Signal Process Control. 2022b;71:103197. https://doi.org/10.1016/j.bspc.2021.103197.
28. Yadav SK, Tiwari K, Pandey HM, Akbar SA. A review of multimodal human activity recognition with special emphasis on classification, applications, challenges and future directions. Knowl Based Syst. 2021;223:106970. https://doi.org/10.1016/j.knosys.2021.106970.
29. Wang Q, Markopoulos P, Yu B, Chen W, Timmermans A. Interactive wearable systems for upper body rehabilitation: a systematic review. J Neuroeng Rehabil. 2017;14(1):20. https://doi.org/10.1186/s12984-017-0229-y.
30. Lu R, Xu Y, Li X, Fan Y, Zeng W, Tan Y, Ren K, Chen W, Cao X. Evaluation of wearable sensor devices in Parkinson's disease: a review of current status and future prospects. Parkinsons Dis. 2020;2020:4693019. https://doi.org/10.1155/2020/4693019.
31. Battista L, Romaniello A. A novel device for continuous monitoring of tremor and other motor symptoms. Neurol Sci. 2018;39(8):1333–43. https://doi.org/10.1007/s10072-018-3414-2.
32. Tsipouras MG, Tzallas AT, Rigas G, Tsouli S, Fotiadis DI, Konitsiotis S. An automated methodology for levodopa-induced dyskinesia: assessment based on gyroscope and accelerometer signals. Artif Intell Med. 2012;55(2):127–35. https://doi.org/10.1016/j.artmed.2012.03.003.
33. Hausdorff JM. Gait dynamics in Parkinson's disease: common and distinct behavior among stride length, gait variability, and fractal-like scaling. Chaos. 2009;19(2):026113. https://doi.org/10.1063/1.3147408.
34. Teshuva I, Hillel I, Gazit E, Giladi N, Mirelman A, Hausdorff JM. Using wearables to assess bradykinesia and rigidity in patients with Parkinson's disease: a focused, narrative review of the literature. J Neural Transm (Vienna). 2019;126(6):699–710. https://doi.org/10.1007/s00702-019-02017-9.
35. Palmerini L, Mellone S, Avanzolini G, Valzania F, Chiari L. Quantification of motor impairment in Parkinson's disease using an instrumented timed up and go test. IEEE Trans Neural Syst Rehabil Eng. 2013;21(4):664–73. https://doi.org/10.1109/tnsre.2012.2236577.
36. Ossig C, Antonini A, Buhmann C, Classen J, Csoti I, Falkenburger B, Schwarz M, Winkler J, Storch A. Wearable sensor-based objective assessment of motor symptoms in Parkinson's disease. J Neural Transm (Vienna). 2016;123(1):57–64. https://doi.org/10.1007/s00702-015-1439-8.
37. Elble RJ, Ondo W. Tremor rating scales and laboratory tools for assessing tremor. J Neurol Sci. 2022;435:120202. https://doi.org/10.1016/j.jns.2022.120202.
38. Bhidayasiri R, Petchrutchatachart S, Pongthornseri R, Anan C, Dumnin S, Thanawattano C. Low-cost, 3-dimension, office-based inertial sensors for automated tremor assessment: technical development and experimental verification. J Parkinsons Dis. 2014;4(2):273–82. https://doi.org/10.3233/jpd-130311.
39. Mari Z, Haubenberger D. Remote measurement and home monitoring of tremor. J Neurol Sci. 2022;435:120201. https://doi.org/10.1016/j.jns.2022.120201.
40. Hssayeni MD, Jimenez-Shahed J, Burack MA, Ghoraani B. Wearable sensors for estimation of parkinsonian tremor severity during free body movements. Sensors (Basel). 2019;19(19):4215. https://doi.org/10.3390/s19194215.
41. Davico G, Pizzolato C, Lloyd DG, Obst SJ, Walsh HPJ, Carty CP. Increasing level of neuromusculoskeletal model personalisation to investigate joint contact forces in cerebral palsy: a twin case study. Clin Biomech (Bristol, Avon). 2020;72:141–9. https://doi.org/10.1016/j.clinbiomech.2019.12.011.

42. Toney ME, Chang YH. The motor and the brake of the trailing leg in human walking: leg force control through ankle modulation and knee covariance. Exp Brain Res. 2016;234(10):3011–23. https://doi.org/10.1007/s00221-016-4703-8.
43. Liew BXW, Rügamer D, Zhai X, Wang Y, Morris S, Netto K. Comparing shallow, deep, and transfer learning in predicting joint moments in running. J Biomech. 2021;129:110820. https://doi.org/10.1016/j.jbiomech.2021.110820.
44. Boukhennoufa I, Altai Z, Zhai X, Utti V, McDonald-Maier KD, Liew BXW. Predicting the internal knee abduction impulse during walking using deep learning. Front Bioeng Biotechnol. 2022a;10:877347. https://doi.org/10.3389/fbioe.2022.877347.
45. Giarmatzis G, Zacharaki EI, Moustakas K. Real-time prediction of joint forces by motion capture and machine learning. Sensors (Basel). 2020;20(23):6933. https://doi.org/10.3390/s20236933.
46. Johnson WR, Alderson J, Lloyd D, Mian A. Predicting athlete ground reaction forces and moments from spatio-temporal driven CNN models. IEEE Trans Biomed Eng. 2019a;66(3):689–94. https://doi.org/10.1109/tbme.2018.2854632.
47. Lee M, Park S. Estimation of three-dimensional lower limb kinetics data during walking using machine learning from a single IMU attached to the sacrum. Sensors (Basel). 2020;20(21):6277. https://doi.org/10.3390/s20216277.
48. Mundt M, Thomsen W, Witter T, Koeppe A, David S, Bamer F, Potthast W, Markert B. Prediction of lower limb joint angles and moments during gait using artificial neural networks. Med Biol Eng Comput. 2020;58(1):211–25. https://doi.org/10.1007/s11517-019-02061-3.
49. Stetter BJ, Krafft FC, Ringhof S, Stein T, Sell S. A machine learning and wearable sensor based approach to estimate external knee flexion and adduction moments during various locomotion tasks. Front Bioeng Biotechnol. 2020;8:9. https://www.frontiersin.org/article/10.3389/fbioe.2020.00009.
50. Wang C, Chan PPK, Lam BMF, Wang S, Zhang JH, Chan ZYS, Chan RHM, Ho KKW, Cheung RTH. Real-time estimation of knee adduction moment for gait retraining in patients with knee osteoarthritis. IEEE Trans Neural Syst Rehabil Eng. 2020a;28(4):888–94. https://doi.org/10.1109/tnsre.2020.2978537.
51. Boswell MA, Uhlrich SD, Kidziński Ł, Thomas K, Kolesar JA, Gold GE, Beaupre GS, Delp SL. A neural network to predict the knee adduction moment in patients with osteoarthritis using anatomical landmarks obtainable from 2D video analysis. Osteoarthritis Cartilage. 2021;29(3):346–56. https://doi.org/10.1016/j.joca.2020.12.017.
52. Camargo J, Molinaro D, Young A. Predicting biological joint moment during multiple ambulation tasks. J Biomech. 2022;134:111020. https://doi.org/10.1016/j.jbiomech.2022.111020.
53. He Z, Liu T, Yi J. A wearable sensing and training system: towards gait rehabilitation for elderly patients with knee osteoarthritis. IEEE Sensors J. 2019;19(14):5936–45. https://doi.org/10.1109/JSEN.2019.2908417.
54. Johnson WR, Mian A, Lloyd DG, Alderson JA. On-field player workload exposure and knee injury risk monitoring via deep learning. J Biomech. 2019b;93:185–93. https://doi.org/10.1016/j.jbiomech.2019.07.002.
55. Favre J, Hayoz M, Erhart-Hledik JC, Andriacchi TP. A neural network model to predict knee adduction moment during walking based on ground reaction force and anthropometric measurements. J Biomech. 2012;45(4):692–8. https://doi.org/10.1016/j.jbiomech.2011.11.057.
56. Xiang L, Wang A, Gu Y, Zhao L, Shim V, Fernandez J. Recent machine learning progress in lower limb running biomechanics with wearable technology: a systematic review. Front Neurorobot. 2022;16:913052. https://doi.org/10.3389/fnbot.2022.913052.
57. Holzinger A, Saranti A, Molnar C, Biecek P, Samek W. Explainable AI methods—A brief overview. In: Holzinger A, Goebel R, Fong R, Moon T, Müller KR, Samek W, editors. xxAI—beyond explainable AI: international workshop, held in conjunction with ICML 2020, July 18, 2020, Vienna, Austria, revised and extended papers. Springer International Publishing. 2022. pp. 13–38. https://doi.org/10.1007/978-3-031-04083-2_2.
58. Molnar C, König G, Bischl B, Casalicchio G. Model-agnostic feature importance and effects with dependent features: a conditional subgroup approach. Data Min Knowl Disc. 2023. https://doi.org/10.1007/s10618-022-00901-9.

59. Brockhaus S, Rügamer D, Greven S. Boosting functional regression models with FDboost. J Stat Softw. 2020;94(10):1–50. https://arxiv.org/abs/1705.10662.
60. Liew BXW, Rugamer D, Abichandani D, De Nunzio AM. Classifying individuals with and without patellofemoral pain syndrome using ground force profiles—development of a method using functional data boosting. Gait Posture. 2020a;80:90–5. https://doi.org/10.1016/j.gaitpost.2020.05.034.
61. Liew BXW, Rugamer D, Stocker A, De Nunzio AM. Classifying neck pain status using scalar and functional biomechanical variables—development of a method using functional data boosting. Gait Posture. 2020b;76:146–50. https://doi.org/10.1016/j.gaitpost.2019.12.008.
62. Rügamer D, Kolb C, Klein N. Semi-structured distributional regression. Am Stat. 2023;1–12. https://doi.org/10.1080/00031305.2022.2164054.
63. Ramchoun H, Idrissi MAJ, Ghanou Y, Ettaouil M. Multilayer perceptron: architecture optimization and training with mixed activation functions. ACM international conference proceeding series, part F129474; 2017. https://doi.org/10.1145/3090354.3090427.
64. Kleinbaum DG, Klein M. Logistic regression. New York: Springer; 2010. https://doi.org/10.1007/978-1-4419-1742-3.
65. Noble WS. What is a support vector machine? Nat Biotechnol. 2006;24(12):1565–7. https://doi.org/10.1038/nbt1206-1565.
66. Guo G, Wang H, Bell D, Bi Y, Greer K. KNN model-based approach in classification. 2003. pp. 986–996. https://doi.org/10.1007/978-3-540-39964-3_62.
67. Rokach L. Ensemble-based classifiers. Artif Intell Rev. 2010;33(1–2):1–39. https://doi.org/10.1007/s10462-009-9124-7.
68. Shaik AB, Srinivasan S. A brief survey on random forest ensembles in classification model. 2019. pp. 253–260. https://doi.org/10.1007/978-981-13-2354-6_27.
69. Swe TT. Analysis of tree based supervised learning algorithms on medical data. Int J Sci Res Publ. 2019;9(4):8817. https://doi.org/10.29322/IJSRP.9.04.2019.p8817.
70. Kalmegh S. Analysis of WEKA data mining algorithm REPTree, simple cart and RandomTree for classification of Indian News. 2015.
71. Bicer M, Phillips ATM, Melis A, McGregor AH, Modenese L. Generative deep learning applied to biomechanics: a new augmentation technique for motion capture datasets. J Biomech. 2022;144:111301. https://doi.org/10.1016/j.jbiomech.2022.111301.
72. Dorschky E, Nitschke M, Martindale CF, van den Bogert AJ, Koelewijn AD, Eskofier BM. CNN-based estimation of sagittal plane walking and running biomechanics from measured and simulated inertial sensor data. Front Bioeng Biotechnol. 2020;8:604. https://doi.org/10.3389/FBIOE.2020.00604.
73. Bringmann O, Ecker W, Feldner I, Frischknecht A, Gerum C, Hämäläinen T, Hanif MA, Klaiber MJ, Mueller-Gritschneder D, Bernardo PP, Prebeck S, Shafique M. Automated HW/SW co-design for edge AI: state, challenges and steps ahead: special session paper. 2021 international conference on hardware/software codesign and system synthesis (CODES+ISSS). 2021. pp. 11–20.
74. Shokri R, Stronati M, Song C, Shmatikov V. Membership inference attacks against machine learning models. In: 2017 IEEE Symposium on Security and Privacy (SP); 2017. pp. 3–18. https://doi.org/10.1109/SP.2017.41.
75. Räuker T, Ho A, Casper S, Hadfield-Menell D. Toward transparent AI: A survey on interpreting the inner structures of deep neural networks; 2022. p. arXiv:2207.13243. https://doi.org/10.48550/arXiv.2207.13243.
76. Zhou J, Zheng L, Chen C, Wang Y, Zheng X, Wu B, Chen C, Wang L, Yin J. Toward scalable and privacy-preserving deep neural network via algorithmic-cryptographic co-design. ACM Trans Intell Syst Technol. 2022;13(4):53. https://doi.org/10.1145/3501809.
77. Wang X, Han Y, Leung VCM, Niyato D, Yan X, Chen X. Convergence of edge computing and deep learning: a comprehensive survey. IEEE Commun Surv Tutor. 2020b;22(2):869–904. https://doi.org/10.1109/COMST.2020.2970550.
78. Elsken T, Metzen JH, Hutter F. Neural architecture search: a survey. J Mach Learn Res. 2019;20(1):1997–2017.

79. Das R, Paul S, Mourya GK, Kumar N, Hussain M. Recent trends and practices toward assessment and rehabilitation of neurodegenerative disorders: insights from human gait. Front Neurosci. 2022;16:379. https://doi.org/10.3389/FNINS.2022.859298/BIBTEX.
80. Wade L, Needham L, McGuigan P, Bilzon J. Applications and limitations of current markerless motion capture methods for clinical gait biomechanics. PeerJ. 2022;10:e12995. https://doi.org/10.7717/PEERJ.12995.
81. Muro-de-la-Herran A, García-Zapirain B, Méndez-Zorrilla A. Gait analysis methods: an overview of wearable and non-wearable systems, highlighting clinical applications. Sensors (Basel). 2014;14(2):3362. https://doi.org/10.3390/S140203362.
82. Wren TAL, Tucker CA, Rethlefsen SA, Gorton GE, Õunpuu S. Clinical efficacy of instrumented gait analysis: systematic review 2020 update. Gait Posture. 2020;80:274–9. https://doi.org/10.1016/J.GAITPOST.2020.05.031.
83. Simon SR. Quantification of human motion: gait analysis—benefits and limitations to its application to clinical problems. J Biomech. 2004;37(12):1869–80. https://doi.org/10.1016/J.JBIOMECH.2004.02.047.
84. Baptista R, Ghorbel E, Shabayek AER, Moissenet F, Aouada D, Douchet A, André M, Pager J, Bouilland S. Home self-training: visual feedback for assisting physical activity for stroke survivors. Comput Methods Prog Biomed. 2019;176:111–20. https://doi.org/10.1016/J.CMPB.2019.04.019.
85. Moro M, Marchesi G, Hesse F, Odone F, Casadio M. Markerless vs. marker-based gait analysis: a proof of concept study. Sensors (Basel). 2022;22(5):2011. https://doi.org/10.3390/S22052011.
86. Colyer SL, Evans M, Cosker DP, Salo AIT. A review of the evolution of vision-based motion analysis and the integration of advanced computer vision methods towards developing a Markerless system. Sports Med Open. 2018;4(1):1–15. https://doi.org/10.1186/S40798-018-0139-Y/FIGURES/7.
87. Mathis A, Mamidanna P, Cury KM, Abe T, Murthy VN, Mathis MW, Bethge M. DeepLabCut: markerless pose estimation of user-defined body parts with deep learning. Nat Neurosci. 2018;21(9):1281–9. https://doi.org/10.1038/s41593-018-0209-y.
88. Clark RA, Mentiplay BF, Hough E, Pua YH. Three-dimensional cameras and skeleton pose tracking for physical function assessment: a review of uses, validity, current developments and Kinect alternatives. Gait Posture. 2019;68:193–200. https://doi.org/10.1016/J.GAITPOST.2018.11.029.
89. Sarbolandi H, Lefloch D, Kolb A. Kinect range sensing: structured-light versus time-of-flight Kinect. Comput Vis Image Underst. 2015;139:1–20. https://doi.org/10.1016/J.CVIU.2015.05.006.
90. Cao Z, Hidalgo G, Simon T, Wei SE, Sheikh Y. OpenPose: realtime multi-person 2D pose estimation using part affinity fields. IEEE Trans Pattern Anal Mach Intell. 2018;43(1):172–86. https://doi.org/10.1109/TPAMI.2019.2929257.
91. Fang HS, Xie S, Tai YW, Lu C. RMPE: regional multi-person pose estimation. In: Proceedings of the IEEE international conference on computer vision, 2017-October. 2016. pp. 2353–2362. https://doi.org/10.1109/ICCV.2017.256.
92. Newell A, Yang K, Deng J. Stacked hourglass networks for human pose estimation. Lecture notes in computer science (including subseries lecture notes in artificial intelligence and lecture notes in bioinformatics), 9912 LNCS; 2016. pp. 483–499. https://doi.org/10.1007/978-3-319-46484-8_29/FIGURES/11.
93. Serrancoli G, Bogatikov P, Huix JP, Barbera AF, Egea AJS, Ribe JT, Kanaan-Izquierdo S, Susin A. Marker-less monitoring protocol to analyze biomechanical joint metrics during pedaling. IEEE Access. 2020;8:122782–90. https://doi.org/10.1109/ACCESS.2020.3006423.
94. Stenum J, Rossi C, Roemmich RT. Two-dimensional video-based analysis of human gait using pose estimation. PLoS Comput Biol. 2021;17(4):e1008935. https://doi.org/10.1371/JOURNAL.PCBI.1008935.

95. Mehta D, Sridhar S, Sotnychenko O, Rhodin H, Shafiei M, Seidel HP, Xu W, Casas D, Theobalt C. VNect: real-time 3D human pose estimation with a single RGB camera. ACM Trans Graph. 2017;36(4):1–14. https://doi.org/10.1145/3072959.3073596.
96. Nakano N, Sakura T, Ueda K, Omura L, Kimura A, Iino Y, Fukashiro S, Yoshioka S. Evaluation of 3D markerless motion capture accuracy using OpenPose with multiple video cameras. Front Sports Active Living. 2020;2:50. https://doi.org/10.3389/FSPOR.2020.00050.
97. Needham L, Evans M, Cosker DP, Colyer SL. Can markerless pose estimation algorithms estimate 3D mass Centre positions and velocities during linear sprinting activities? Sensors. 2021a;21(8):2889. https://doi.org/10.3390/S21082889.
98. Slembrouck M, Luong H, Gerlo J, Schütte K, Van Cauwelaert D, De Clercq D, Vanwanseele B, Veelaert P, Philips W. Multiview 3D markerless human pose estimation from OpenPose skeletons. In: Lecture notes in computer science (including subseries lecture notes in artificial intelligence and lecture notes in bioinformatics), 12002 LNCS; 2020. pp. 166–178. https://doi.org/10.1007/978-3-030-40605-9_15.
99. Zago M, Luzzago M, Marangoni T, De Cecco M, Tarabini M, Galli M. 3D tracking of human motion using visual skeletonization and stereoscopic vision. Front Bioeng Biotechnol. 2020;8:181. https://doi.org/10.3389/FBIOE.2020.00181/BIBTEX.
100. Miranda DL, Rainbow MJ, Crisco JJ, Fleming BC. Kinematic differences between optical motion capture and biplanar videoradiography during a jump–cut maneuver. J Biomech. 2013;46(3):567–73. https://doi.org/10.1016/J.JBIOMECH.2012.09.023.
101. Seethapathi N, Wang S, Saluja R, Blohm G, Kording KP. Movement science needs different pose tracking algorithms. 2019. https://arxiv.org/abs/1907.10226v1.
102. Baldewijns G, Claes V, Debard L, Mertens M, Devriendt E, Milisen K, Tournoy J, Croonenborghs T, Vanrumste B. Automated in-home gait transfer time analysis using video cameras. J Ambient Intell Smart Environ. 2016;8(3):273–86. https://doi.org/10.3233/AIS-160379.
103. Cunningham R, Sánchez MB, Butler PB, Southgate MJ, Loram ID. Fully automated image-based estimation of postural point-features in children with cerebral palsy using deep learning. R Soc Open Sci. 2019;6(11):191011. https://doi.org/10.1098/RSOS.191011.
104. Eichler N, Hel-Or H, Shimshoni I, Itah D, Gross B, Raz S. 3D motion capture system for assessing patient motion during Fugl-Meyer stroke rehabilitation testing. IET Comput Vis. 2018;12(7):963–75. https://doi.org/10.1049/IET-CVI.2018.5274.
105. Kidziński Ł, Yang B, Hicks JL, Rajagopal A, Delp SL, Schwartz MH. Deep neural networks enable quantitative movement analysis using single-camera videos. Nat Commun. 2020;11(1):4054. https://doi.org/10.1038/s41467-020-17807-z.
106. Martinez HR, Garcia-Sarreon A, Camara-Lemarroy C, Salazar F, Guerrero-González ML. Accuracy of markerless 3D motion capture evaluation to differentiate between on/off status in Parkinson's disease after deep brain stimulation. Parkinsons Dis. 2018;2018:5830364. https://doi.org/10.1155/2018/5830364.
107. Moro M, Marchesi G, Odone F, Casadio M. Markerless gait analysis in stroke survivors based on computer vision and deep learning: a pilot study. Proceedings of the ACM Symposium on Applied Computing; 2020. pp. 2097–2104. https://doi.org/10.1145/3341105.3373963.
108. Shin JH, Yu R, Ong JN, Lee CY, Jeon SH, Park H, Kim HJ, Lee J, Jeon B. Quantitative gait analysis using a pose-estimation algorithm with a single 2D-video of Parkinson's disease patients. J Parkinsons Dis. 2021;11(3):1271–83. https://doi.org/10.3233/JPD-212544.
109. Fugl-Meyer AR, Jääskö L, Leyman I, Olsson S, Steglind S. The post-stroke hemiplegic patient. 1. A method for evaluation of physical performance. Scand J Rehabil Med. 1975;7(1):13–31.
110. Michaelsen SM, Jacobs S, Roby-Brami A, Levin MF. Compensation for distal impairments of grasping in adults with hemiparesis. Exp Brain Res. 2004;157(2):162–73. https://doi.org/10.1007/S00221-004-1829-X.
111. iPi Soft—Markerless Motion Capture. 2023. https://www.ipisoft.com/.
112. Kotsifaki A, Whiteley R, Hansen C. Dual Kinect v2 system can capture lower limb kinematics reasonably well in a clinical setting: concurrent validity of a dual camera marker-

less motion capture system in professional football players. BMJ Open Sport Exerc Med. 2018;4(1):e000441. https://doi.org/10.1136/BMJSEM-2018-000441.
113. Captury—Markerless motion capture technology. 2023. https://captury.com/.
114. Harsted S, Holsgaard-Larsen A, Hestbæk L, Boyle E, Lauridsen HH. Concurrent validity of lower extremity kinematics and jump characteristics captured in pre-school children by a markerless 3D motion capture system. Chiropr Man Therap. 2019;27(1):1–16. https://doi.org/10.1186/S12998-019-0261-Z/FIGURES/4.
115. Vicon | Award Winning Motion Capture Systems. 2023. https://www.vicon.com/.
116. Harsted S, Holsgaard-Larsen A, Hestbæk L, Andreasen DL, Lauridsen HH. Test-retest reliability and agreement of lower-extremity kinematics captured in squatting and jumping preschool children using markerless motion capture technology. Front Digit Health. 2022;4:1027647. https://doi.org/10.3389/FDGTH.2022.1027647/FULL.
117. SIMI—Reality Motion Systems. 2023. http://www.simi.com/en/home.html.
118. Becker L. Evaluation of joint angle accuracy using markerless silhouette-based tracking and hybrid tracking against traditional marker tracking. 2016.
119. Theia Markerless—Markerless Motion Capture Redefined. 2023. https://www.theiamarkerless.ca/.
120. McGuirk TE, Perry ES, Sihanath WB, Riazati S, Patten C. Feasibility of markerless motion capture for three-dimensional gait assessment in community settings. Front Hum Neurosci. 2022;16:325. https://doi.org/10.3389/FNHUM.2022.867485/BIBTEX.
121. Kanko RM, Laende EK, Davis EM, Selbie WS, Deluzio KJ. Concurrent assessment of gait kinematics using marker-based and markerless motion capture. J Biomech. 2021a;127:110665. https://doi.org/10.1016/J.JBIOMECH.2021.110665.
122. Kanko RM, Laende EK, Strutzenberger G, Brown M, Selbie WS, DePaul V, Scott SH, Deluzio KJ. Assessment of spatiotemporal gait parameters using a deep learning algorithm-based markerless motion capture system. J Biomech. 2021b;122:110414. https://doi.org/10.1016/J.JBIOMECH.2021.110414.
123. Kanko RM, Laende E, Selbie WS, Deluzio KJ. Inter-session repeatability of markerless motion capture gait kinematics. J Biomech. 2021c;121:110422. https://doi.org/10.1016/J.JBIOMECH.2021.110422.
124. Riazati S, McGuirk TE, Perry ES, Sihanath WB, Patten C. Absolute reliability of gait parameters acquired with markerless motion capture in living domains. Front Hum Neurosci. 2022;16:867474. https://doi.org/10.3389/FNHUM.2022.867474.
125. Merletti R. Controllo motorio ed elettromiografia di superficie. 2023. https://www.robertomerletti.it/it/emg/material/teaching/.
126. Wu J, Li X, Liu W, Jane Wang Z. SEMG signal processing methods: a review. J Phys Conf Ser. 2019;1237(3):032008. https://doi.org/10.1088/1742-6596/1237/3/032008.
127. Ahmed T, Islam MK. EMG signal classification for detecting neuromuscular disorders. J Phys Conf Ser. 2021;1921(1):012043. https://doi.org/10.1088/1742-6596/1921/1/012043.
128. Belkhou A, Achmamad A, Jbari A. Classification and diagnosis of myopathy EMG signals using the continuous wavelet transform. 2019 Scientific meeting on electrical-electronics and biomedical engineering and computer science, EBBT 2019. 2019. https://doi.org/10.1109/EBBT.2019.8742051.
129. Kehri V, Awale RN. EMG signal analysis for diagnosis of muscular dystrophy using wavelet transform, SVM and ANN. Biomed Pharmacol J. 2018;11(3):1583–91. https://doi.org/10.13005/BPJ/1525.
130. Torres-Castilllo JR, López-López CO, Padilla-Castañeda MA. Neuromuscular disorders detection through time-frequency analysis and classification of multi-muscular EMG signals using Hilbert-Huang transform. Biomed Signal Process Control. 2022;71:103037. https://doi.org/10.1016/J.BSPC.2021.103037.
131. Jafarzadeh M, Hussey DC, Tadesse Y. Deep learning approach to control of prosthetic hands with electromyography signals. In: 2019 22nd IEEE international symposium on measurement and control in robotics: robotics for the benefit of humanity, ISMCR 2019. 2019. https://doi.org/10.1109/ISMCR47492.2019.8955725.

132. Dost Sürücü G, Tezen Ö. The effect of EMG biofeedback on lower extremity functions in hemiplegic patients. Acta Neurol Belg. 2021;121(1):113–8. https://doi.org/10.1007/S13760-019-01261-W/TABLES/6.
133. Lipinski CL, Donovan L, McLoughlin TJ, Armstrong CW, Norte GE. Surface electromyography of the forearm musculature during an overhead throwing rehabilitation progression program. Phys Ther Sport. 2018;33:109–16. https://doi.org/10.1016/J.PTSP.2018.07.006.
134. Zebis MK, Sørensen MH, Lauridsen HB, Bencke J, Andersen CH, Carlsbæk JB, Jespersen P, Kallehauge AH, Andersen LL. Electromyography evaluation of bodyweight exercise progression in a validated anterior cruciate ligament injury rehabilitation program: a cross-sectional study. Am J Phys Med Rehabil. 2019;98(11):998. https://doi.org/10.1097/PHM.0000000000001232.
135. Zory R, Molinari F, Knaflitz M, Schena F, Rouard A. Muscle fatigue during cross country sprint assessed by activation patterns and electromyographic signals time–frequency analysis. Scand J Med Sci Sports. 2011;21(6):783–90. https://doi.org/10.1111/J.1600-0838.2010.01124.X.
136. Varrecchia T, De Marchis C, Rinaldi M, Draicchio F, Serrao M, Schmid M, Conforto S, Ranavolo A. Lifting activity assessment using surface electromyographic features and neural networks. Int J Ind Ergon. 2018;66:1–9. https://doi.org/10.1016/J.ERGON.2018.02.003.
137. Papagiannis GI, Triantafyllou AI, Roumpelakis IM, Zampeli F, Garyfallia Eleni P, Koulouvaris P, Papadopoulos EC, Papagelopoulos PJ, Babis GC. Methodology of surface electromyography in gait analysis: review of the literature. J Med Eng Technol. 2019;43(1):59–65. Taylor and Francis Ltd. https://doi.org/10.1080/03091902.2019.1609610.
138. Wang W, De Stefano A, Allen R. A simulation model of the surface EMG signal for analysis of muscle activity during the gait cycle. Comput Biol Med. 2006;36(6):601–18. https://doi.org/10.1016/J.COMPBIOMED.2005.04.002.
139. Caliandro P, Insola A, Scarnati E, Padua L, Russo G, Granieri E, Mazzone P. Effects of unilateral pedunculopontine stimulation on electromyographic activation patterns during gait in individual patients with Parkinson's disease. J Neural Transm. 2011;118(10):1477–86. https://doi.org/10.1007/S00702-011-0705-7/METRICS.
140. Papagiannis GI, Roumpelakis IM, Triantafyllou AI, Makris IN, Babis GC. No differences identified in transverse plane biomechanics between medial pivot and rotating platform total knee implant designs. J Arthroplasty. 2016a;31(8):1814–20. https://doi.org/10.1016/j.arth.2016.01.050.
141. Papagiannis GI, Roumpelakis IM, Triantafyllou AI, Makris IN, Babis GC. Response to Letter to the Editor on "No differences identified in transverse plane biomechanics between medial pivot and rotating platform total knee implant designs". J Arthroplasty. 2016b;31(10):2373. https://doi.org/10.1016/j.arth.2016.03.055.
142. Rodríguez-Tapia B, Soto I, Marínez DM, Arballo NC. Myoelectric interfaces and related applications: current state of EMG signal processing—A systematic review. In: IEEE Access. 2020;8:7792–7805. Institute of Electrical and Electronics Engineers Inc. https://doi.org/10.1109/ACCESS.2019.2963881.
143. Merletti R, Parker PA. Electromyography—physiology, engineering, and noninvasive applications. IEEE. 2004.
144. McManus L, Lowery M, Merletti R, Søgaard K, Besomi M, Clancy EA, van Dieën JH, Hug F, Wrigley T, Besier T, Carson RG, Disselhorst-Klug C, Enoka RM, Falla D, Farina D, Gandevia S, Holobar A, Kiernan MC, McGill K, et al. Consensus for experimental design in electromyography (CEDE) project: terminology matrix. J Electromyogr Kinesiol. 2021;59:102565. https://doi.org/10.1016/j.jelekin.2021.102565.
145. De Luca CJ. DELSYS. Surface electromyography: detection and recording. 2002. https://delsys.com/downloads/TUTORIAL/semg-detection-and-recording.pdf.
146. De Luca CJ. DELSYS. A practicum on the use of sEMG signals in movement sciences. 2008. https://delsyseurope.com/downloads/TUTORIAL/a-practicum-on-the-use-of-semg-signals-in-movement-sciences.pdf.

147. Balbinot G, Joner Wiest M, Li G, Pakosh M, Cesar Furlan J, Kalsi-Ryan S, Zariffa J. The use of surface EMG in neurorehabilitation following traumatic spinal cord injury: a scoping review. Clin Neurophysiol. 2022;138:61–73. https://doi.org/10.1016/j.clinph.2022.02.028.
148. Baptista R, Salvaggio F, Cavallo C, Pizzocaro S, Galasso S, Schmid M, De Nunzio AM. Training-induced muscle fatigue with a powered lower-limb exoskeleton: a preliminary study on healthy subjects. Med Sci. 2022;10(4):55. https://doi.org/10.3390/medsci10040055.
149. Liu SH, Lin CB, Chen Y, Chen W, Huang TS, Hsu CY. An EMG patch for the real-time monitoring of muscle-fatigue conditions during exercise. Sensors. 2019;19(14):3108. https://doi.org/10.3390/s19143108.
150. Twomey R, Aboodarda SJ, Kruger R, Culos-Reed SN, Temesi J, Millet GY. Neuromuscular fatigue during exercise: methodological considerations, etiology and potential role in chronic fatigue. Neurophysiol Clin. 2017;47(2):95–110. Elsevier Masson SAS. https://doi.org/10.1016/j.neucli.2017.03.002.
151. Makaram N, Karthick PA, Swaminathan R. Analysis of dynamics of EMG signal variations in fatiguing contractions of muscles using transition network approach. IEEE Trans Instrum Meas. 2021;70. https://doi.org/10.1109/TIM.2021.3063777.
152. Piper H. Elektrophysiologie Menschlicher Muskeln. 1912. https://scholar.google.com/scholar?hl=it&as_sdt=0,5&q=H.+Piper,+Elektrophysiologie+Menschlicher+Muskeln,+p.126,+Springer,+Berlin+1912.
153. Cifrek M, Medved V, Srhoj-Egekher V. The application of Hilbert–Huang transform in the analysis of muscle fatigue during cyclic dynamic contractions. Med Biol Eng Comput. 2011;49(6):659–69.
154. Cifrek M, Tonkovic S, Medved V. Measurement and analysis of surface myoelectric signals during fatigued cyclic dynamic contractions. Measurement. 2000;27:85–92. www.elsevier.com/locate/measurement.
155. Royer TD. Electromyography and muscle force. In: Athletic therapy today, human kinetics; 2005. pp. 43–45.
156. Yousif HA, Ammar Z, Norasmadi AR. Assessment of muscles fatigue based on surface EMG signals using machine learning and statistical approaches: a review. IOP Conf Ser Mater Sci Eng. 2019;705:012010. https://doi.org/10.1088/1757-899X/705/1/012010.
157. Schepens P. Systematic evaluation of EMG data. Ghent University. 2017.
158. Kamaruddin NA, Khalid PI, Shaameri AZ. The use of surface electromyography in muscle fatigue assessments—A review. Teknologi J. 2015;74(6):2180–3722. www.jurnalteknologi.utm.my
159. Nussbaum MA. Static and dynamic myoelectric measures of shoulder muscle fatigue during intermittent dynamic exertions of low to moderate intensity. Eur J Appl Physiol. 2001;85(3–4):299–309. https://doi.org/10.1007/s004210100454.
160. González-Izal M, Malanda A, Navarro-Amézqueta I, Gorostiaga EM, Mallor F, Ibañez J, Izquierdo M. EMG spectral indices and muscle power fatigue during dynamic contractions. J Electromyogr Kinesiol. 2010;20(2):233–40. https://doi.org/10.1016/j.jelekin.2009.03.011.
161. Bonato P, Roy SH, Knaflitz M, De Luca CJ. Time frequency parameters of the surface myoelectric signal for assessing muscle fatigue during cyclic dynamic contractions. IEEE Trans Biomed Eng. 2001;48(7):745–53. https://doi.org/10.1109/10.930899.
162. Karlsson S, Yu J, Akay M. Time-frequency analysis of myoelectric signals during dynamic contractions: a comparative study. IEEE Trans Biomed Eng. 2000;47(2):228–38. https://doi.org/10.1109/10.821766.
163. Georgakis A, Stergioulas LK, Giakas G. Fatigue analysis of the surface EMG signal in isometric constant force contractions using the averaged instantaneous frequency. IEEE Trans Biomed Eng. 2003;50(2):262–5. https://doi.org/10.1109/TBME.2002.807641.
164. Huang NE. The empirical mode decomposition and the Hilbert spectrum for nonlinear and non-stationary time series analysis. Proc R Soc Lond Ser A Math Phys Eng Sci. 1998;454(1971):903–95.

165. Farina D, Fattorini L, Felici F, Filligoi G. Nonlinear surface EMG analysis to detect changes of motor unit conduction velocity and synchronization. J Appl Physiol. 2002;93(5):1753–63. https://doi.org/10.1152/japplphysiol.00314.2002.
166. Bramson A, Baland A, Iriki A. Measuring dynamical uncertainty with revealed dynamics Markov models. Front Appl Math Stat. 2019;5:7. https://doi.org/10.3389/fams.2019.00007.
167. McCullough M, Small M, Iu HHC, Stemler T. Multiscale ordinal network analysis of human cardiac dynamics. Philos Trans R Soc A Math Phys Eng Sci. 2017;375(2096):20160292. https://doi.org/10.1098/rsta.2016.0292.
168. Sakellariou K, Stemler T, Small M. Markov modeling via ordinal partitions: an alternative paradigm for network-based time-series analysis. Phys Rev E. 2019;100(6):062307. https://doi.org/10.1103/PHYSREVE.100.062307/FIGURES/27/MEDIUM.
169. Wu Y, Chen P, Luo X, Huang H, Liao L, Yao Y, Wu M, Rangayyan RM. Quantification of knee vibroarthrographic signal irregularity associated with patellofemoral joint cartilage pathology based on entropy and envelope amplitude measures. Comput Methods Prog Biomed. 2016;130:1–12. https://doi.org/10.1016/j.cmpb.2016.03.021.
170. Korupolu R, Stampas A, Singh M, Zhou P, Francisco G. Electrophysiological outcome measures in spinal cord injury clinical trials: a systematic review. Top Spinal Cord Inj Rehabil. 2019;25(4):340–54. https://doi.org/10.1310/SCI2504-340.
171. Balbinot G, Li G, Wiest MJ, Pakosh M, Furlan JC, Kalsi-Ryan S, Zariffa J. Properties of the surface electromyogram following traumatic spinal cord injury: a scoping review. J Neuroeng Rehabil. 2021;18(1):105. BioMed Central Ltd. https://doi.org/10.1186/s12984-021-00888-2.
172. Morecognition | Technology for an efficient, simple and fun rehabilitation. 2023. https://morecognition.com/en/more-2/.
173. Pregnolato G, Rimini D, Baldan F, Maistrello L, Salvalaggio S, Celadon N, Ariano P, Pirri CF, Turolla A. Clinical features to predict the use of a sEMG wearable device (REMO®) for hand motor training of stroke patients: a cross-sectional cohort study. Int J Environ Res Public Health. 2023;20(6):5082. https://doi.org/10.3390/ijerph20065082.
174. Castiblanco JC, Ortmann S, Mondragon IF, Alvarado-Rojas C, Jöbges M, Colorado JD. Myoelectric pattern recognition of hand motions for stroke rehabilitation. Biomed Signal Process Control. 2020;57:101737. https://doi.org/10.1016/j.bspc.2019.101737.
175. Sathiyanarayanan M, Rajan S. MYO armband for physiotherapy healthcare: a case study using gesture recognition application. In: 2016 8th international conference on communication systems and networks, COMSNETS 2016. 2016. https://doi.org/10.1109/COMSNETS.2016.7439933.
176. Geryes M, Charara J, Skaiky A, Mcheick A, Girault JM. A novel biomedical application for the myogesture control armband. 2017.
177. Merletti R, Temporiti F, Gatti R, Gupta S, Sandrini G, Serrao M. Translation of surface electromyography to clinical and motor rehabilitation applications: the need for new clinical figures. Transl Neurosci. 2023;14(1):20220279. https://doi.org/10.1515/tnsci-2022-0279.
178. Merletti R, Campanini I, Rymer WZ, Disselhorst-Klug C. Editorial: surface electromyography: barriers limiting widespread use of sEMG in clinical assessment and neurorehabilitation. Front Neurol. 2021;12:642257. https://doi.org/10.3389/fneur.2021.642257.
179. Campanini I, Cosma M, Manca M, Merlo A. Added value of dynamic EMG in the assessment of the equinus and the equinovarus foot deviation in stroke patients and barriers limiting its usage. Front Neurol. 2020a;11:583399. https://doi.org/10.3389/fneur.2020.583399.
180. Manca A, Cereatti A, Bar-On L, Botter A, Della Croce U, Knaflitz M, Maffiuletti NA, Mazzoli D, Merlo A, Roatta S, Turolla A, Deriu F. A survey on the use and barriers of surface electromyography in neurorehabilitation. Front Neurol. 2020;11:573616. https://doi.org/10.3389/fneur.2020.573616.
181. CEDE Project | International Society of Electrophysiology and Kinesiology (ISEK). 2023. https://isek.org/cede-project/.
182. International Society of Electrophysiology and Kinesiology. 2023. https://isek.org/.

183. Feldner HA, Papazian C, Peters K, Steele KM. "It's all sort of cool and interesting…but what do I do with it?" A qualitative study of stroke survivors' perceptions of surface electromyography. Front Neurol. 2020;11:1037. https://doi.org/10.3389/fneur.2020.01037.
184. Martin BJ, Acosta-Sojo Y. sEMG: a window into muscle work, but not easy to teach and delicate to practice—A perspective on the difficult path to a clinical tool. Front Neurol. 2021;11:588451. https://doi.org/10.3389/fneur.2020.588451.
185. Campanini I, Disselhorst-Klug C, Rymer WZ, Merletti R. Surface EMG in clinical assessment and neurorehabilitation: barriers limiting its use. Front Neurol. 2020b;11:934. https://doi.org/10.3389/fneur.2020.00934.
186. Goetz CG, Tilley BC, Shaftman SR, Stebbins GT, Fahn S, Martinez-Martin P, Poewe W, Sampaio C, Stern MB, Dodel R, Dubois B, Holloway R, Jankovic J, Kulisevsky J, Lang AE, Lees A, Leurgans S, LeWitt PA, Nyenhuis D, et al. Movement Disorder Society-sponsored revision of the unified Parkinson's disease rating scale (MDS-UPDRS): scale presentation and clinimetric testing results. Mov Disord. 2008;23(15):2129–70. https://doi.org/10.1002/MDS.22340.
187. McGuirk TE, Perry ES, Sihanath WB, Riazati S, Patten C. Feasibility of markerless motion capture for three-dimensional gait assessment in community settings. Front Hum Neurosci. 2022;16:867485. https://doi.org/10.3389/fnhum.2022.867485.

# Robotics in Neurorehabilitation: From Research to Clinical Practice

**8**

Mirjam Bonanno, Giovanni Pioggia, Andrea Santamato, and Rocco Salvatore Calabrò

## 8.1 The Robotic Era as the Golden Age of Neurorehabilitation

In the early 2000s, the robotics in the neurorehabilitation began to be increasingly studied by researchers. Initially, these particularly systems were strictly used by soldiers for military purposes, then they were translated for rehabilitation, especially in those patients with neurological disorders [1, 2]. In these 23 years of clinical research in the neurorehabilitation field, robotics purchased a pivotal role next to the conventional physical therapies. Research on these new rehabilitative devices goes hand in hand with discoveries on neuroplasticity. The term of "neuroplasticity" must be intended as the neurophysiological process involving molecular changes in cell signals and neurotransmission, linked to functional and structural modifications in neurons [3]. In fact, it has been widely accepted that, following brain injury, both connectivity maps and behavioural skills can at least be partially restored through intense motor repetitions, task-oriented activities, and motivation, boosting newly neural connections [4]. In this vein, robot-mediated learning is based on four bases: robotic assistance, robotic perturbation, the adjunct of virtual reality, and the interface within brain and robotic device.

M. Bonanno (✉) · R. S. Calabrò
IRCCS Centro Neurolesi Bonino-Pulejo, Messina, Italy
e-mail: mirjam.bonanno@irccsme.it; roccos.calabro@irccsme.it

G. Pioggia
Institute for Biomedical Research and Innovation, National Research Council of Italy (IRIB-CNR), Messina, Italy
e-mail: giovanni.pioggia@irib.cnr.it

A. Santamato
Rehabilitation Centre-Physical Medicine and Rehabilitation Section, OORR-Hospital-University of Foggia, Foggia, Italy
e-mail: andrea.santamato@unifg.it


R. S. Calabrò (ed.), *Translational Neurorehabilitation*,
https://doi.org/10.1007/978-3-031-63604-2_8

First, the robotic assistance consists in the haptic interaction provided by the device. It has been suggested that robot assistance could promote motivation linked to better motor performance during training than without assistance. Indeed, the use of robot-mediated assistance during motor activities can modulate neural functioning compared to un-assisted or active voluntary movements [5].

The robotic perturbation is related to the application of an external force which renders movement more difficult to perform. The increased difficulty, in addition to intensity of motor training, represents stronger learning stimulus [3].

The adjunct of virtual reality to robotic training represents an important stimulus thanks to its audio-video feedback and the different scenarios that engage the attention of the subject involving him/her in the rehabilitation sessions. In details, virtual reality (VR) with its graphics and scenarios can amplify the patients' compliance, promoting long-lasting training session and more intensive sessions [6].

The brain interface with robotic device is related to bridge the gap between the movement intention and the actual movement of an orthosis or robot device without the need of a limb [7]. Then, the robotic devices can be considered as a milestone in the neurorehabilitation field, thus further potentiating neural plasticity processes, especially in those patients affected by brain injury [8]. Recovering from a stroke or brain injury events comprises both spontaneous and motor-learning mediated mechanisms. It is recognized that spontaneous behavioural recovery, due to plastic process, occur in the first 3 months after the brain lesion [4]. This important issue should be addressed during the rehabilitation path to better understand when to expect recovery, to choose the most appropriate treatment, and determine the timing of rehabilitation.

## 8.2 Classification of Robotic Rehabilitation Systems and Description of Main Technical Aspects

Some authors distinguish robotic systems into two different types, according to their motion assistance in active and passive [9]. Firstly, active devices are equipped with one actuator and allow active movement assistance. This assistance improves when patient is too weak to perform the specific exercises. However, even with active robotic systems, some exercises can be considered passive when the patient's effort is not required. On the other hand, passive devices are characterized by actuators providing resistive force only. These kinds of actuators are cheaper than those of active robotic systems. In this context, upper limb robotic devices present other two types of motion assistance: haptic and coaching systems. The haptic devices are classified as either active or passive, depending on their type of actuator. These systems provide a tactile sensation through touch by the patient. While coaching devices are sensorized tools included in the non-actuated device group, in which the user interacts with a virtual environment [10].

From biomechanical point of view, robotic devices can be distinguished into two main categories: (1) end-effectors, which are stationary devices with a distal moveable part requiring an active participation; (2) exoskeletons which are wearable devices that are able to reproduce passive movements [11]. These two categories principally differ from each other for the modality of interaction between patients' movement and robotic system. In particular, the end-effectors are suitable tools that can be adapted to different shapes and sizes. However, the end-effectors for upper limb could present a disadvantage related to the lack of rotation movement which also limits pronation and supination movements [9]. While exoskeleton robots present a splint or a bionic structure in which the limb is allocated, in this way they require a smaller work setting than end-effectors. Indeed, they can also calculate the torque for each joint, controlling limb movements [12]. Therefore exoskeletons can be used to train patients with severe motor alterations, such as spinal cord injury, cerebral palsy, and severe acquired brain injury. Notably, exoskeletons in the lower limb can be differentiated into (1) tethered exoskeletons equipped with a treadmill, such as Lokomat; (2) untethered exoskeletons which are used to promote overground gait, such as Ekso (Fig. 8.1) [13].

In the upper limb robotic training, clinicians should consider that exoskeletons as well as end-effectors are distinguished also in systems for proximal movements training (i.e. shoulder, elbow, and wrist) and in systems for distal hand movements (i.e. hand and fingers) [14, 15], as showed in Fig. 8.2. Other authors [16] further distinguished exoskeletons for upper limb in two categories: fixed or added. The fixed systems are allocated in specific rooms, although some of these systems presents wheels facilizing the transport. Otherwise, the added systems consist of only adjustable arm tool.

In the following part, we described briefly the most used robotic devices either for lower and upper limb, dividing them into end-effectors and exoskeletons.

### 8.2.1 Exoskeletons for Lower Limb

*Lokomat* is a tethered exoskeleton, comprising an external gait orthosis supported by a body weight support system (BWSS), in addition to computer-controlled linear actuators at each hip and knee joint. Generally, Lokomat devices are often equipped

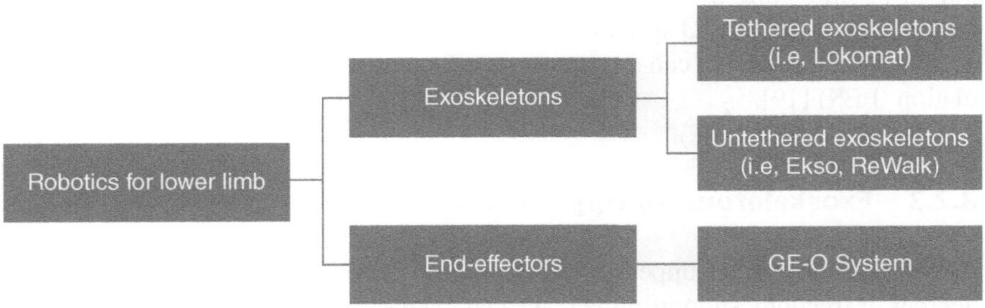


**Fig. 8.1** The scheme of the subdivision of robotics for lower limb neurological rehabilitation according to their biomechanical characteristics and the most used systems

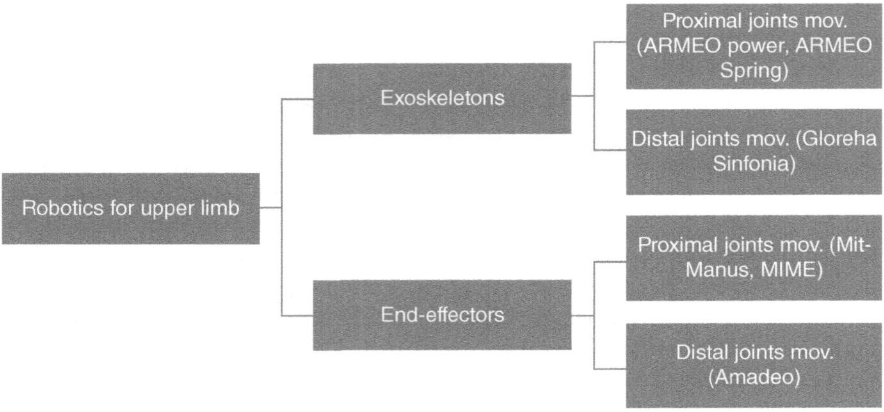


**Fig. 8.2** The scheme of robotics for upper limb neurological rehabilitation according to their biomechanical and anatomical characteristics, in addition to the most used systems

with a virtual reality (VR) screen to get patients more involved in the gait training [17].

Ekso and ReWalk consist in bilateral robots characterized by actuators moving patient's leg during gait. These tools can be used in the home context promoting independence in the activity of daily living (assistive devices), or in the rehabilitative setting to improve gait and postural stability (rehabilitative devices). These wearable exoskeletons contain 2 joints actuated at hip and knee, while the ankle is mechanically controlled. Notably, the ReWalk needs also crutches to move around easily [18].

### 8.2.2 End-Effectors for Lower Limb

GE-O system provides propulsive motion of legs thanks to 2 footplates that generates an alternation of stance and swing gait phases. The main difference between end-effectors and exoskeletons is that the feet are always in contact with the moving platform, simulating the gait phases but not necessarily generating real swing and stance phases. The GE-O system not only allow walk on the flat, but also ascending and descending stairs, it can be adjunct to VR exercises and functional electric stimulation (FES) [19].

### 8.2.3 Exoskeletons for Upper Limb

ARMEO Power is an upper limb rehabilitative exoskeleton, equipped with six degrees of freedom, supporting 1D, 2D, and 3D movements, especially for proximal joints. Therapists can control the provided assistance to the patients, based on their level of injury and can also evaluate objectively ranges of movement and forces.

Like most robotic devices, the ARMEO power is equipped with a screen for VR exercises [20].

ARMEO Spring is an arm and hand exoskeleton with 3D workspace and six degrees of freedom, as well as ARMEO power. The main difference between these two exoskeletons is that ARMEO spring does not provide any assistance, while ARMEO power guides all patient's movements. In fact, the ARMEO spring is based on all activity on the patients' movements [20].

Gloreha Sinfonia is a powered exoskeleton glove designed for rehabilitation of hand and fingers, mobilizing them in flexion and extension. This robotic device can detect the voluntary movements based on the residual motor skills. The system can be used also in a bilateral modality to promote bimanual motricity for activity of daily living [21].

### 8.2.4 End-Effectors for Upper Limb

Mit-Manus consists in a robotic workstation in which patients can interact with a specific screen for computer-based games. In particular, this device has five degrees of freedom, two for shoulder (abduction-adduction) and elbow (flexion-extension; pronation-supination), while wrist presents three degrees of freedom, including radial-ulnar inclination, pronation-supination, and flexion-extension. The Mit-Manus also allows to register some spatial-temporal parameters of motor performances such as speed, force, and position [14].

MIME robot manipulator (PUMA 560, Staubli Unimation Inc., Duncan, South Carolina) comprises a splint in which mechanical forces are applied to the limb, as therapists usually do. This device presents six degrees of freedom in 3D and sensors to measure forces and joint torques between the robot and the affected limb [22].

AMADEO is a powered end-effector for hand and fingers rehabilitation, both for adults and children. It is a stationary and fixed robotic device, which connects directly with the patient's fingers [23].

## 8.3 Clinical Implications for Robotic Training in Neurorehabilitation

From a clinical point of view, gait recovery in patients affected by brain/spinal cord injury is one of the most important goals in physiotherapy. In this way, robotic systems meet the needs of both patients and therapists, providing an intensive, task-oriented, and long-lasting training [8]. In fact, the robotic-assisted gait training (RAGT) safely helps overcome the initial phases of rehabilitation, when patients need more support to perform motor activity [24]. The RAGT promotes the trunk stability by using a BWSS with more difficult standing levels, from completely supported to standing up actively. The BWSS of robotic systems guarantees the body alignment during gait weight-shifting to facilitate postural reactions [25]. Indeed, postural reactions and trunk stability are fundamental to ensure the correct steps during gait and to avoid falls. In this way, robotic devices such as Lokomat can used also to train balance on the hypothesis that intensive and repetitive walking training can improve non-walking tasks including static and dynamic balance [26]. Moreover, RAGT has positive influences on cardio-pulmonary systems, increasing endurance, as confirmed by some studies [24, 27]. However, the presence of an assistive guide during movement inevitable reduces the active contribution provided by the patient. Since compensation mechanisms are the most used by people affected by brain injury, Lokomat could discourage these neural processes when it is set at 50% of guidance [28]. To overcome these concerns, clinicians should consider increasing gait speed or provide more motivating feedback, also through VR. On the other hand, GE-O system as an end-effector is a promising tool in those patients with residual motor abilities including incomplete spinal cord injury, as suggested by Calabrò RS et al. [29]. It seems that the GE-O system promoted motor coordination, visual and spatial attention, in addition to looking and pointing function, which are all connected to gait performance. It is noteworthy that there is no evidence about the superiority of one of these robotic systems on each other in patients affected by

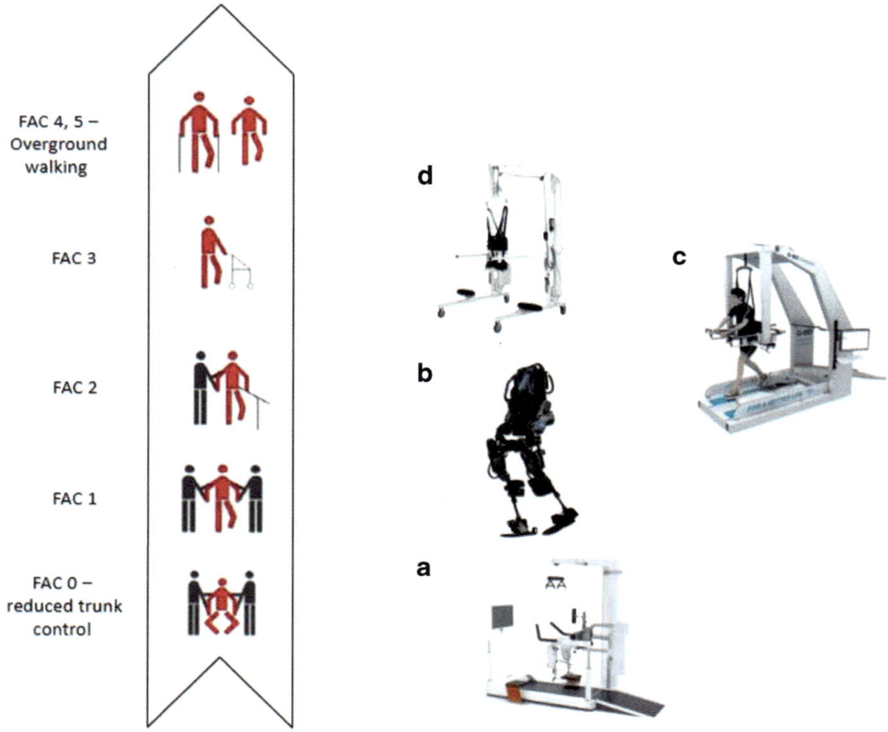


**Fig. 8.3** Theoretical scheme based on the functional ambulation classification (FAC) (on the left) and the lower limb robotic device most used (on the right), according to the FAC's level, adapted from study by Morone G. et al. [19]. (**a**) Lokomat; (**b**) wearable exoskeleton; (**c**) GE-O system; (**d**) BWSS for gait

neurological disorders. Therefore, the extension and time of injury and age of patients play a key role in choosing the right therapeutic robotic tool, in addition to standard physiotherapy (Fig. 8.3).

Even more controversial is the robotic upper limb rehabilitation because several guidelines recommend it, especially after stroke but the variation of training characteristics, type of training, and the absence of specific outcome measures limit its application in clinical practice [20]. Regarding the type of robot, as well as for robotic lower limb devices, depends on the residual motor skills of patients. In fact, the use of end-effectors (i.e., MIME, Mit-Manus, and Amadeo) should be considered when patients are able to control their movements, due to the lack of intersegmental control provided by these devices [30]. While exoskeletons are more suitable in the first rehabilitation steps because of their 3D workspace and the supported guidance during movements, improving motivation [20, 30]. Otherwise, the upper extremities limitation represents one of the most functional impairments in patients with cervical SCI and robotic devices could be promising tools to promote arm strength and functional ability [31]. However, it seems that they are particularly effective in incomplete SCI or in chronic lesions, as confirmed by some authors [32].

## 8.4 Combined Approach: Is This the Right Future Direction?

Nowadays, the combined approach using non-invasive brain stimulation (NIBS) in addition to robotic rehabilitation is getting popular in the neurorehabilitation settings. In details, the transcranial magnetic stimulation (TMS) and transcranial direct current stimulation (tDCS) are the most common types of NIBS used in clinical practice to modulate cortical excitability, inducing plastic changes [33, 34]. It has been already demonstrated that the use of NIBS is effective in improving gait recovery for chronic stroke patients [35]. In this sense, NIBS and neurorobotics could further enhance cortical plasticity by providing cortical stimuli (top-down approaches—NIBS) that acts on the produced effects of robotics (bottom-down approaches), improving patients' motor skills [36]. Despite these promising results in the recovery of gait, the evidence about of NIBS and robotic-assisted therapy for upper limb is still lacking and the approach is not superior to the only RAT for upper limb [36, 37]. These discrepancies in literature are supposed to be linked to the heterogeneity of intervention protocols both for motor treatment and NIBS. Another unsolved question is when clinicians should apply NIBS, as a preparation for brain before RAT, during the RAT or post-RAT to maintain the effects.

Thus, to answer to the initial question: the combined approach could be a promising future direction for research and then for clinical practice in the neurorehabilitation field, always taking the patient's needs into account with a person-centred care perspective.

## 8.5 Conclusion

This book chapter addressed an important and under debate issue. We started from the research basis of robotics in neurorehabilitation, describing briefly the most common robotic devices commercially available and arrived at their current use in clinical practice. It is clear that the robotic devices represent a valid approach to be added to standard physiotherapy, even if remains a lack of evidence about RAT for upper limb effectiveness as well as for the combined approach (NIBS plus RAT for upper limb) in patients affected by traumatic brain/spinal cord injury. However, this does not mean that they are not effective in inducing those use-dependent neuroplasticity. Therefore, we recommend their clinical use, alongside standard physiotherapy, in order to increasingly motivate patients and improve their motor outcomes.

## References


1. Hidler J, Sainburg R. Role of robotics in neurorehabilitation. Top Spinal Cord Inj Rehabil. 2011;17(1):42–9. https://doi.org/10.1310/sci1701-42.
2. Oña ED, Cano-de la Cuerda R, Sánchez-Herrera P, Balaguer C, Jardón A. A review of robotics in neurorehabilitation: towards an automated process for upper limb. J Healthc Eng. 2018;2018:9758939. https://doi.org/10.1155/2018/9758939.
3. Turner DL, Ramos-Murguialday A, Birbaumer N, Hoffmann U, Luft A. Neurophysiology of robot-mediated training and therapy: a perspective for future use in clinical populations. Front Neurol. 2013;4:184. https://doi.org/10.3389/fneur.2013.00184.
4. Alia C, Spalletti C, Lai S, Panarese A, Lamola G, Bertolucci F, Vallone F, Di Garbo A, Chisari C, Micera S, Caleo M. Neuroplastic changes following brain ischemia and their contribution to stroke recovery: novel approaches in neurorehabilitation. Front Cell Neurosci. 2017;11:76. https://doi.org/10.3389/fncel.2017.00076. PMID: 28360842; PMCID: PMC5352696.
5. Özen Ö, Buetler KA, Marchal-Crespo L. Promoting motor variability during robotic assistance enhances motor learning of dynamic tasks. Front Neurosci. 2021;14:600059. https://doi.org/10.3389/fnins.2020.600059.
6. Zanatta F, Giardini A, Pierobon A, D'Addario M, Steca P. A systematic review on the usability of robotic and virtual reality devices in neuromotor rehabilitation: patients' and healthcare professionals' perspective. BMC Health Serv Res. 2022;22(1):523. https://doi.org/10.1186/s12913-022-07821-w. PMID: 35443710; PMCID: PMC9020115.
7. Baniqued PDE, Stanyer EC, Awais M, Alazmani A, Jackson AE, Mon-Williams MA, Mushtaq F, Holt RJ. Brain-computer interface robotics for hand rehabilitation after stroke: a systematic review. J Neuroeng Rehabil. 2021;18(1):15. https://doi.org/10.1186/s12984-021-00820-8. PMID: 33485365; PMCID: PMC7825186.
8. Bonanno M, De Luca R, De Nunzio AM, Quartarone A, Calabrò RS. Innovative technologies in the neurorehabilitation of traumatic brain injury: a systematic review. Brain Sci. 2022;12(12):1678. https://doi.org/10.3390/brainsci12121678.
9. Maciejasz P, Eschweiler J, Gerlach-Hahn K, Jansen-Troy A, Leonhardt S. A survey on robotic devices for upper limb rehabilitation. J Neuroeng Rehabil. 2014;11:3. https://doi.org/10.1186/1743-0003-11-3. PMID: 24401110; PMCID: PMC4029785.
10. Kabir R, Sunny MSH, Ahmed HU, Rahman MH. Hand rehabilitation devices: a comprehensive systematic review. Micromachines (Basel). 2022;13(7):1033. https://doi.org/10.3390/mi13071033. PMID: 35888850; PMCID: PMC9325203.

11. Molteni F, Gasperini G, Cannaviello G, Guanziroli E. Exoskeleton and end-effector robots for upper and lower limbs rehabilitation: narrative review. PM R. 2018;10(9 Suppl 2):S174–88. https://doi.org/10.1016/j.pmrj.2018.06.005. PMID: 30269804.
12. Qassim HM, Wan Hasan WZ. A review on upper limb rehabilitation robots. Appl Sci. 2020;10:6976. https://doi.org/10.3390/app10196976.
13. Baud R, Manzoori AR, Ijspeert A, Bouri M. Review of control strategies for lower-limb exoskeletons to assist gait. J Neuroeng Rehabil. 2021;18(1):119. https://doi.org/10.1186/s12984-021-00906-3.
14. Sobiech M, Wolański W, Karpiel I. Brief overview upper limb rehabilitation robots/devices. In: Biele C, Kacprzyk J, Kopeć W, Owsiński JW, Romanowski A, Sikorski M, editors. Digital interaction and machine intelligence. MIDI 2021. Lecture notes in networks and systems, vol 440. Cham: Springer; 2022. https://doi.org/10.1007/978-3-031-11432-8_29.
15. de-la-Torre R, Oña ED, Balaguer C, Jardón A. Robot-aided systems for improving the assessment of upper limb spasticity: a systematic review. Sensors (Basel). 2020;20(18):5251. https://doi.org/10.3390/s20185251.
16. Ayad S, Ayad M, Megueni A, Spaich EG, Struijk LNSA. Toward standardizing the classification of robotic gait rehabilitation systems. IEEE Rev Biomed Eng. 2019;12:138–53. https://doi.org/10.1109/RBME.2018.2886228. Epub 2018 Dec 11. PMID: 30561350.
17. Calafiore D, Negrini F, Tottoli N, Ferraro F, Ozyemisci-Taskiran O, de Sire A. Efficacy of robotic exoskeleton for gait rehabilitation in patients with subacute stroke: a systematic review. Eur J Phys Rehabil Med. 2022;58(1):1–8. https://doi.org/10.23736/S1973-9087.21.06846-5. Epub 2021 Jul 12. PMID: 34247470; PMCID: PMC9980569.
18. de Miguel-Fernández J, Lobo-Prat J, Prinsen E, Font-Llagunes JM, Marchal-Crespo L. Control strategies used in lower limb exoskeletons for gait rehabilitation after brain injury: a systematic review and analysis of clinical effectiveness. J Neuroeng Rehabil. 2023;20(1):23. Published 2023 Feb 19. https://doi.org/10.1186/s12984-023-01144-5.
19. Morone G, Paolucci S, Cherubini A, De Angelis D, Venturiero V, Coiro P, Iosa M. Robot-assisted gait training for stroke patients: current state of the art and perspectives of robotics. Neuropsychiatr Dis Treat. 2017;13:1303–11. https://doi.org/10.2147/NDT.S114102. PMID: 28553117; PMCID: PMC5440028.
20. Calabrò RS, Morone G, Naro A, Gandolfi M, Liotti V, D'aurizio C, Straudi S, Focacci A, Pournajaf S, Aprile I, Filoni S, Zanetti C, Leo MR, Tedesco L, Spina V, Chisari C, Taveggia G, Mazzoleni S, Smania N, Paolucci S, Franceschini M, Bonaiuti D. Robot-assisted training for upper limb in stroke (ROBOTAS): an observational, multicenter study to identify determinants of efficacy. J Clin Med. 2021;10:5245. https://doi.org/10.3390/jcm10225245.
21. Vanoglio F, Luisa A, Garofali F, Mora C. Evaluation of the effectiveness of Gloreha (hand rehabilitation glove) on hemiplegic patients. Pilot study. In: XIII congress of Italian Society of Neurorehabilitation, 18–20 April. Italy: Bari; 2013.
22. Lum PS, Burgar CG, Van der Loos M, Shor PC, Majmundar M, Yap R. MIME robotic device for upper-limb neurorehabilitation in subacute stroke subjects: a follow-up study. J Rehabil Res Dev. 2006;43(5):631–42. https://doi.org/10.1682/jrrd.2005.02.0044.
23. Serrano-López Terradas PA, Criado Ferrer T, Jakob I, Calvo-Arenillas JI. Quo Vadis, Amadeo hand robot? A randomized study with a hand recovery predictive model in subacute stroke. Int J Environ Res Public Health. 2022;20(1):690. https://doi.org/10.3390/ijerph20010690. PMID: 36613027; PMCID: PMC9820043.
24. Esquenazi A, Talaty M. Robotics for lower limb rehabilitation. Phys Med Rehabil Clin N Am. 2019;30(2):385–97. https://doi.org/10.1016/j.pmr.2018.12.012.
25. Moucheboeuf G, Griffier R, Gasq D, Glize B, Bouyer L, Dehail P, Cassoudesalle H. Effects of robotic gait training after stroke: a meta-analysis. Ann Phys Rehabil Med. 2020;63(6):518–34. https://doi.org/10.1016/j.rehab.2020.02.008. Epub 2020 Mar 27. PMID: 32229177.
26. Baronchelli F, Zucchella C, Serrao M, Intiso D, Bartolo M. The effect of robotic assisted gait training with Lokomat® on balance control after stroke: systematic review and meta-analysis. Front Neurol. 2021;12:661815. https://doi.org/10.3389/fneur.2021.661815. PMID: 34295298; PMCID: PMC8289887.

27. Bruni MF, Melegari C, De Cola MC, Bramanti A, Bramanti P, Calabrò RS. What does best evidence tell us about robotic gait rehabilitation in stroke patients: a systematic review and meta-analysis. J Clin Neurosci. 2018;48:11–7. https://doi.org/10.1016/j.jocn.2017.10.048. Epub 2017 Dec 6. PMID: 29208476.
28. van Kammen K, Boonstra AM, van der Woude LHV, Reinders-Messelink HA, den Otter R. Differences in muscle activity and temporal step parameters between Lokomat guided walking and treadmill walking in post-stroke hemiparetic patients and healthy walkers. J Neuroeng Rehabil. 2017;14(1):32. https://doi.org/10.1186/s12984-017-0244-z. PMID: 28427422; PMCID: PMC5397709.
29. Calabrò RS, Filoni S, Billeri L, et al. Robotic rehabilitation in spinal cord injury: a pilot study on end-effectors and neurophysiological outcomes. Ann Biomed Eng. 2021;49(2):732–45. https://doi.org/10.1007/s10439-020-02611-z.
30. Morone G, Palomba A, Cinnera AM, Agostini M, Aprile I, Arienti C, Paci M, Casanova E, Marino D, La Rosa G, et al.; "CICERONE" Italian Consensus Conference on Robotic in Neurorehabilitation. Systematic review of guidelines to identify recommendations for upper limb robotic rehabilitation after stroke. Eur J Phys Rehabil Med. 2021;57:238–245. https://doi.org/10.23736/S1973-9087.21.06625-9.
31. Morone G, de Sire A, Martino Cinnera A, Paci M, Perrero L, Invernizzi M, Lippi L, Agostini M, Aprile I, Casanova E, Marino D, La Rosa G, Bressi F, Sterzi S, Giansanti D, Battistini A, Miccinilli S, Filoni S, Sicari M, Petrozzino S, Solaro CM, Gargano S, Benanti P, Boldrini P, Bonaiuti D, Castelli E, Draicchio F, Falabella V, Galeri S, Gimigliano F, Grigioni M, Mazzoleni S, Mazzon S, Molteni F, Petrarca M, Picelli A, Gandolfi M, Posteraro F, Senatore M, Turchetti G, Straudi S. Upper limb robotic rehabilitation for patients with cervical spinal cord injury: a comprehensive review. Brain Sci. 2021;11(12):1630. https://doi.org/10.3390/brainsci11121630. PMID: 34942935; PMCID: PMC8699455.
32. Zariffa J, Kapadia N, Kramer JLK, Taylor P, Alizadeh-Meghrazi M, Zivanovic V, Willms R, Townson A, Curt A, Popovic MR, et al. Feasibility and efficacy of upper limb robotic rehabilitation in a subacute cervical spinal cord injury population. Spinal Cord. 2012;50:220–6. https://doi.org/10.1038/sc.2011.104.
33. Di Pino G, Pellegrino G, Assenza G, Capone F, Ferreri F, Formica D, et al. Modulation of brain plasticity in stroke: a novel model for neurorehabilitation. Nat Rev Neurol. 2014;10:597–608. https://doi.org/10.1038/nrneurol.2014.162.
34. Bressi F, Cinnera AM, Morone G, et al. Combining robot-assisted gait training and non-invasive brain stimulation in chronic stroke patients: a systematic review. Front Neurol. 2022;13:795788. Published 2022 May 2. https://doi.org/10.3389/fneur.2022.795788.
35. Bai X, Guo Z, He L, Ren L, McClure MA, Mu Q. Different therapeutic effects of transcranial direct current stimulation on upper and lower limb recovery of stroke patients with motor dysfunction: a meta-analysis. Neural Plast. 2019;2019:1372138. https://doi.org/10.1155/2019/1372138.
36. Naro A, Calabrò RS. Improving upper limb and gait rehabilitation outcomes in post-stroke patients: a scoping review on the additional effects of non-invasive brain stimulation when combined with robot-aided rehabilitation. Brain Sci. 2022;12(11):1511. https://doi.org/10.3390/brainsci12111511.
37. Reis SB, Bernardo WM, Oshiro CA, Krebs HI, Conforto AB. Effects of robotic therapy associated with noninvasive brain stimulation on upper-limb rehabilitation after stroke: systematic review and meta-analysis of randomized clinical trials. Neurorehabil Neural Repair. 2021;35:256–66. https://doi.org/10.1177/1545968321989353.

# Virtual Reality in Neurorehabilitation: Is It Time to Go Beyond Research?

9

Maria Grazia Maggio and Rocco Salvatore Calabrò

## 9.1 Introduction

Cognitive Rehabilitation (CR) is a non-pharmacological treatment aimed at subjects affected by cognitive difficulties due to brain damage or neurodegenerative disorders [1]. CR is based on the learning of compensatory strategies (compensatory CR) and the strengthening of residual abilities (reparative CR) in order to promote the best possible quality of life in the patient. In particular, the reparative CR is based on specific cognitive training to recover deficient cognitive abilities. In the compensatory CR, the use of tools/helpers is promoted to allow the deficit to be overcome [1, 2].

CR is based on the principle of neuroplasticity, i.e., the ability of the brain, if suitably stimulated, to modify itself to increase its functional capabilities through the neo-generation of neurons and the increase in interneuronal dendritic connections. The brain can therefore be adequately stimulated through exercise, the study of new languages or musical instruments, active life, and relationships, the execution of intellectual games, and the use of specific technological tools and cognitive enhancement programs [1, 2].

In fact, one aspect of CR is the way it can be used. CR techniques can be conventional (i.e., paper/pencil exercises) or innovative, such as computerized CR, robotics techniques, or virtual reality (VR). The latter has shown promising results in recent years, becoming a relevant tool in CR. VR consists of interactive simulations

M. G. Maggio (✉)
Department of Biomedical and Biotechnological Science, University of Catania, Catania, Italy

R. S. Calabrò
IRCCS Centro Neurolesi Bonino-Pulejo, Messina, Italy
e-mail: roccos.calabro@irccsme.it





R. S. Calabrò (ed.), *Translational Neurorehabilitation*,
https://doi.org/10.1007/978-3-031-63604-2_9

created with hardware and software in order to allow the patient to interact in highly stimulating environments, similar to the real world [1, 3–5].

VR systems are based on immersion and presence characteristics [6, 7]. Immersion is the objective perception of being sensory absorbed in a three-dimensional, computer-generated environment. Presence, on the other hand, consists in the subjective psychological perception, whereby the user is aware of being involved in the virtual context. Based on the degree and type of immersion and presence aroused in the user, three types of VR are distinguished: immersive VR, non-immersive VR, and semi-immersive VR [8]. Immersive virtual reality allows to provide the sensation and illusion of being inside the virtual world; then, it absorbs the user internally from a sensory point of view. It is usually based on some tools that produce audio and visual diffusion, such as a helmet, to allow the user to concentrate on the stimuli presented on the 3D screen. An example is "the Computer Assisted Rehabilitation Environment (CAREN—Motek Medical B.V.)." This instrument is a dome simulator with up to 360° VR screen and various software and peripherals, including a sensorized treadmill (with plantar pressure sensors), a mobile platform, and high-speed infrared cameras that map the position of the placed markers on the patient. The CAREN is used for the improvement of motor and cognitive performance in neurological and orthopedic patients [9].

Semi-immersive virtual reality is based on a three-dimensional modality, similar to reality seen through a window, in which there is less immersion but the user's presence sensation is maintained. An example is BTs-Nirvana (BTS Bioengineering, Milan). BTs-N is a motor and cognitive rehabilitation system, based on interactive virtual scenarios, which offer audiovisual feedback, activated by the patient's movement, and guided by the therapist [10].

Finally, non-immersive virtual reality is characterized by reduced immersion and presence, as the user has the role of external observer of a scenario presented on a monitor, such as PC-based software [11].

The VR systems, thanks to the characteristics of immersion, interaction, and imagination allow to stimulate patient interaction and involvement in therapeutic sessions, favoring functional recovery [6, 7]. Indeed, VR allows the therapist to customize the exercises of patients, increasing the motivation and active participation of patients. In the literature, it has been observed that VR can favor the reactivation and/or improvement of various functions connected to the cortex [8] and strengthen the efficiency of the sensory cortex. VR through audio/visual feedback promotes motor awareness and emotional involvement, improving functional execution and control of complex behavioral sequences and cognitive processes [8–10, 12–14].

## 9.2 Dementia and VR

Dementia is a cognitive impairment that causes a deterioration in social functioning, work skills, and daily life activities. Worldwide, dementia affects around 35.6 million people [15]. The most common form of dementia is Alzheimer's disease, characterized by a cognitive decline, in particular memory dysfunctions [16, 17]. Cognitive decline occurs insidiously with a gradual and steady progression. The application of virtual reality (VR) can be helpful in the rehabilitation of different types of cognitive disorders [16–20]. It has been shown that VR can improve the reactivation of some areas of the cortex by enhancing neuroplasticity processes [16–19] and promoting a better cognitive outcome [20]. In particular, growing evidence has highlighted that VR can enhance spatial memory, episodic memory, daily life autonomy, language, executive function, short-term and working memory, attention, movement, and balance [20–22]. Additionally, recent applications of VR in patients with dementia highlighted that VR can promote patient recovery in a shorter time than traditional methods, as confirmed by neuropsychological and neuroimaging tools [23].

## 9.3 Stroke and VR

Stroke is characterized by a sudden onset of signs and symptoms due to focal damage, caused by impaired cerebral circulation. It represents the third cause of death and the main cause of adult disability [24]. On the basis of the vascular mechanisms that cause stroke, it is possible to distinguish "ischemic" stroke (80%), due to the occlusion of a blood vessel, or "hemorrhagic" (20%), due to a ruptured blood vessel. Stroke causes sensory, motor, and cognitive deficits, which differ according to the area affected. Furthermore, it affects social activities and daily life causing dependence and reduced autonomy [25–27]. Neuropsychological rehabilitation can be very useful to increase recovery through neuroplasticity processes, especially through VR. Various authors have observed that VR using oriented and repetitive tasks can promote effective recovery of neurological disorders [25–27]. Jang et al. have implemented a study on mice after a cortical infarction, observing that rehabilitation in stimulating environments promotes problem-solving skills and stimulates optimal performance [28]. In fact, VR allows the creation of stimulating and engaging environments for patients, encouraging their motivation to follow the training sessions for extended times [28]. Studies conducted with instrumental examinations have observed that the use of VR, thanks to the augmented feedback, can stimulate a sensorimotor activation of the neural representation from contralesional to ipsilesional in post-stroke patients [25–28]. VR allows a high degree of customization of programs, defining exercises calibrated to the patient's needs, and allowing for a long-term improvement of functional skills [28].

## 9.4 Traumatic Brain Injury (TBI) and VR

Traumatic brain injury (TBI) consists of damage to the brain caused by an external physical force. It can occur from a sudden blow, shock, or trauma to the head or an object penetrating the brain. Other common events that can lead to TBI are falls, automobile accidents, violence, sports injuries, explosions, or other combat injuries.

TBI can cause an alteration of consciousness that can vary from mild to serious disorientation and also a coma. In some cases, the person may experience memory dysfunctions after the traumatic event [29]. In the general population, TBI is estimated to have a prevalence of 12–16.7% in males and 8.5% in females [30]. Usually, following TBI, diffuse syndromes occur due to the involvement of multiple brain areas. In these cases, patients can experience cognitive, physiological, and psychosocial deficits with a negative impact on the patient's life. Various studies have shown that rehabilitation performed through innovative and highly motivating tools, including VR training, can be an effective way to improve cognitive functions in patients with TBI [31–33]. Virtual environments simulate real and interactive contexts that encourage the training of brain functions, such as memory and other cognitive skills [32, 33]. In particular, the VR environment allows sending increased feedback to the central nervous system, inducing changes in neural plasticity, and resulting in the activation of motor activity and/or cognitive function [32]. For this reason, the repetitive virtual task is effective in TBI patients for both motor skills, such as improving walking speed, gait, and balance, as well as cognitive function and functional ability (e.g., crossing a road, driving, preparing food) [33]. In recent years, interest in VR has increased, where in TBI patients it has been observed that this innovative tool can be an alternative and highly functional rehabilitation for the recovery of TBI patients [32].

## 9.5 Multiple Sclerosis and VR

Multiple sclerosis is a neurodegenerative disease that affects the central nervous system. The inflammation triggered by the immune system can damage both myelin (the sheath that surrounds and insulates nerve fibers) and the cells specialized in its production (oligodendrocytes) and the nerve fibers themselves. This process, called demyelination, can cause areas of myelin loss or damage, which are called plaques. They can occur anywhere in the central nervous system, especially in the optic nerves, cerebellum, and spinal cord. Plaques can evolve from an initial inflammatory phase to a chronic phase, in which they take on scar-like characteristics (called sclerosis).

Multiple sclerosis presents different symptoms, which depend on the extent and location of the lesion in the central nervous system. MS onset at any age in life causes deficits in various cognitive domains. Cognitive deficits affect 70% of MS patients, and 10% of them have severe symptoms, while 50–60% of patients have mild or moderate symptoms [34]. The most affected areas are attention, short- and long-term memory, speed of information processing, and the processes of planning, reasoning, and language difficulties. SM can affect the quality of life of patients and their families [35].

The rehabilitation of MS patients consists of strengthening residual capacities and learning new strategies. For these purposes, various studies have shown VR as a promising tool [36–41]. Studies related to VR in MS initially focused on motor disorders and then extended to cognitive functions, highlighting the efficacy of VR for the functional recovery of patients [36, 39–41]. In fact, VR is feasible and safe in patients with moderate disabilities and can positively influence both walking and cognitive abilities. Consequently, VR is a useful tool for assessment and rehabilitation in MS patients, with long-term maintenance of the positive results obtained.

## 9.6 Parkinson's Disease and VR

Parkinson's disease is an extrapyramidal syndrome due to dopaminergic neuronal degeneration of the substantia nigra (substantia nigra pars compacta, SNPC), with a gradual progression and a prolonged course. The prevalence of Parkinson's disease is 0.3% in the general population (1–3% in persons older than 65 years) [42]. PD is characterized by motor symptoms (muscle stiffness, tremor at rest, bradykinesia, and postural instability) as well as non-motor symptoms, particularly cognitive dysfunction, such as deficits in executive function and visuospatial memory. Cognitive function is often comorbid with depression and anxiety, which negatively impact the quality of life of patients and their caregivers [43]. PD is managed by a combination of dopaminergic drugs, physiotherapy, and cognitive rehabilitation (CR), which can improve patient management and promote adequate therapeutic and rehabilitative treatments [43]. In recent decades, various authors have underlined that the use of innovative rehabilitation techniques can be effective for the recovery of PD patients and the improvement of quality of life [44–47]. In particular, various studies have highlighted that VR can increase motor skills in PD. To date, few studies have considered the rehabilitation of cognitive functions in PD using VR, underlining the potential role of this tool in PD. VR can allow for a global involvement of the patient, favoring the development of motor and cognitive skills and general well-being [42–47]. In particular, VR can stimulate executive, attentional, and visual-spatial skills, which are mostly impaired in patients with PD.

## 9.7 Discussion: Accessibility and Healthcare Allocation of New Technology

VR represents a promising approach to neurorehabilitation, enhancing the effects of conventional therapies, and encouraging longer training sessions and a reduction in overall hospitalization time [3–6]. Virtual interactive training can stimulate overall cognitive abilities, by stimulating multisensory activity that affects the recovery of memory function, visuospatial cognition, executive processes, and behavioral skills in patients with neurological disorders. In particular, one of the main benefits of VR is to provide the patient with a positive, fun, and motivating learning experience [2–6]. VR is based on an interactive sensory experience through information technology, which allows for greater patient compliance with treatment [36]. Furthermore, various authors have underlined that the repeated execution of cognitive and physical tasks can allow improvements at various levels, especially in the level of attention and concentration, with positive effects in the cognitive domains

[36, 48]. Thus, VR can facilitate the rehabilitation of neurological patients by enhancing cognitive domains, with a significant improvement in global cognitive functioning. This is possible through brain plasticity processes through complex mechanisms [36, 49–51].

A disadvantage of new technologies may be related to the cost of these tools in healthcare [52–56]. Indeed, evidence shows that low- and middle-income countries may have more difficulties in accessing resources [52–56]. Hence, the evaluation of medical devices requires further research to establish the appropriateness of these tools, through a better evaluation of the clinical effectiveness and cost-effectiveness of medical devices.

## 9.8 Translation of VR in Daily Life: The Metaverse

It seems useful to understand how VR can be useful in the translation to daily life, especially given the cost of the VR tools used up to now in rehabilitation. There is a new frontier that has emerged in recent years: the metaverse, a new way of conceiving reality in virtual terms.

The metaverse is a virtual expansion of the real world, where it is possible to interact through an avatar and through technological and wearable devices, such as VR helmets and viewers, gloves, and tactile suits. It is a digital world where VR is connected to the web and where people can live a life parallel to that of the real world. In fact, the term metaverse literally means "universe beyond" and was coined by science fiction writer Neal Stephenson in his 1992 novel "Snow Crash." Stephenson describes the metaverse as a virtual world in which people connect via augmented reality technology devices and can create whatever they want in 3D: shops, offices, and clubs where everyone can meet and interact. Lately, several virtual world moguls aspire to the metaverse, such as Mark Zuckerberg, owner of Facebook, who describes the metaverse as a virtual environment that you can enter instead of just watching it on a screen, a kind of "internet brought to life." The metaverse is different from VR in that it uses VR to allow you to assume your digital identity, connect with other people and things, and engage in social activities. It allows you to live a second parallel life, like in a huge video game. The metaverse can expand the possibilities of technology in reducing the limits of time and space.

Some authors highlighted that the metaverse could be helpful in the mental health field, especially for the therapy of psychological disorders. Indeed, the possibility of managing the virtual world through an avatar, which involves an illusion of the whole body of the patient, could promote personal well-being, especially for patients suffering from symptoms of body dysmorphia or social deficits [57–60]. According to Calabro et al., the metaverse allows for a multisensory and multiuser experience and could improve neural plasticity, boosting functional recovery in neurological patients [58]. Furthermore, Yang and colleagues noted that the metaverse could improve medical care, as the whole process could take place in a computer-generated 3D virtual environment where doctors and patients can meet [61]. In agreement with these authors, Mescko hypothesized that the metaverse could replace telemedicine and online forums, allowing for patient monitoring [62]. Furthermore, you proposed the possibility of creating virtual universities, with lessons in 3D virtual classrooms. Moreover, Wiederhold and Riva have underlined that the metaverse could allow innovative treatment methods that reduce healthcare costs [63]. A recent article by Cerasa et al. showed the importance of predictive coding: the ability of the brain to actively build a simulation (or internal model) of the body and its environment [57]. This embodied simulation of the body is able to predict various sensory inputs to avoid prediction errors (or "surprise") and interact efficiently with the world. These simulations use sensorimotor experiences derived from visceral/autonomic, motor, and sensory input, activating multimodal brain networks that can simulate the expected outcome [64, 65]. This multisensory processing is augmented by virtual tools [66, 67]. In fact, according to Paulus and colleagues, mental illnesses are characterized by incorrect representations of the world and the body [68]. In practice, when the subject receives a result, there is an error in interpreting it, this causes anxiety, tension, and sadness, for a distorted misinterpretation of what will happen. Thus, the metaverse, through augmented and virtual feedback, could modify, correct, or augment the bodily experience, with important clinical implications in the treatment of mental health disorders.

To sum up, the metaverse could stimulate the effect of innovative VR therapies, allowing total VR immersion which could have a greater impact on rehabilitation outcomes.

## 9.9 Conclusions

In conclusion, VR can be used as a CR method to improve the cognitive and behavioral functions of patients, thanks to global sensory involvement. VR has been used in the rehabilitation of various neurological pathologies, resulting in effectiveness in activating and training the cognitive and behavioral abilities of patients.

## References


1. De Luca R, Calabrò RS, Bramanti P. Cognitive rehabilitation after severe acquired brain injury: current evidence and future directions. Neuropsychol Rehabil. 2018;28(6):879–98.
2. Salazar AM, Warden DL, Schwab K, Spector J, Braverman S, Walter J, Cole R, Rosner MM, Martin EM, Ecklund J, Ellenbogen RG. Cognitive rehabilitation for traumatic brain injury: a randomized trial. Defense and Veterans Head Injury Program (DVHIP) Study Group. JAMA. 2000;283(23):3075–81.
3. Maggio MG, Luca A, D'Agate C, Italia M, Calabrò RS, Nicoletti A. Feasibility and usability of a non-immersive virtual reality tele-cognitive app in cognitive rehabilitation of patients affected by Parkinson's disease. Psychogeriatrics. 2022;22(6):775–9.
4. Leonardi S, Maggio MG, Russo M, Bramanti A, Arcadi FA, Naro A, Calabrò RS, De Luca R. Cognitive recovery in people with relapsing/remitting multiple sclerosis: a randomized clinical trial on virtual reality-based neurorehabilitation. Clin Neurol Neurosurg. 2021;208:106828.
5. Maggio MG, Torrisi M, Buda A, De Luca R, Piazzitta D, Cannavò A, Leo A, Milardi D, Manuli A, Calabro RS. Effects of robotic neurorehabilitation through lokomat plus virtual reality on

cognitive function in patients with traumatic brain injury: a retrospective case-control study. Int J Neurosci. 2020;130(2):117–23.
6. Bohil CJ, Alicea B, Biocca FA. Virtual reality in neuroscience research and therapy. Nat Rev Neurosci. 2011;12(12):752–62.
7. Riva G. Virtual reality in psychotherapy: review. Cyberpsychol Behav. 2005;8(3):220–30; discussion 231–40.
8. Maggio MG, De Luca R, Molonia F, Porcari B, Destro M, Casella C, Salvati R, Bramanti P, Calabro RS. Cognitive rehabilitation in patients with traumatic brain injury: a narrative review on the emerging use of virtual reality. J Clin Neurosci. 2019;61:1–4.
9. Calabrò RS, Naro A, Cimino V, Buda A, Paladina G, Di Lorenzo G, Manuli A, Milardi D, Bramanti P, Bramanti A. Improving motor performance in Parkinson's disease: a preliminary study on the promising use of the computer-assisted virtual reality environment (CAREN). Neurol Sci. 2020;41(4):933–41.
10. De Luca R, Maggio MG, Maresca G, Latella D, Cannavò A, Sciarrone F, Lo Voi E, Accorinti M, Bramanti P, Calabrò RS. Improving cognitive function after traumatic brain injury: a clinical trial on the potential use of the semi-immersive virtual reality. Behav Neurol. 2019;2019:9268179.
11. De Luca R, Russo M, Gasparini S, Leonardi S, Foti Cuzzola M, Sciarrone F, Zichittella C, Sessa E, Maggio MG, De Cola MC, Calabrò RS. Do people with multiple sclerosis benefit from PC-based neurorehabilitation? A pilot study. Appl Neuropsychol Adult. 2021;28(4):427–35.
12. Aida J, Chau B, Dunn J. Immersive virtual reality in traumatic brain injury rehabilitation: a literature review. NeuroRehabilitation. 2018;42(4):441–8.
13. Manivannan S, Al-Amri M, Postans M, Westacott LJ, Gray W, Zaben M. The effectiveness of virtual reality interventions for improvement of neurocognitive performance after traumatic brain injury: a systematic review. J Head Trauma Rehabil. 2019;34(2):E52–65.
14. Zanier ER, Zoerle T, Di Lernia D, Riva G. Virtual reality for traumatic brain injury. Front Neurol. 2018;9:345.
15. Prince M, Bryce R, Albanese E, Wimo A, Ribeiro W, Ferri CP. The global prevalence of dementia: a systematic review and metaanalysis. Alzheimers Dement. 2013;9(1):63–75.e2.
16. Allison SL, Fagan AM, Morris JC, Head D. Spatial navigation in preclinical Alzheimer's disease. J Alzheimers Dis. 2016;52(1):77–90.
17. Valladares-Rodriguez S, Perez-Rodriguez R, Facal D, Fernandez-Iglesias MJ, Anido-Rifon L, Mouriño-Garcia M. Design process and preliminary psychometric study of a video game to detect cognitive impairment in senior adults. PeerJ. 2017;5:e3508.
18. Schedlbauer AM, Copara MS, Watrous AJ, Ekstrom AD. Multiple interacting brain areas underlie successful spatiotemporal memory retrieval in humans. Sci Rep. 2014;4:6431.
19. Carrieri M, Petracca A, Lancia S, Basso Moro S, Brigadoi S, Spezialetti M, Ferrari M, Placidi G, Quaresima V. Prefrontal cortex activation upon a demanding virtual hand-controlled task: a new frontier for neuroergonomics. Front Hum Neurosci. 2016;10:53.
20. Bus B. Virtual reality training system for patients with dementia. Zürich: ETH: Institute of Neuroinformatics, Swiss Federal Institute of Technology; 2009.
21. Seo K, Kim JK, Oh DH, Ryu H, Choi H. Virtual daily living test to screen for mild cognitive impairment using kinematic movement analysis. PLoS One. 2017;12(7):e0181883.
22. Fernandez Montenegro JM, Argyriou V. Cognitive evaluation for the diagnosis of Alzheimer's disease based on Turing test and virtual environments. Physiol Behav. 2017;173:42–51.
23. Wen D, Lan X, Zhou Y, Li G, Hsu SH, Jung TP. The study of evaluation and rehabilitation of patients with different cognitive impairment phases based on virtual reality and EEG. Front Aging Neurosci. 2018;10:88.
24. Jones KM, Bhattacharjee R, Krishnamurthi R, Blanton S, Theadom A, Barker-Collo S, Thrift A, Parmar P, Maujean A, Ranta A, Sanya E, Feigin VL, SMART Study Group. Methodology of the stroke self-management rehabilitation trial: an international, multisite pilot trial. J Stroke Cerebrovasc Dis. 2015;24(2):297–303.

25. Saleh S, Adamovich SV, Tunik E. Mirrored feedback in chronic stroke: recruitment and effective connectivity of ipsilesional sensorimotor networks. Neurorehabil Neural Repair. 2014;28(4):344–54.
26. Pedreira da Fonseca E, da Silva Ribeiro NM, Pinto EB. Therapeutic effect of virtual reality on post-stroke patients: randomized clinical trial. J Stroke Cerebrovasc Dis. 2017;26(1):94–100.
27. Laver KE, Lange B, George S, Deutsch JE, Saposnik G, Crotty M. Virtual reality for stroke rehabilitation. Cochrane Database Syst Rev. 2017;11(11):CD008349.
28. Jang SH, You SH, Hallett M, et al. Cortical reorganization and associated functional motor recovery after virtual reality in patients with chronic stroke: an experimenter blind preliminary study. Arch Phys Med Rehabil. 2005;86:2218–23.
29. Adams H, Donnelly J, Czosnyka M, et al. Temporal profile of intracranial pressure and cerebrovascular reactivity in severe traumatic brain injury and association with fatal outcome: an observational study. PLoS Med. 2017;14(7):e1002353.
30. Saatian M, Ahmadpoor J, Mohammadi Y, Mazloumi E. Epidemiology and pattern of traumatic brain injury in a developing country regional trauma center. Bull Emerg Trauma. 2018;6(1):45–53.
31. Fasczewski KS, Gill DL. A model of motivation for physical activity in individuals diagnosed with multiple sclerosis. Disabil Rehabil. 2019;41(18):2143–50.
32. Gamito P, Oliveira J, Coelho C, Morais D, Lopes P, Pacheco J, Brito R, Soares F, Santos N, Barata AF. Cognitive training on stroke patients via virtual reality-based serious games. Disabil Rehabil. 2017;39(4):385–8.
33. Imam B, Jarus T. Virtual reality rehabilitation from social cognitive and motor learning theoretical perspectives in stroke population. Rehabil Res Pract. 2014;2014:594540.
34. Sumowski JF, Benedict R, Enzinger C, et al. Cognition in multiple sclerosis: state of the field and priorities for the future. Neurology. 2018;90(6):278–88.
35. Al-Sharman A, Khalil H, Nazzal M, et al. Living with multiple sclerosis: a Jordanian perspective. Physiother Res Int. 2018;23(2):e1709.
36. Calabrò RS, Russo M, Naro A, De Luca R, et al. Robotic gait training in multiple sclerosis rehabilitation: can virtual reality make the difference? Findings from a randomized controlled trial. J Neurol Sci. 2017;377:25–30.
37. Lozano-Quilis JA, Gil-Gómez H, Gil-Gómez JA, et al. Virtual rehabilitation for multiple sclerosis using a kinect-based system: randomized controlled trial. JMIR Serious Games. 2014;2(2):e12.
38. Jonsdottir J, Bertoni R, Lawo M, et al. Serious games for arm rehabilitation of persons with multiple sclerosis. A randomized controlled pilot study. Mult Scler Relat Disord. 2017;19:25–9.
39. Lamargue-Hamel D, Deloire M, Saubusse A, et al. Cognitive evaluation by tasks in a virtual reality environment in multiple sclerosis. J Neurol Sci. 2015;359(1–2):94–9.
40. Leocani L, Comi E, Annovazzi P, et al. Impaired short-term motor learning in multiple sclerosis: evidence from virtual reality. Neurorehabil Neural Repair. 2007;21(3):273–8.
41. Peruzzi A, Cereatti A, Della Croce U, Mirelman A. Effects of a virtual reality and treadmill training on gait of subjects with multiple sclerosis: a pilot study. Mult Scler Relat Disord. 2016;5:91–6.
42. Lee A, Gilbert RM. Epidemiology of Parkinson disease. Neurol Clin. 2016;34(4):955–65.
43. Maggio MG, De Luca R, Maresca G, Di Lorenzo G, Latella D, Calabro RS, Bramanti A. Personal computer-based cognitive training in Parkinson's disease: a case study. Psychogeriatrics. 2018;18(5):427–9.
44. Albiol-Pérez S, Gil-Gómez JA, Muñoz-Tomás MT, et al. The effect of balance training on postural control in patients with Parkinson's disease using a virtual rehabilitation system. Methods Inf Med. 2017;56(2):138–44.
45. Dockx K, Bekkers EM, Van den Bergh V, et al. Virtual reality for rehabilitation in Parkinson's disease. Cochrane Database Syst Rev. 2016;12:CD010760.
46. Morales-Gomez S, Elizagaray-Garcia I, Yepes-Rojas O, et al. Effectiveness of virtual immersion programmes in patients with Parkinson's disease. A systematic rev. Rev Neurol. 2018;66(3):69–80.

47. Maidan I, Rosenberg-Katz K, Jacob Y, et al. Disparate effects of training on brain activation in Parkinson disease. Neurology. 2017;89(17):1804–10.
48. Calabrò RS, Naro A, Russo M, et al. The role of virtual reality in improving motor performance as revealed by EEG: a randomized clinical trial. J Neuroeng Rehabil. 2017;14:53.
49. Kizony R, Katz N. Adapting an immersive virtual reality system for rehabilitation. Comput Animat Virtual Worlds. 2003;14:261–8.
50. Subramanian SK, Chilingaryan G, Levin MF, Sveistrup H. Influence of training environment and cognitive deficits on use of feedback for motor learning in chronic stroke. In: Virtual rehabilitation proceedings (ICVR). New York: International Conference on IEEE; 2015. pp. 38–43.
51. Varabyova Y, Blankart CR, Greer AL, Schreyögg J. The determinant of medical technology in different decisional systems: a systematic literature review. Health Policy. 2017;121:230–42.
52. Gray J. The shift to personalised and population medicine. Lancet. 2013;382(9888):200–1.
53. Elshaug AG, Rosenthal MB, Lavis JN, et al. Right care 4. Levers for addressing medical underuse and overuse: achieving high-value health care. Lancet. 2017;390(10090):191–202.
54. Gooch RA, Kahn JM. ICU bed supply, utilization, and health care spending: an example of demand elasticity. JAMA. 2014;311(6):567–8.
55. Bhuiya A, Hanifi S, Urni F, Mahmood SS. Three methods to monitor utilization of healthcare services by the poor. Int J Equity Health. 2009;8:29.
56. Ciani O, Wilcher B, Blankart CR, et al. Health technology assessment of medical devices: a survey of non-European union agencies. Int J Technol Assess Health Care. 2015;31(3):154–65.
57. Cerasa A, Gaggioli A, Marino F, Riva G, Pioggia G. The promise of the metaverse in mental health: the new era of MEDverse. Heliyon. 2022;8(11):e11762.
58. Calabrò RS, Cerasa A, Ciancarelli I, Pignolo L, Tonin P, Iosa M, Morone G. The arrival of the metaverse in neurorehabilitation: fact, fake or vision? Biomedicines. 2022;10(10):2602.
59. Koohsari MJ, McCormack GR, Nakaya T, Yasunaga A, Fuller D, Nagai Y, Oka K. The metaverse, the built environment, and public health: opportunities and uncertainties. J Med Internet Res. 2023;25:e43549.
60. Checcucci E, Verri P, Amparore D, Cacciamani GE, Rivas JG, Autorino R, Mottrie A, Breda A, Porpiglia F. The future of robotic surgery in urology: from augmented reality to the advent of metaverse. Ther Adv Urol. 2023;15:17562872231151853.
61. Yang D, Zhou J, Chen RC, Song YL, Song ZJ, Zhang XJ, Wang Q, Zhou CZ, Sun JY, Zhang LC, et al. Expert consensus on the metaverse in medicine. Clin eHealth. 2022;5:1–9.
62. Mesko B. The promise of the metaverse in cardiovascular health. Eur Heart J. 2022;22:ehac231.
63. Wiederhold BK, Riva G. Metaverse creates new opportunities in healthcare. Annu Rev Cyber Telemed. 2022;20:3–7.
64. Riva G, Di Lernia D, Sajno E, Sansoni M, Bartolotta S, Serino S, Gaggioli A, Wiederhold BK. Virtual reality therapy in the metaverse: merging VR for the outside with VR for the inside. Annu Rev Cyber Telemed. 2021;19:3–8.
65. Riva G, Serino S, Di Lernia D, Pagnini F. Regenerative virtual therapy: the use of multisensory technologies and mindful attention for updating the altered representations of the bodily self. Front Syst Neurosci. 2021;15:749268.
66. Blanke O, Slater M, Serino A. Behavioral, neural, and computational principles of bodily self-consciousness. Neuron. 2015;88(1):145–66.
67. Riva G, Wiederhold B, Chirico A, Di Lernia D, Mantovani F, Gaggioli A. Brain and virtual reality: what do they have in common and how to exploit their potential. Annu Rev Cyber Ther Telemed. 2018;16:3–7.
68. Paulus MP, Feinstein JS, Khalsa SS. An active inference approach to interoceptive psychopathology. Annu Rev Clin Psychol. 2019;15:97–122.

# Lower Limb Exoskeletons for Gait Rehabilitation

# 10

David Rodriguez-Cianca, Juan C. Moreno, and Diego Torricelli

## 10.1 A Brief History of Robotic Exoskeletons

A robotic exoskeleton is a mechanical apparatus that is worn externally to the human body and operates alongside and in parallel with the human limbs to extend, complement, substitute, or enhance some specific function or activity of its wearer. The first patent describing an exoskeleton system dates from 1890 [1]. The inventor presented a passive device using bow springs worn in parallel with the lower limbs and described it as an 'exercising apparatus specially adapted for conditioning the cardio-vascular system, for training agility or co-ordination of movements'.

Formal scientific and technological work on robotic exoskeletons research started no earlier than in the 1960s [2]. Around 1965, General Electric started the development of Hardiman (Human Augmentation Research and Development Investigation) [3], a large full-body exoskeleton designed to augment the user's strength to enable the lifting of heavy objects. The first lower limb exoskeletons for gait assistance were developed at the end of the 1960s at the Mihajlo Pupin Institute Serbia [4]. This was the predecessor of more complex exoskeletal devices developed in 1972 and 1974 by the same institution. Indeed, the last one is considered as the first example in the world of an active exoskeleton that used electric motors as actuators, and the predecessor of contemporary humanoid robots driven by electric motors [5]. Nevertheless, because of the existing technical limitations of that time, such as low computational and mechanical power, high weight, bulkiness, low sensor reliability as well as the lack of experience and knowledge on the topic, it took several decades until the robotic exoskeleton technology matured and the first exoskeletons were ready for the market. It wasn't until the beginning of the twenty-first

D. Rodriguez-Cianca (✉) · J. C. Moreno · D. Torricelli
BioRobotics Group, Centre for Automation and Robotics (CAR-UPM), Spanish National Research Council (CSIC), Madrid, Spain
e-mail: david.rodriguez.cianca@csic.es; jc.moreno@csic.es; diego.torricelli@csic.es



R. S. Calabrò (ed.), *Translational Neurorehabilitation*,
https://doi.org/10.1007/978-3-031-63604-2_10

century that the interest in robotic exoskeletons flourished again. In the last two decades, robotic exoskeleton research has experienced a renaissance, reaching a much higher level of maturity thanks to the new advances in the fields of computation, robotics, and electronics. In this period, as a result of the extensive research, new applications and more advanced prototypes with improved performance have been developed, including some that have reached the market.

The existing types of exoskeleton devices cover a wide range of application domains, end-users, and goals. For these reasons, exoskeleton technology has recently attracted the attention of the public and the industry and is currently considered a technology area with the potential to have a massive impact on people's lives in the coming years.

## 10.2 Types of Exoskeleton Devices

Nowadays, many types of exoskeletons exist. Multiple categories of robotic exoskeletons can be established according to different criteria. We have chosen three main classification categories, as illustrated in Fig. 10.1: (1) the intended use of the exoskeleton in a given application domain, (2) the body parts covered, and (3) the technical characteristics of the device itself.

### 10.2.1 Exoskeletons by Application Domain

With respect to their intended use of application, exoskeletons can be further categorized into: (1) rehabilitation, (2) assistance, and (3) performance augmentation.

Rehabilitation of neurologically injured patients is a key application domain for robotic exoskeletons, and historically, one of their first application fields. These

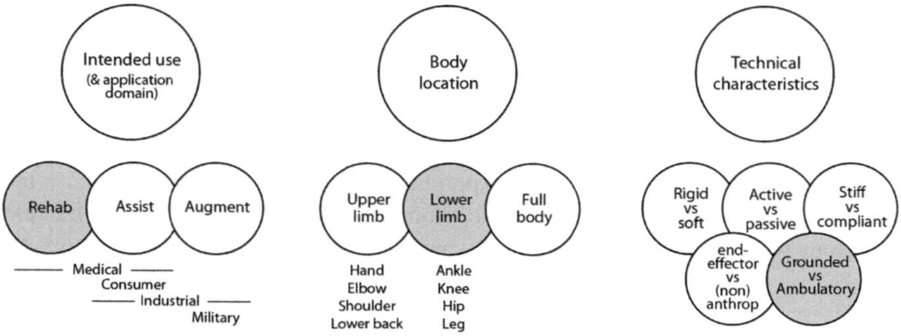


**Fig. 10.1** Different taxonomies to classify exoskeletons. An exoskeleton can be characterized using a combination of multiple categories. For example, an exoskeleton can be non-anthropomorphic, static, and active using electric actuators and rigid parts for the upper limbs aimed for assistance; or it can be anthropomorphic, mobile, soft, passive, and compliant and used in the lower limbs for rehabilitation purposes, etc. Highlighted categories indicate the focus of this book chapter

robots aim at facilitating the recovery of patients after a stroke, traumatic brain injury, or spinal cord injury (SCI) accident by improving the quality and efficacy of the patient's rehabilitation process. Using exoskeletons during these therapies allows increased training durations, more reproducible and biologically accurate movements with the active involvement of the user, a closer monitoring of patient's progress during the therapy process and, finally, the release of burden of therapists, allowing them to focus on the clinical decisions and other (e.g. cognitive) interactions with the patient.

Exoskeletons can also be used as assistive devices. These assistive devices allow users to complete movements they could not complete on their own, or to complete them with a reduced effort or risk of injury. These consumer exoskeletons can be used, for instance, by the elderly population to support activities of daily living, injured subjects or even by workers to reduce their long-term risk of injury due to repetitive movements and non-ergonomic positions.

Exoskeletons for performance augmentation are built mostly for: (a) workers conducting physically demanding labour in industrial applications and (b) military purposes. Their main goal is to increase the performance of the user while doing a certain activity or reduce his/her fatigue while performing a task. This type of exoskeleton can be used, for instance, for lifting heavy objects, carrying heavy loads over large distances, or working with heavy tools.

### 10.2.2 Exoskeletons by Body Location

Exoskeletons can also be classified according to the human limb onto which the external framework couples to the human body, leading to three main categories. Thus, exoskeletons can be classified as: upper limb exoskeletons, lower limb exoskeletons and full-body exoskeletons. Upper limb exoskeletons enhance the human manipulation function and normally include the shoulder, elbow, and wrist articulations together with the upper arm, forearm, and hand segments. Lower limb exoskeletons are generally designed to provide support, stability, and/or mobility to the user's legs during walking and standing. Finally, full-body exoskeletons are built to provide a combination of the advantages and assistance of the two previously mentioned types.

### 10.2.3 Exoskeletons by Technical Characteristics

In general, robotic exoskeletons resemble many of the technical characteristics of "traditional" robots: mechanical joints are connected in series by means of rigid links, which are driven by actuators that provide active power to the device. The exoskeleton is then connected to the human through some type of mechanical interface, which transmits the forces/torques created by the exoskeleton to provide support to its wearer. The joint arrangement and the length of the links determine the exoskeleton's kinematic configuration, which generally maps on to the human limb anatomy resulting in a one-to-one correspondence between human and robot joints. Although this general definition can be applied to most existing devices, in the last year new technical developments have led to variations on this basic design, in terms of new types and configurations of actuators, joints, and links, as well as on new materials with improved mechanical properties. This has led to many additional categories based on certain characteristics and technical aspects of the design and working principle of each robot.

**Rigidity**. Rigid exoskeletons follow a similar approach as traditional robots, including rigid links driven by actuators that drive each of the human joints. Nevertheless, robotic exoskeletons are not necessary constructed with rigid parts, but can also be made from textiles and soft materials. These non-rigid devices are generally called exosuits or soft exoskeletons. Exosuits include garments and flexible transmissions, such as cables or artificial tendons, driven by motors that can produce rotations about the human joints thanks to specially designed routings that translate pulling forces into torques about a specific joint, thereby producing motion at the human joint level.

**Kinematic layout**. Depending on the kinematic configuration, exoskeletons can also be classified as: end-effector, non-anthropomorphic, or anthropomorphic. End-effector exoskeletons generally interact with the user at a single location, and their kinematic configuration does not necessarily resemble the human one. Nevertheless, exoskeletons generally have multiple connection points with their users. In the case of one-to-one map between the joints and degrees of freedom (DOFs) of the exoskeleton and those of the user, the exoskeleton is considered to be anthropomorphic. Otherwise, exoskeletons can be classified as non-anthropomorphic (e.g., the exoskeleton having more DOFs than the corresponding human limb).

**Power source**. Exoskeletons can also be subdivided into active and passive, depending on whether they use an external power source that provides additional energy to the wearer, or if they only re-distribute the user's own energy. Active systems in general use actuators driven by some form of power source to assist and reduce the load on human muscles, whereas passive systems generally implement springs that store energy of the user's own joints during phases of negative power to release it in phases where active power is required, or to dissipate energy by means of mechanical dampers in order to provide some biomechanical benefit. Active exoskeletons can be classified as electric, hydraulic, or pneumatic depending on the energy source required to drive the exoskeleton's joints.

**Interaction mode**. Depending on the type of interaction with the user, exoskeletons can be classified as compliant or stiff. Most of current robots include stiff actuators that consist of motors coupled with high transmission ratios of gearboxes that provide high torque, bandwidth, and precision control in expense of making them highly non-backdrivable. This approach is very suitable for applications that require position control, but lacks transparency in applications that require a safer interaction with the user, such as the case of exoskeletons. An alternative approach consists on the use of compliant actuators. They place a compliant element, generally a spring, between the motor and the human so that the interaction between both systems is made in a compliant manner as determined by the compliance of the spring. This decouples the dynamics of the motor from the human, allows for more accurate and safe force control, and improves energy efficiency. However, despite the theoretical advantages of these systems, their increased complexity has so far prevented their implementation into commercial exoskeleton systems, and their use has been constrained to research prototypes. In most cases, compliance between the exoskeleton and the human is implemented by control using impedance/admittance control schemes.

**Portability**. Depending on their portability, exoskeletons can be divided into grounded, also called treadmill-based, or static; and ambulatory, portable, or overground. More information about these systems is provided in the next section.

## 10.3 Exoskeletons for Gait Rehabilitation

There is strong evidence that early, intensive, and repetitive task- and goal-oriented gait training, if progressively adapted to the patient's level of impairment and rehabilitation stage, can improve clinical outcomes [6]. Traditional therapies for gait rehabilitation require physical therapists to manually move the feet and legs of patients to facilitate the re-learning of movement patterns. However, due to the limited human resources and the heterogeneity of the impairment, it is challenging for physiotherapists to provide the required intensity and dose of training. The use of robotic exoskeletons in the field of rehabilitation started in 1994 as a method to automate and optimize traditional gait rehabilitation therapies [7]. The device offloads the burden from physical therapists, allowing to deliver precise interventions and training at optimal intensities while being unconstrained by the limits of manual assistance. Lower limb exoskeletons can promote task-oriented repetitive movements, muscle strengthening, and movement coordination, which have been shown to positively impact energy efficiency, gait speed, and balance control [8, 9].

The interest in the use of lower limb exoskeletons for gait rehabilitation has increased exponentially in the last years, both at clinical and research levels. As a result, many robotic exoskeletons for lower limb rehabilitation have been developed in these years. This trend is also reflected in the considerable number of reviews on the topic currently available [10–20].

### 10.3.1 Grounded Vs Ambulatory Exoskeletons

According to their portability and function, exoskeletons for gait rehabilitation can be grouped into two main categories: grounded, also referred to as treadmill-based or fixed, and ambulatory, also referred to as overground or portable [21].

A grounded exoskeleton typically consists of a lower limb exoskeleton mounted on top of a treadmill, so that the gait therapy can take place in a permanent and static allocation. This way, the patient can walk on the treadmill with the support of the exoskeleton in a controlled and supervised environment. Usually, a body weight support (BWS) system is used to reduce gravitational forces acting on the legs, to ensure safety and help the patient maintain balance. This kind of robotic medical device allows therapists to design customized exercises that work towards improving patient balance, ambulation, coordination, stamina, and posture through an intensive, repetitive re-training of functional movements of the lower limbs, in order to reproduce reciprocal, synchronized, and natural gait patterns [22]. Besides, since the robot body is generally connected to a fixed platform, an important advantage of treadmill-based exoskeletons is that the patient does not need to carry the entire weight of the robot. However, there are some important disadvantages of these kinds of exoskeletons. First, it has been reported that there are differences between walking on a treadmill and natural gait walking overground [23, 24], which is an important aspect to consider in clinical evaluations [25]. Second, these devices are not affordable for personal use, being their use confined to rehabilitation centres or hospitals, making the entire procedure not self-oriented, and not always accessible for most patients. Another argument against treadmill-based gait training is the lack of variability in visuospatial flow, which is considered to be an essential element during overground walking [26].

Ambulatory exoskeletons emerged to address the limitations of grounded exoskeletons. The main difference consists in their ability to allow the user to move overground, i.e., without being constrained to a fixed location, and is usually compatible with conventional assistive elements such as walkers or crutches. The main advantages of using ambulatory exoskeletons are the possibility to overcome the typical intermittent short sessions available in hospitals and rehabilitation centres by allowing practising the rehabilitation exercises at home, and to allow more extra activities besides walking, such as standing up or other daily-life tasks. Therefore, these devices have greater potential to promote an active participation and physical activity of the user. At the same time, these systems are subject to important challenges and limitations in terms of autonomy and balance capabilities. Since the power supply of the exoskeleton must be worn by the user, similarly as with the rest of the components of the device, this adds extra weight that must be carried out by the patient, reducing the autonomy of the system. Besides, most available exoskeletons of this type do not self-maintain balance, so generally the patient requires crutches for retaining balance while walking.

### 10.3.2 State of the Art of Grounded Exoskeletons

Grounded exoskeletons are currently the most common types of exoskeletons for clinical use. Many robotic systems of this type, both for commercial purposes and for research, have been developed in the past years. Three of these systems are currently on the market: the Lokomat [27] (Hocoma AG, Switzerland), the Walkbot [28], (P&S Mechanics, Korea), and the ReoAmbulator (Motorika Ltd). The Lokomat, which leads the sector with more than 1000 devices sold up to date, is the most clinically evaluated system and adopted rehabilitation robot in the world [29], and one of the first of its type being currently the benchmark for the rest of the available commercial and research platforms. This robot actuates the hip and knee joints of both legs in the sagittal plane by means of direct current motors and ball screws. The implemented BWS system can deliver a tuneable uplifting constant force, while a pivoting and spring-loaded parallelogram provides a semi-passive DOF to allow pelvis vertical motion during walking. In a new version of Lokomat, the system includes an optional module to improve the quality of the walking therapy by adding two extra active DOFs to the pelvis that guide the user's pelvic lateral translation and transverse rotation in a coupled way. Similar design concepts as the original Lokomat system are found in the Walkbot and ReoAmbulator devices. Interestingly, in any of these commercial exoskeletons, the ankle joint is actuated.

Other non-commercial projects have been looking for new solutions to deliver more freedom of movement, improve wearability of the device, and to stimulate active participation of the patient by means of the development of new robotic systems. Researchers at the University of Twente developed the LOPES (Lower Extremity Powered Exoskeleton), a treadmill-based robotic system for gait rehabilitation and motor function assessment in 2007 [30]. LOPES combines an actuated pelvis module with three DOFs: two active (forward and sideways) and one passive (vertical); with two exoskeleton legs containing each three actuated rotational joints: two at the hip (flexion/extension and ad/abduction) and one at the knee (flexion/extension). In the newest version of LOPES (LOPES II [31], researchers implemented the same amount of DOFs, but developed a new actuation system called "shadow-leg approach", in which a structure of push-pull rods was used to drive the user's lower limbs and the pelvis. Another noteworthy treadmill-based exoskeleton is ALEX (Active Leg Exoskeleton) [22]. Designed by researchers at the University of Delaware in 2006, ALEX possesses two lineal DOFs (forward and sideways) plus an extra rotational DOF (rotation about the vertical axis) at the pelvis level, two active DOFs on the exoskeleton for the hip (flexion/extension and ad/abduction), and one DOF at the knee joint for flexion/extension. In 2011, a modified version (ALEX II) was proposed [32]. In this version, the single exoskeleton leg could be switched between each side, and rotary motors were applied to actuate the hip and knee joints directly. In 2013, ALEX III was developed [33], which included a total of four active DOFs at the pelvis level (forward, sideways, vertical displacement, and vertical rotation) and four DOFs at each leg (hip adduction/abduction, hip and knee flexion/extension, ankle plantar/dorsiflexion). Finally, in 2015, the same research group released the C-ALEX exoskeleton [34], a unilateral cable-driven active leg exoskeleton able to drive the hip and knee flexion/extension. In 2008, the Vrije Universiteit Brussel (VUB) developed the ALTACRO (Automated Locomotion Training using an Actuated Compliant Robotic Orthosis) exoskeleton [35]. ALTACRO consisted of a treadmill-based exoskeleton with a total of 11 active DOFs: 5 at the pelvis (forward, sideways and vertical displacement, plus rotation about the vertical and sagittal axes) and three additional DOFs for the legs (flexion/extension of the hip, knee, and ankle joints) using variable compliance actuators. Additionally, the system included a passive DOF for hip ad/abduction.

### 10.3.3 State of the Art of Ambulatory Exoskeletons

In the last years, the number of commercial ambulatory exoskeletons has increased notably in comparison to their grounded counterparts. A recent review by Rodriguez-Fernandez et al. [11] identified a total of 25 exoskeleton systems for gait rehabilitation, with more being developed or at a prototype stage. Nowadays, nine of these devices have FDA approval: HAL [36], Ekso [37], Rewalk [38], Indego [39], ExoAtlet [40], Atalante [41], Keeogo [42], Honda walking assist device [43], and GEMS-H [44], while seven have CE mark: HANK [45], ExoAtlet, HAL, Ekso, Indego, REX [46], and Keoogo.

Ekso, HAL, and ReWalk have been tested in a considerably higher number of clinical studies, and together with the Indego exoskeleton, they have been the most evaluated exoskeletons up to date. Indeed, the first clinical study using an ambulatory wearable exoskeleton was published in 2009 using the HAL exoskeleton.

The big majority of current exoskeletons only assist the sagittal plane, with a number of active degrees of freedom (DOF) ranging from one to three per leg. Most systems assist two or more joints, while only a few of them assist a single joint. The most frequent number of active DOFs in commercial systems is two, corresponding to the hip and knee joints: HAL, Ekso, ReWalk, Indego, ExoAtlet. Only two commercial lower limb exoskeletons also provide assistance to the ankle joints: HANK and Atalante; whereas others present passive joints to allow free ankle motion.

Current exoskeletons can usually adapt to the user's height with a range of approximately 1.45–1.95 m, and most systems have a maximum allowed weight of around 100 kg. Despite the new and promising advances, current systems are perceived as heavy and bulky due to their rigid structures, big actuators, and heavy batteries. The average weight of hip-knee exoskeletons is 14.28 kg (7.14 kg/leg), although the weight varies depending on the number of active joints. Regarding their autonomy, systems are generally only capable of sustaining 2–4 h of continuous use by means of portable batteries that increase the weight as compared to their grounded counterparts. Most devices use electric motors placed close or very close to the joint level, which also increase the inertia of the systems as weight is placed in distal locations.

Despite the advantages of these systems in terms of portability when compared to grounded exoskeletons, only the ReWalk Personal and Indego are currently approved for a home or personal use, whereas the remaining systems are meant to be used in a rehabilitation centre environment under the supervision of a trained healthcare professional. This is in part due to the lack of balance capabilities of these systems, which requires the use of supportive devices such as crutches, walkers, and canes, which can limit the independence of the user and lead to shoulder pain and risk of falls. Only two exoskeletons provide self-balancing capabilities: REX and Atalante. These systems, however, are heavier, more expensive and more difficult to use in a clinical environment and so far they have not been approved for home-based rehabilitation.

## 10.4 Control Methods for Gait Rehabilitation Exoskeletons

Besides the mechanical design and characteristics of the used device, the implemented control strategy, i.e., how the exoskeleton processes the sensory information and derives commands to its actuated joints to fulfill a specific goal, plays a key role on the effectiveness of the robotic treatment [47].

There are multiple literature reviews focused on control strategies used in lower extremity exoskeletons [48–51]. In the following section, we provide a simplified overview of the most common control strategies implemented on current powered lower limb exoskeletons for gait rehabilitation of individuals with brain injuries based on a recent review [52].

### 10.4.1 Exoskeleton's Control Taxonomy

A common taxonomy to classify exoskeleton's controllers consists of a hierarchical representation based on three levels (Fig. 10.2): high-level control, mid-level control, and low-level control. High-level controllers are defined as control strategies that identify the human's intention and select a suitable exoskeleton response behaviour. The mid-level control reacts to the user state and defines the reference position or force that the robot should follow based on the control aim and the state estimated by the human–robot synchronization algorithm and the sensor measurements. Finally, the low-level control tries to reach the desired state, or setpoint, at the joint level, generally in terms of a reference position, velocity, force, or torque, as determined by the mid-level controller.

In the following, we focus on high- and mid-level controllers for exoskeletons, given that the low-level controller, being related to the hardware system, involves more technical aspects of the device that are considered out of the scope of this chapter.

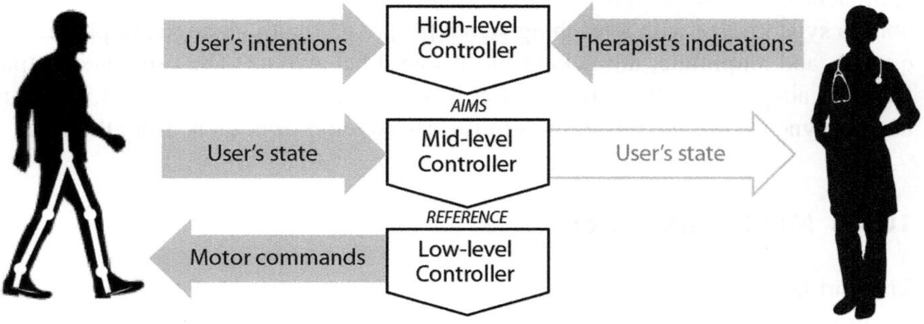


**Fig. 10.2** Simplified scheme of the control structure used to command a lower limb exoskeleton

### 10.4.2 High-Level Control

A high-level control system modifies the state of the actuation system according to the control aim, which may vary depending on the desired treatment. Three main types can be distinguished:

- *Assistive* high-level controllers aim at supporting the patients' movements to complete a certain task. Assistance can be provided: (1) while patients are fully guided by the exoskeleton and remain passive during the training or (2) while patients actively execute the task while they are guided/corrected by the robot [53]. This high-level controller is currently the most widely implemented on lower limb exoskeletons for individuals with brain injuries [52].
- *Challenge-based* high-level controllers aim at perturbating and opposing to the movements of the patient to train him/her under challenging conditions to improve physical performance and movement control [54].
- *Adaptive control strategies* consist of a combination of the previous two, whose aim is to modify the control parameters based on the patients' specific needs to either support their movements or resist them, when necessary. Although they are still not very common in the field given their challenging and time-consuming parameter tuning procedure [55], they are considered the most promising techniques to improve functional outcomes of individuals with brain injury [52].

Another important task of the high-level controller is to synchronize the robot joints with the user motion. This is usually achieved by estimating the state of the patient based on kinematics, kinetics, and/or bioelectric data. Three main methods are currently used: threshold-based, stochastic and bio-inspired. Threshold-based algorithms define the motor activity of the patient, e.g., gait phase, falling, and/or start-stop walking, by means of a state machine that allows the transition between states depending on logical rules driven by a specific signal, e.g., joint kinematics, ground reaction forces, human–robot interaction forces, muscular activity, and brain activity, among others. Stochastic methods, instead, infer the state using statistical models, e.g., Linear Discriminant Analysis (LDA) [56] or neural networks [57]. Bio-inspired models provide a more recent alternative to the previous two approaches. The most common type consists of one or more adaptive oscillators able to synchronize with a teaching signal [58], e.g., the thigh angle, in phase, frequency, and amplitude, mimicking bio-inspired behaviours. The estimated output from the adaptive oscillator is then used to estimate the phase of the gait, which is used to synchronize the movements of the exoskeleton from those from the user.

### 10.4.3 Mid-Level Control

The mid-level control uses the information coming from exoskeleton's sensors and the user to generate reference control commands to be followed by the low-level controller for each joint. Three different families of mid-level control strategies can be distinguished depending on the control inputs/outputs and controllers employed.

- *Trajectory-tracking* controllers generate predefined position or force/torque trajectories as reference commands for the low-level controller. These trajectories are determined based on pre-recordings of unimpaired individuals, information from the non-paretic limb, or pre-recorded trajectories during therapist-guided assistance [59] and are currently the most common control strategy used in commercial lower limb exoskeletons [52].
- *Neuromuscular control* strategies use biosignal recordings as control signals to decode the actions of the patient, which are later transformed into reference values for the low-level control [60]. Approaches like myoelectric and Brain–Computer Interface (BCI) control use muscular, electromyography (EMG), or brain electroencephalography (EEG) signals, respectively, to handle the control objective [52].
- Finally, *compliant controllers* regulate the impedance [61] or admittance [62] of the exoskeleton to make the joint behave as a spring-mass-damper system. Impedance controllers provide force/torque commands to the low-level controller based on readings of position/velocity according to a desired dynamic interaction, whereas admittance controllers modulate joint's speed/velocity according to measured forces/torques. An usual approach in the field consists on combining trajectory-tracking control with compliant control to allow for patient's movement variability around a desired predefined trajectory.

## 10.5 Clinical Evidence

High repeatability and intensity of training, together with patients' engagement, have been listed as fundamental factors to induce neural plasticity and motor learning [63–65]. Therefore, clinical evidence suggests that combining robotic and conventional rehabilitation training could positively impact the ability to walk independently and with increased speed. However, up to date, there is still no solid evidence about the superiority of robotic rehabilitation over conventional therapy [66–69].

In the following, we provide a summary of the currently available clinical evidence on the use of exoskeletons for gait rehabilitation.

### 10.5.1 Grounded Exoskeletons

There is currently conflicting evidence regarding the efficacy of treadmill-based exoskeletons for gait training compared to conventional therapy. Several studies have illustrated the effectiveness of stationary robot-assisted gait training to improve the ambulation of individuals with spinal cord injury (SCI). Duan et al. [70], through a pairwise meta-analysis, concluded that robotic-based training is more effective than conventional training in improving ambulation. Furthermore, through a pairwise meta-analysis, Fang et al. [71] reported that the use of stationary exoskeletons can improve spasticity and walking ability in people with SCI better than conventional training can. In a systematic review of 13 RCTs [25], authors compared different training strategies to improve gait in people with SCI and concluded that body weight supported gait training and robot-assisted gait training do not increase walking speed more than overground gait training and other forms of physiotherapy, and their effects on walking distance remain unclear. Although these studies have provided diverse outcomes on different training strategies for patients with SCI, they used only pairwise meta-analysis without direct comparisons. A more recent meta-analysis by Yang et al. [72], which includes direct and indirect comparisons, assessed the training efficacy of different body weight-supported gait-training strategies for improving ambulation in individuals with spinal cord injury (SCI). This study considered body weight-supported overground training (BWSOGT), body weight-supported treadmill training (BWSTT), and robot-assisted gait training (RAGT). Results showed that RAGT had significantly more favourable performance than manual-based therapy. The ranking probabilities indicated that the most effective approach for SCI ambulation rehabilitation was RAGT, followed by BWSOGT, BWSTT, and finally manual-based therapy. These recent findings suggest clear benefits of the use of stationary exoskeletons.

In stroke survivors, the use of exoskeleton-based gait training for rehabilitation is also expanding. Nonetheless, its competence is still being argued due, in part, to the lack of systematic and randomized control studies [73]. In an early study in 2008, therapist-assisted locomotor training showed superior to robotic-assisted locomotor training amongst ambulatory survivors with chronic stroke [74]. However, more recent results of randomized control trials (RCT) focusing on the efficacy of exoskeleton-assisted BWSTT for chronic stroke have showed positive improvements in gait [75–77], indicating that there is a potential value in further research and important advances in the field during the last decades. A Cochrane review by Mehrholz et al. [69] reported that individuals who received electromechanical-assisted gait training in combination with physiotherapy after stroke were more likely to achieve independent walking than people who received gait training without these devices. Besides, results showed that people who are non-ambulatory at the start of the intervention may benefit, but ambulatory people may not benefit from this type of training. Specifically, people in the first 3 months after stroke and those who are not able to walk seemed to benefit most from this type of intervention. However, the role of the type of device is still not clear. A more recent review by Yamamoto et al. [73] suggests that exoskeleton-assisted treadmill walking may be effective in improving walking function in chronic stroke patients, although they could not conclude whether the difference with respect to traditional therapies is significant. Authors concluded that further studies are required to verify the efficacy of robot-based therapies and strengthen evidence on intervention protocols. Interestingly, according to this review, the Lokomat was the most clinically evaluated device in this field, with seven randomized clinical trials evaluating its efficacy.

### 10.5.2 Ambulatory Exoskeletons

Robotic therapy seems to be progressing towards ambulatory exoskeletons since, in theory, they offer the advantages of grounded exoskeletons, as well as providing more active participation of the user. However, clinical validation studies of these devices are currently in their early stages. Indeed, current evidence is still limited to short intervention trials with few participants [78], and study designs are primarily focused on observational and pilot with a clear lack of randomized-controlled trials (RCTs). For instance, in a multisession case study involving three participants with complete SCI, training with the Ekso resulted in linear increases in the number of steps and distances in 2 out of 3 participants, and a 2- to 3-fold increase in walking speed for all participants [79]. In [11], authors summarized the current clinical evidence on the efficacy of ambulatory exoskeletons. They found promising results on the use of this technology on SCI patients, although results were based mainly on pilot studies, as also concluded by [25, 80]. For instance, training with the Indego device resulted in an average walking speed of 0.37 $ms^{-1}$ in 32 individuals with SCI who were otherwise non-ambulatory without the device after 26 visits [81], which is about half the median dose of body-weight-supported treadmill-walking regimens (50 sessions) [82]. Authors from [11] also noted that, despite the higher potential use of ambulatory exoskeletons in this population given their advantages in terms of independence, most of the current studies focus on stroke patients.

Authors from [11] identified only 5 RCTs to date including ambulatory exoskeletons for gait rehabilitation, involving a total of 183 patients. Some examples include the work of Calabro et al. [83], who compared the combination of robotic training with Ekso and conventional gait training with conventional gait training alone. Authors showed several significant improvements such as gait speed, corticospinal excitability, and muscle activation. Jayaraman et al. [84] found significant differences in gait variables such as improvements in gait speed, step length, and spatial symmetry when using the SMA exoskeleton. Watanabe et al. [85], in contrast, did not find significant improvements in either walking speed or stride length when comparing the HAL exoskeleton with conventional gait therapy. However, the HAL group showed a significant improvement in the Functional Ambulation Categories (FAC) test that was maintained at the 2-month follow-up evaluation. Lastly, authors identified one randomized cross over trial in which researchers evaluated the effects of the Keeogo exoskeleton on the physical performance of people with multiple sclerosis, both in a clinical setting and in a home setting [42], being this one the only study that measured the benefits of using an overground exoskeleton at home. Contrary to what was expected, wearing the Keoogo did not improve physical performance, and participants were slower both in walking functional tests and stair climbing.

### 10.5.3 Control Strategies

Although the number of studies that evaluated the effectiveness of robotic-assisted gait rehabilitation has increased exponentially in the last decade, there are still critical limitations in the clinical comparison of the effectiveness of different control strategies of lower limb exoskeletons for gait rehabilitation that allow determining the most effective control strategies. A recent review [52] tried to answer this question but could only extract moderate conclusions in post-stroke participants given the current limitations and the lack of comparable studies and metrics. For instance, they could not identify a control strategy that was clearly better than others to improve the patient's functional status in acute stroke. Nonetheless, assistive control strategies that implemented a combination of trajectory-tracking and compliant control seemed to provide the highest clinical effectiveness. In subacute stroke, assistive control strategies that followed a threshold-based algorithm with EMG as detection metric and control signal provided the highest improvements. Finally, in chronic stroke, assistive control strategies, which followed threshold-based or adaptive oscillator algorithms together with trajectory-tracking control, resulted in the highest improvements. For other brain injuries, such as cerebral palsy and traumatic brain injury, authors could not analyse the effect of the control strategies alone given the lack of standardization on the available clinical studies.

Despite the current lack of evidence on the best type of controllers for gait rehabilitation, the current knowledge on neurorehabilitation suggests that the best controller will be the one that can provide user-dependent, intensive, symbiotic and goal-oriented training while requiring an active participation of the patient to promote brain plasticity.

## 10.6 Conclusions

This chapter provided an overview in Layman's terms to the field of exoskeleton systems with a focus on lower limb exoskeleton systems for gait rehabilitation. The wide number of applications for exoskeletons that exist nowadays and the highly increasing number of commercially available systems show the high level of technical maturity and progress the field has achieved in the last years. Nevertheless, in the field of gait rehabilitation, exoskeletons are still challenged by complex technical problems and acceptability issues. From the user perspective, the main drawbacks reside in the size and weight of the device, the ease of adjustment to different subjects, and the ease of use [86]. Size and weight of wearable exoskeletons should be reduced by means of lighter components placed in more proximal locations to users' centre of mass to reduce system's inertia. Another relevant aspect to be improved is the way the exoskeleton is connected to the human limbs to guarantee an efficient and safe physical human-exoskeleton interaction [87]. Structures should also be simplified and include features to allow fast and independent donning/doffing. On the control side, future generations of exoskeletons should implement robust control methods able to understand the state and intention of the patient to efficiently implement control algorithms able to conveniently adapt the assistance to the user needs. Similarly, devices should be more intuitive to use, both for the user and the clinician, and designed to adapt to the different states of the therapy process according to the users' and therapeutic needs along the recovery process. Balance capabilities of exoskeletons should also be improved to reduce the use of supportive devices, such as walkers and crutches, which increase the risk of falls and injuries and limit their use in home-based scenarios. Finally, new regulations and standard test methods are strongly needed to quantify the performance of wearable robots in diverse contexts, to have a better understanding of the possible adverse events and risks that may occur during the use of exoskeletons in real-world applications.

Besides the aforementioned technical limitations, the lack of a solid scientific evidence on their superiority with respect to traditional manual therapies, together with the high purchasing costs, are hampering the inclusion of exoskeletons in rehabilitation centres and hospitals on a wider scale. With respect to the clinical evidence, there is some sentiment that grounded exoskeletons for gait rehabilitation have not yet lived up to the expectations originally predicted [26]. One possible reason is that a pre-set belt speed, combined with body weight support, create an environment where the patient has less control over the initiation of each step when compared to overground walking. Another possible limitation is the lack of variability in visuospatial flow, which represents an essential element of overground walking. Nevertheless, as most authors conclude, while these grounded powered exoskeletons hold promise, the literature surrounding their use for gait training is still limited and more systematic and randomized controlled trials are needed to determine their real efficacy. This limitation is even more evident in the ambulatory exoskeleton subcategory, where more efforts should be made in conducting experimental studies with control groups to obtain stronger evidence on clinical effectiveness of these devices. Literature comparing ambulatory exoskeletons with other types of gait therapy is still scarce, especially in people with SCI. Current results seem to indicate positive user satisfaction in terms of safety, efficacy, and comfort [86] despite some known occurrences of non-injurious falls [88] and skin issues, such as redness and minor abrasion, owing to a poor device fit [89]. Nevertheless, further studies comparing ambulatory exoskeletons with other types of robotic gait therapy or conventional gait therapy are needed to demonstrate their effectiveness as a rehabilitation device.

In conclusion, the current findings suggest the potential of robot-assisted gait rehabilitation. Nevertheless, the optimal type of rehabilitation robot and control strategy for a specific user and neuromuscular impairment remains unclear. Larger trials with control groups are needed to fully understand the generalizability of the current results, the effectiveness of these technologies in the clinic, the optimal frequency and duration of training, and how long benefits last before this technology can be effectively adopted as a standard and valuable clinical tool for gait rehabilitation. Efforts should also be invested in identifying which patient population would benefit the most from each robotic device, rather than assessing a robot's generalized benefit for all users. The device, training, and control paradigms and patient-inclusion criteria will need to be optimized in the future to maximize the benefits of exoskeletons for gait rehabilitation.

## References


1. Yagn N. Apparatus for facilitating walking. Google Patents; 1890.
2. Pons JL. Wearable robots: biomechatronic exoskeletons. Hoboken: Wiley; 2008.
3. Fick BR, Makinson JB. Hardiman I prototype for machine augmentation of human strength and endurance; 1971.
4. Vukobratovic M, Hristic D, Stojiljkovic Z. Development of active anthropomorphic exoskeletons. Med Biol Eng. 1974;12(1):66–80.
5. Vukobratovic MK. When were active exoskeletons actually born? Int J Human Robot. 2007;4(3):459–86.
6. Schroeder J, Truijen S, Van Criekinge T, Saeys W. Feasibility and effectiveness of repetitive gait training early after stroke: a systematic review and meta-analysis. J Rehabil Med. 2019;51(2):78–88.
7. Gassert R, Dietz V. Rehabilitation robots for the treatment of sensorimotor deficits: a neurophysiological perspective. J Neuroeng Rehabil. 2018;15(1):1–15.
8. Trushkova N, Cochran O, Ermolina N, Zelano G. Is training with a focus on motor learning effective in improving body coordination in chronic post stroke patients? J Neurol Sci. 2021;429:118583.
9. Roth EJ, Merbitz C, Mroczek K, Dugan SA, Suh WW. HEMIPLEGIC GAIT: relationships between walking speed and other temporal parameters: 1. Am J Phys Med Rehabil. 1997;76(2):128–33.
10. Young AJ, Ferris DP. State-of-the-art and future directions for robotic lower limb exoskeletons. IEEE Trans Neural Syst Rehab Eng. 2016;99:1–1. https://doi.org/10.1109/TNSRE.2016.2521160.
11. Rodríguez-Fernández A, Lobo-Prat J, Font-Llagunes JM. Systematic review on wearable lower-limb exoskeletons for gait training in neuromuscular impairments. J Neuroeng Rehabil. 2021;18(1):1–21.
12. Hobbs B, Artemiadis P. A review of robot-assisted lower-limb stroke therapy: unexplored paths and future directions in gait rehabilitation. Front Neurorobot. 2020;14:19.
13. Louie DR, Eng JJ. Powered robotic exoskeletons in post-stroke rehabilitation of gait: a scoping review. J Neuroeng Rehabil. 2016;13:1–10.
14. Morone G, et al. Robot-assisted gait training for stroke patients: current state of the art and perspectives of robotics. Neuropsychiatr Dis Treat. 2017;13:1303–11.
15. del C. Sanchez-Villamañan M, Gonzalez-Vargas J, Torricelli D, Moreno JC, Pons JL. Compliant lower limb exoskeletons: a comprehensive review on mechanical design principles. J Neuroeng Rehabil. 2019;16(1):1–16.
16. Chen B, et al. Recent developments and challenges of lower extremity exoskeletons. J Orthop Transl. 2016;5:26–37.
17. Esquenazi A, Talaty M. Robotics for lower limb rehabilitation. Phys Med Rehab Clin. 2019;30(2):385–97.
18. Huo W, Mohammed S, Moreno JC, Amirat Y. Lower limb wearable robots for assistance and rehabilitation: a state of the art. IEEE Syst J. 2014;10(3):1068–81.
19. Contreras-Vidal JL, et al. Powered exoskeletons for bipedal locomotion after spinal cord injury. J Neural Eng. 2016;13(3):31001.
20. Meng W, Liu Q, Zhou Z, Ai Q, Sheng B, Xie SS. Recent development of mechanisms and control strategies for robot-assisted lower limb rehabilitation. Mechatronics. 2015;31:132–45.
21. Chen G, Chan CK, Guo Z, Yu H. A review of lower extremity assistive robotic exoskeletons in rehabilitation therapy. Crit Rev Biomed Eng. 2013;41(4–5):343.

22. Banala SK, Agrawal SK, Scholz JP. Active Leg Exoskeleton (ALEX) for gait rehabilitation of motor-impaired patients. In: 2007 IEEE 10th international conference on rehabilitation robotics; 2007. pp. 401–407.
23. Hollman JH, Watkins MK, Imhoff AC, Braun CE, Akervik KA, Ness DK. A comparison of variability in spatiotemporal gait parameters between treadmill and overground walking conditions. Gait Posture. 2016;43:204–9.
24. Lee SJ, Hidler J. Biomechanics of overground vs. treadmill walking in healthy individuals. J Appl Physiol. 2008;104(3):747–55.
25. Mehrholz J, Harvey LA, Thomas S, Elsner B. Is body-weight-supported treadmill training or robotic-assisted gait training superior to overground gait training and other forms of physiotherapy in people with spinal cord injury? A systematic review. Spinal Cord. 2017;55(8):722–9.
26. Dobkin BH, Duncan PW. Should body weight–supported treadmill training and robotic-assistive steppers for locomotor training trot back to the starting gate? Neurorehabil Neural Repair. 2012;26(4):308–17.
27. Riener R, Lünenburger L, Maier IC, Colombo G, Dietz V. Locomotor training in subjects with sensori-motor deficits: an overview of the robotic gait orthosis lokomat. J Healthc Eng. 2010;1(2):197–216.
28. Kim DH, Shin Y-I, Joa K-L, Shin YK, Lee JJ, You SJH. Immediate effect of Walkbot robotic gait training on neuromechanical knee stiffness in spastic hemiplegia: a case report. NeuroRehabilitation. 2013;32(4):833–8. https://doi.org/10.3233/NRE-130907.
29. Tefertiller C, Pharo B, Evans N, Winchester P. Efficacy of rehabilitation robotics for walking training in neurological disorders: a review. J Rehabil Res Dev. 2011;48(4):387.
30. Veneman JF, Kruidhof R, Hekman EEG, Ekkelenkamp R, Van Asseldonk EHF, Van Der Kooij H. Design and evaluation of the LOPES exoskeleton robot for interactive gait rehabilitation. IEEE Trans Neural Syst Rehabil Eng. 2007;15(3):379–86.
31. Meuleman J, Van Asseldonk E, Van Oort G, Rietman H, Van Der Kooij H. LOPES II—design and evaluation of an admittance controlled gait training robot with shadow-leg approach. IEEE Trans Neural Syst Rehabil Eng. 2016;24:352. https://doi.org/10.1109/TNSRE.2015.2511448.
32. Winfree KN, Stegall P, Agrawal SK. Design of a minimally constraining, passively supported gait training exoskeleton: ALEX II. In: 2011 IEEE international conference on rehabilitation robotics; 2011. pp. 1–6.
33. Zanotto D, Stegall P, Agrawal SK. ALEX III: A novel robotic platform with 12 DOFs for human gait training. In: 2013 IEEE international conference on robotics and automation; 2013. pp. 3914–3919.
34. Jin X, Cui X, Agrawal SK. Design of a cable-driven active leg exoskeleton (c-alex) and gait training experiments with human subjects. In: 2015 IEEE international conference on robotics and automation (ICRA); 2015. pp. 5578–5583.
35. Grosu V, Rodriguez Guerrero C, Grosu S, Vanderborght B, Lefeber D. Design of smart modular variable stiffness actuators for robotic assistive devices. IEEE/ASME Trans Mechatron. 2017;22:1.
36. Kawamoto H, Hayashi T, Sakurai T, Eguchi K, Sankai Y. Development of single leg version of HAL for hemiplegia. In: 2009 annual international conference of the IEEE engineering in medicine and biology society; 2009. pp. 5038–5043.
37. Strausser KA, Swift TA, Zoss AB, Kazerooni H, Bennett BC. Mobile exoskeleton for spinal cord injury: development and testing. In: Dynamic systems and control conference; 2011. pp. 419–425.
38. Esquenazi A, Talaty M, Packel A, Saulino M. The ReWalk powered exoskeleton to restore ambulatory function to individuals with thoracic-level motor-complete spinal cord injury. Am J Phys Med Rehabil. 2012;91(11):911–21.
39. Hartigan C, et al. Mobility outcomes following five training sessions with a powered exoskeleton. Top Spinal Cord Inj Rehabil. 2015;21(2):93–9.
40. Pais-Vieira C, et al. Method for positioning and rehabilitation training with the ExoAtlet® powered exoskeleton. MethodsX. 2020;7:100849.

41. Harib O, et al. Feedback control of an exoskeleton for paraplegics: toward robustly stable, hands-free dynamic walking. IEEE Control Syst Mag. 2018;38(6):61–87.
42. McGibbon CA, et al. Evaluation of the Keeogo exoskeleton for assisting ambulatory activities in people with multiple sclerosis: an open-label, randomized, cross-over trial. J Neuroeng Rehabil. 2018;15:1–14.
43. Firouzi M, et al. Immediate effects of the Honda walking assist on spatiotemporal gait characteristics in individuals after stroke. Med Nov Technol Devices. 2022;16:100173.
44. Lee J, Huber ME, Hogan N. Gait entrainment to torque pulses from a hip exoskeleton robot. IEEE Trans Neural Syst Rehabil Eng. 2022;30:656–67.
45. Gil-Agudo Á, et al. Exoskeleton-based training improves walking independence in incomplete spinal cord injury patients: results from a randomized controlled trial. J Neuroeng Rehabil. 2023;20(1):1–11.
46. Barbareschi G, Richards R, Thornton M, Carlson T, Holloway C. Statically vs dynamically balanced gait: analysis of a robotic exoskeleton compared with a human. In: 2015 37th annual international conference of the IEEE engineering in medicine and biology society (EMBC); 2015. pp. 6728–6731.
47. Xiloyannis M, et al. Soft robotic suits: state of the art, core technologies, and open challenges. IEEE Trans Robot. 2021;38(3):1343–62.
48. Baud R, Manzoori AR, Ijspeert A, Bouri M. Review of control strategies for lower-limb exoskeletons to assist gait. J Neuroeng Rehabil. 2021;18(1):1–34.
49. Li W-Z, Cao G-Z, Zhu A-B. Review on control strategies for lower limb rehabilitation exoskeletons. IEEE Access. 2021;9:123040–60.
50. Tucker MR, et al. Control strategies for active lower extremity prosthetics and orthotics: a review. J Neuroeng Rehabil. 2015;12(1):1–30.
51. Yan T, Cempini M, Oddo CM, Vitiello N. Review of assistive strategies in powered lower-limb orthoses and exoskeletons. Robot Auton Syst. 2015;64:120–36.
52. de Miguel-Fernández J, Lobo-Prat J, Prinsen E, Font-Llagunes JM, Marchal-Crespo L. Control strategies used in lower limb exoskeletons for gait rehabilitation after brain injury: a systematic review and analysis of clinical effectiveness. J Neuroeng Rehabil. 2023;20(1):23.
53. Basalp E, Wolf P, Marchal-Crespo L. Haptic training: which types facilitate (re) learning of which motor task and for whom? Answers by a review. IEEE Trans Haptics. 2021;14(4):722–39.
54. Veldema J, Jansen P. Resistance training in stroke rehabilitation: systematic review and meta-analysis. Clin Rehabil. 2020;34(9):1173–97.
55. Fricke SS, Bayón C, Van Der Kooij H, Edwin EH. Automatic versus manual tuning of robot-assisted gait training in people with neurological disorders. J Neuroeng Rehabil. 2020;17(1) https://doi.org/10.1186/S12984-019-0630-9.
56. Gui K, Liu H, Zhang D. Toward multimodal human–robot interaction to enhance active participation of users in gait rehabilitation. IEEE Trans Neural Syst Rehabil Eng. 2017;25(11):2054–66.
57. Gurriet T, Tucker M, Duburcq A, Boeris G, Ames AD. Towards variable assistance for lower body exoskeletons. IEEE Robot Autom Lett. 2019;5(1):266–73.
58. Ronsse R, et al. Oscillator-based assistance of cyclical movements: model-based and model-free approaches. Med Biol Eng Comput. 2011;49:1173–85.
59. Lora-Millan JS, Sanchez-Cuesta FJ, Romero JP, Moreno JC, Rocon E. A unilateral robotic knee exoskeleton to assess the role of natural gait assistance in hemiparetic patients. J Neuroeng Rehabil. 2022;19(1):109.
60. Gordleeva SY, et al. Real-time EEG–EMG human–machine interface-based control system for a lower-limb exoskeleton. IEEE Access. 2020;8:84070–81.
61. Aguirre-Ollinger G, Colgate JE, Peshkin MA, Goswami A. Active-impedance control of a lower-limb assistive exoskeleton. In: 2007 IEEE 10th international conference on rehabilitation robotics; 2007. pp. 188–195.
62. Calanca A, Muradore R, Fiorini P. A review of algorithms for compliant control of stiff and fixed-compliance robots. IEEE/ASME Trans Mechatronics. 2015;21(2):613–24.

63. Fisher BE, Sullivan KJ. Activity-dependent factors affecting poststroke functional outcomes. Top Stroke Rehabil. 2001;8(3):31–44.
64. Winter DA. Biomechanics and motor control of human gait: normal, elderly and pathological. 1991.
65. Krakauer JW. Motor learning: its relevance to stroke recovery and neurorehabilitation. Curr Opin Neurol. 2006;19(1):84–90.
66. Moucheboeuf G, et al. Effects of robotic gait training after stroke: a meta-analysis. Ann Phys Rehabil Med. 2020;63(6):518–34.
67. Tedla JS, Dixit S, Gular K, Abohashrh M. Robotic-assisted gait training effect on function and gait speed in subacute and chronic stroke population: a systematic review and meta-analysis of randomized controlled trials. Eur Neurol. 2019;81(3–4):103–11.
68. Goffredo M, et al. Stroke gait rehabilitation: a comparison of end-effector, overground exoskeleton, and conventional gait training. Appl Sci. 2019;9(13):2627.
69. Mehrholz J, Thomas S, Kugler J, Pohl M, Elsner B. Electromechanical-assisted training for walking after stroke. Cochrane Database Syst Rev. 2020;(10):CD006185.
70. Duan R, et al. Clinical benefit of rehabilitation training in spinal cord injury: a systematic review and meta-analysis. Spine (Phila Pa 1976). 2021;46(6):E398–410.
71. Fang C-Y, Tsai J-L, Li G-S, Lien AS-Y, Chang Y-J. Effects of robot-assisted gait training in individuals with spinal cord injury: a meta-analysis. Biomed Res Int. 2020;2020:2102785.
72. Yang F-A, et al. Body weight-supported gait training for patients with spinal cord injury: a network meta-analysis of randomised controlled trials. Sci Rep. 2022;12(1):19262.
73. Yamamoto R, Sasaki S, Kuwahara W, Kawakami M, Kaneko F. Effect of exoskeleton-assisted body weight-supported treadmill training on gait function for patients with chronic stroke: a scoping review. J Neuroeng Rehabil. 2022;19(1):143.
74. Hornby TG, Campbell DD, Kahn JH, Demott T, Moore JL, Roth HR. Enhanced gait-related improvements after therapist-versus robotic-assisted locomotor training in subjects with chronic stroke: a randomized controlled study. Stroke. 2008;39(6):1786–92.
75. Westlake KP, Patten C. Pilot study of Lokomat versus manual-assisted treadmill training for locomotor recovery post-stroke. J Neuroeng Rehabil. 2009;6(1):18.
76. Erbil D, et al. Effects of robot-assisted gait training in chronic stroke patients treated by botulinum toxin-A: a pivotal study. Physiother Res Int. 2018;23(3):e1718.
77. Bae Y, et al. Effects of robot-assisted gait training combined with functional electrical stimulation on recovery of locomotor mobility in chronic stroke patients: a randomized controlled trial. J Phys Ther Sci. 2014;26(12):1949–53.
78. Mekki M, Delgado AD, Fry A, Putrino D, Huang V. Robotic rehabilitation and spinal cord injury: a narrative review. Neurotherapeutics. 2018;15(3):604–17. https://doi.org/10.1007/S13311-018-0642-3.
79. Kressler J, et al. Understanding therapeutic benefits of overground bionic ambulation: exploratory case series in persons with chronic, complete spinal cord injury. Arch Phys Med Rehabil. 2014;95(10):1878–87.
80. Fisahn C, et al. The effectiveness and safety of exoskeletons as assistive and rehabilitation devices in the treatment of neurologic gait disorders in patients with spinal cord injury: a systematic review. Global Spine J. 2016;6:822. https://doi.org/10.1055/s-0036-1593805.
81. Tefertiller C, et al. Initial outcomes from a multicenter study utilizing the indego powered exoskeleton in spinal cord injury. Top Spinal Cord Inj Rehab. 2018;24(1):78–85.
82. Siviy C, et al. Opportunities and challenges in the development of exoskeletons for locomotor assistance. Nat Biomed Eng. 2023;7:456–72.
83. Calabrò RS, et al. Shaping neuroplasticity by using powered exoskeletons in patients with stroke: a randomized clinical trial. J Neuroeng Rehabil. 2018;15(1):1–16.
84. Jayaraman A, et al. Stride management assist exoskeleton vs functional gait training in stroke: a randomized trial. Neurology. 2019;92(3):e263–73.
85. Watanabe H, Goto R, Tanaka N, Matsumura A, Yanagi H. Effects of gait training using the hybrid assistive limb® in recovery-phase stroke patients: a 2-month follow-up, randomized, controlled study. NeuroRehabilitation. 2017;40(3):363–7.

86. Cumplido-Trasmonte C, et al. Satisfaction analysis of overground gait exoskeletons in people with neurological pathology. A systematic review. J Neuroeng Rehabil. 2023;20(1):47. https://doi.org/10.1186/S12984-023-01161-4.
87. Massardi S, Rodriguez-Cianca D, Pinto-Fernandez D, Moreno JC, Lancini M, Torricelli D. Characterization and evaluation of human–exoskeleton interaction dynamics: a review. Sensors. 2022;22(11):3993.
88. Kolakowsky-Hayner SA, Crew J, Moran S, Shah A. Safety and feasibility of using the EksoTM bionic exoskeleton to aid ambulation after spinal cord injury. J Spine. 2013;S4:003. https://doi.org/10.4172/2165-7939.S4-003.
89. Awad LN, Esquenazi A, Francisco GE, Nolan KJ, Jayaraman A. The ReWalk ReStore™ soft robotic exosuit: a multi-site clinical trial of the safety, reliability, and feasibility of exosuit-augmented post-stroke gait rehabilitation. J Neuroeng Rehabil. 2020;17(1):1–11. https://doi.org/10.1186/S12984-020-00702-5/FIGURES/4.

# Wearable Sensors and Soft Robotics 11

Franco Molteni, Alessandro Robustelli, Rachele Penati,
Giovanni Cannaviello, Alessandro Specchia,
Giulio Gasperini, and Eleonora Guanziroli

## 11.1 From Robotic to Soft Robotic Neurorehabilitation

Neurorehabilitation aims to restore or enhance motor function in individuals with neurological disorders or injuries. The use of robotic devices is a useful treatment option in rehabilitation to assist patients with neurological impairments in regaining movement and function [1–4].

Nowadays, both traditional and robotic rehabilitation treatments are considered essential in improving motor outcome and the quality of life for patients [5]. It is important to tailor treatment to each patient's needs and goals, using a combination of traditional and robotic therapies as appropriate, and providing the best tailored customization of the robot settings [6]. Rehabilitation robots can be used either as therapeutic or assistive robots. Assistive robots are actually used as compensation for a loss of function, whereas therapeutic robots provide task-specific training to address the central nervous system toward the recovery of function. Robotic devices have been introduced in rehabilitation since they can maximize afferent input from peripheral joints and provide task-specific stimulation to the central nervous system (CNS) promoting neuroplasticity, the capacity to modify the organization of neural components during human life [7]. According to the principles of neuroplasticity and motor learning, robotic devices, by increasing therapy dosage, intensity, number of repetition, execution of task-oriented exercises, and combining top-down and bottom-up approaches, can promote plasticity and functional recovery [8–10] and have been shown to improve movement and functional outcomes in patients, particularly in those with severe impairments.

F. Molteni (✉) · A. Robustelli · R. Penati · G. Cannaviello · A. Specchia · G. Gasperini · E. Guanziroli
Villa Beretta Rehabilitation Center, Valduce Hospital, Lecco, Italy
e-mail: arobustelli@valduce.it; rpenati@valduce.it; gcannaviello@valduce.it; aspecchia@valduce.it; ggasperini@valduce.it


R. S. Calabrò (ed.), *Translational Neurorehabilitation*, https://doi.org/10.1007/978-3-031-63604-2_11

Robots can be divided into end-effector-type devices and exoskeletons. The end-effector systems, in which movements are generated from the most distal segment of the extremity and no alignment between patient-robot joints is required, use foot-plates or handles to generate a motion of the limb in space [11–13]. In the exoskeleton systems, there is a one-to-one correspondence between robots and human joints, and each single joint is guided along a pre-programmed trajectory. The exoskeletons can be classified according to the number of joints controlled and can be divided into unilateral and bilateral robots.

For what concerns the upper limb, there are robots for the hand and fingers, which can provide support and movement assistance during exercise; exoskeletons for the shoulder and the arm, which can provide support to the unstable and weak muscles and help them to regain strength [14]; virtual reality (VR) devices, which create simulated environments to sustain patient's engagement in challenging and rewarding activities that simulate the upper body movements; these devices can provide visual, auditory, and haptic feedback to enhance the experience and increase motivation.

Instead, for what concerns the lower limb, in recent years clinical and commercial interests have been focused on gait-assistive devices. Initially, robots have been developed for gait training on the treadmill, as it happens during Lokomat training, where the robot guides the patient's leg movements while walking on a treadmill through an electrically powered, computer-controlled lower limb exoskeleton, with an integrated dynamic body weight support system allowing constant unloading of the patient's body weight throughout the gait cycle and enabling upright posture for gait training in a safe environment. For what concerns overground walking, different types of robots have been developed to allow gait training after brain injuries or spinal cord lesions, such as Ekso, which also allows patients with no muscle activity in the lower limbs to walk with the support of therapists. One last development for this type of robots is represented by Atalante X, a hands-free exoskeleton, which enables patients with severe gait impairment to stand up and walk without hands utilization and without the need of the therapist's support.

The main problem encountered with available robotic devices is the limited portability outside the clinical setting. This reduces the opportunity to use robotics after the rehabilitation stay for further therapy as well as assistance during daily life. In the last decades, robotic neurorehabilitation relied on conventional rigid robotics, but recent advancements in materials engineering have opened up new possibilities for more effective and patient-centric approaches. For this reason, in recent years, a growing interest has developed for robotic devices used outside the clinical settings, such as soft robotics that could provide to patients both assistance with activities of daily living (ADLs) and customized rehabilitation programs in different settings.

## 11.2 Soft Robotics

The evolution of technology and the miniaturization of components have given the technical possibility of reducing the weight and dimensions of robotic devices, and the clinical possibility of easing the patient setup for the robotic rehabilitation treatment, while ensuring a continuous monitoring during exercise.

This new area of robotic rehabilitation research is focused on the development of "soft robotics," a field of robotics that deals with the production of devices made of flexible structures and materials, adaptable to any shape and segment of the human body. These can also better adapt to the environment and are more suitable for interactions with humans than traditional rigid robots.

Soft robotics offers several advantages over conventional rigid robotics in the context of neurorehabilitation. First, the properties of soft materials reduce the risk of accidental injuries or tissue damage during therapy sessions, which makes soft robots inherently safer to interact with, compared to their rigid counterparts. Moreover, they can conform to the anatomical shape of the patient's body, allowing for a customized and personalized approach to rehabilitation, adapting to the patient's range of motion and accommodating variations in muscle tone and joint stiffness.

Soft robotics can be integrated into wearable devices, exoskeletons, or assistive devices, making them suitable for a wide range of neurorehabilitation applications. These devices can be lightweight, compact, and portable, enabling patients to undergo therapy in different settings.

In particular, soft robots for the upper limb have been used to help patients to regain their hand and arm movements after an injury. These robots use soft, flexible pneumatic actuators to guide the movement of the patient's arm or hand [14]. The device fits like a glove, and the pneumatic actuators are controlled by a computer program that creates an exercise routine tailored to the patient's needs [15]. One example is the Gloreha device [16], a robotic glove which allows customizable and task-oriented therapies. This device allows paretic hand rehabilitation also in a master/slave mode, guiding the paretic hand through movements executed with the contralateral hand, exploiting the interaction between perception and muscle activation.

Soft robotics for lower limb rehabilitation can be designed in the form of soft exosuits that are worn over the patient's legs. These exosuits use fabric-based actuators that provide assistance to the patient's legs during gait training. The actuators can sense the patient's movements and provide assistance when needed to help the patient move more efficiently.

For example, Myosuit, an FDA and CE approved wearable soft robot for the proximal lower limb, was designed to provide hip and knee extension torque in UMNS patients, to improve the activation of the lower limb muscles during walking, reducing the force demand in the load acceptance phase and guaranteeing adequate proximal support to the hips during the stance phase. Robotics for gait rehabilitation can help to activate appropriate muscle groups during physiological movement patterns, facilitating motor learning.

Another application of soft robotics is foot and ankle rehabilitation. The devices fit over the patient's foot and ankle and are controlled by a computer program that creates an exercise routine tailored to the patient's needs. The soft device uses air pressure to provide resistance and support, helping the patient regain foot and ankle function.

In this way, soft robotic exosuits and wearable devices can support gait training and enhance mobility in patients with lower limb impairments. The compliance of these devices allows for natural joint movement and assists in restoring proper walking patterns.

Soft robotics has the potential to revolutionize the field of rehabilitation by providing more flexible, adaptable, and patient-friendly devices, providing a more personalized and effective rehabilitation program.

## 11.3 Wearable Sensors

Due to the heterogeneity of robotic devices available on the market, both in terms of power transferred to the patient and in terms of joint kinematics, the quantification of rehabilitation outcomes is an important factor in clinical practice, because there is no agreement on the most appropriate modality to select outcome measures [17–20]. The main limitations of outcome measures in rehabilitation can be identified in a large number of instruments available, but with poor psychometric properties; no shared consensus on specific clinical outcome measures that should be used to assess the effects of robotic treatment. Also, the outcome measures regarding the recovery of function and compensation processes, which strongly affect the patient's independence in ADLs, are often unclear and misinterpreted.

The most used clinical tests and evaluations, intended as outcome measures of a robotic neurorehabilitation treatment, have several limitations, such as:

1. Lack of specificity: Many outcome measures are not specific to the area of the brain that has been affected. For example, a measure of hand function may not differentiate between a stroke that has affected the motor cortex versus one that has affected the sensory cortex. The pure evaluation of the performance fails to distinguish between a damage related to the selective neuromotor control from a neuropsychological problem related to the proprioception and the perception of a limb.
2. Ceiling and floor effects: Outcome measures may not be able to measure change for patients who are already at their maximum or minimum level of function.
3. Lack of sensitivity: Some outcome measures may not be sensitive enough to detect small changes in function, making it difficult to measure progress over time, above all in patients who display only minimal impairments.
4. Lack of standardization: There is often variability in how outcome measures are administered and scored, which can make it difficult to compare results across studies or clinics.
5. Inability to capture subjective experiences: Outcome measures typically focus on objective measures of function and may not fully capture the patient's experience of their condition or the impact it has on their quality of life.

For the above-mentioned reasons, researchers have tried to find new instruments to properly characterize each patient.

To understand the complexity of the factors determining mobility and treat the related dysfunctions, it is necessary to perform a multidimensional evaluation of patient's needs and disabilities, both in a healthcare setting and, more importantly, in an ecological care perspective. In this sense, the use of relatively small and cheap wearable sensors could allow a quantitative and qualitative assessment of patient's movement while performing ADLs.

Wearable sensors have been increasingly used in neurorehabilitation for their ability to measure physical and behavioral data in real time, providing insights into patient functional status and progress and allowing a more accurate customization of rehabilitation programs. The main applications of wearable sensors in neurorehabilitation are as follows:

1. Monitoring mobility and movements: Wearable sensors such as accelerometers [21], gyroscopes, and inertial measurement units (IMUs) [22] can track movements like walking speed, gait patterns, upper limb kinematics during functional tasks, and postural stability to provide insights into the patient's mobility progress.
2. Assessing muscle activity: Wearable electromyography (EMG) sensors can measure the electrical signals generated by muscles during movement, providing information on muscle activation timing and fatigue to guide rehabilitation interventions [23].
3. Evaluating sensorimotor function: Wearable sensors can assess aspects of sensorimotor function like coordination, balance, and proprioception, helping to identify impairments and track improvements over time [24].
4. Monitoring physiological responses: Devices such as heart rate monitors [25], pulse oximeters, and skin conductance sensors can track physiological responses to rehabilitation exercises to provide insights into patient progress and overall wellness.
5. Enhancing neurofeedback and biofeedback: Wearable sensors can be incorporated into therapeutic interventions such as biofeedback and neurofeedback to provide real-time feedback on brain activity, muscle tension, or other physiological parameters, facilitating patient engagement and motivation [26].

Overall, wearable sensors have tremendous potential to improve the effectiveness and efficiency of neurorehabilitation interventions, allowing clinicians to tailor interventions to patients' needs and track progress more accurately.

Wearable soft robotics relies heavily on sensors to enable the robotic device to interact with the wearer's body and environment. These sensors can help to control the movement and pressure exerted by the device as well as monitor the device's impact on the wearer. Some common types of sensors used in wearable soft robotics include:

1. Strain sensors: These sensors measure the deformation or elongation of the device's materials while in use. They can detect changes in the device's shape, which provide information on how it is moving and applying force to the wearer.
2. Pressure sensors: These sensors measure the pressure or force being applied by the device on the wearer. They can help to ensure that the device is not causing discomfort or injury to the wearer.

3. Accelerometers: These sensors measure the acceleration of the device in different directions. They can help to monitor the speed and movement of the device and adjust its behavior accordingly.
4. Gyroscopes: These sensors measure the orientation and rotation of the device. They can help to control the stability and balance of the device on the wearer's body.
5. Temperature sensors [27]: These sensors can help to monitor the temperature of the device and the wearer's skin to ensure that the device is not causing overheating or discomfort.
6. Flex sensors: Used to measure the amount of bending or flexing of the robot's limbs or joints.
7. Inertial sensors: Used to track motion, orientation, and acceleration of the robot's body.

Overall, sensors are an essential component of wearable soft robotics, enabling the device to interact with the body and environment in a safe and effective manner.

## 11.4 Electromyography

A further useful component in the management, programming, and use of soft robots is undoubtedly electromyography (EMG), a method used to measure the electrical activity of skeletal muscles. It is commonly used in neurorehabilitation to assess muscle function and physical interaction between patients and rehabilitation devices and can provide real-time feedback to patients and therapists. EMG is also used to determine the effectiveness of different therapies, such as robotic devices or electrical stimulation, in improving muscle function and movement control. In addition, it can be used to monitor muscle activity during exercises, allowing therapists to adjust the intensity and duration of rehabilitation programs to meet the specific needs of each patient. By using EMG, therapists can track improvements in muscle function over time and adjust therapy accordingly. Furthermore, the EMG feedback can motivate patients to fully engage in rehabilitation programs, helping them to achieve better outcomes in shorter periods of time.

Surface electromyography (sEMG) [28] has been widely used in neurorehabilitation as a tool for evaluating muscle activity during movement and as a biofeedback intervention for improving motor control in people with neurological disorders such as stroke, spinal cord injury, and cerebral palsy.

In stroke rehabilitation [29], sEMG is used to assess muscle activity in both affected and non-affected limbs and provide feedback to patients during rehabilitation exercises. This feedback can help patients to learn how to activate and control their muscles more effectively, which can improve their ability to perform functional activities such as walking and reaching.

In spinal cord injury rehabilitation [30], sEMG is used to assess muscle activity in residual or implanted muscles for functional purposes. By detecting and amplifying the electrical activity generated by the skeletal muscles, sEMG can provide important information about the level of muscle activation and the motor function in paralyzed limbs. This information can be used to design targeted interventions such as electrical stimulation, biofeedback, and adaptive devices.

As technology advances, sEMG systems are becoming more portable, wearable and affordable making it a more accessible option for people who require neurorehabilitation. Moreover, they are increasingly integrated into soft robotics devices not only as activity monitoring but also as a trigger to initiate or terminate a distal movement with proximal impulses or vice versa.

Surface EMG has some limitations:

1. Limited specificity: Surface electromyography (sEMG) does not exclusively measure the activity of a specific muscle but signals, if not properly analyzed, can be influenced by the activity of the nearby muscles or by specific properties of the skin, adipose tissue, and other tissues.
2. Limited ability to detect deep muscles: sEMG can only measure activity in superficial muscles, and it cannot measure deep muscles that are critical to movement and postural control. These limitations can though be overcome by specific techniques which insert a needle into the deep muscle and connect the wire into the surface EMG.
3. Subset of movements: sEMG can only measure muscle activity during a subset of movements. This limits its use in detecting muscle imbalances and identifying functional deficits that occur during a broader range of activities.
4. Limited assessment of muscle force: Although sEMG can provide information about muscle activity, it cannot directly measure muscle force output. This can limit the ability to assess changes in strength and power.
5. Inability to differentiate between voluntary and compensatory activity: sEMG cannot differentiate between voluntary and reflexive muscle activity. This can make it difficult to assess the effects of training or therapy on individual muscle groups.
6. Cost: sEMG is a relatively expensive technology, which can limit its availability in research and clinical settings.
7. Complexity: sEMG analysis requires advanced technical knowledge to properly collect, process, and interpret the data. This can make it challenging for clinicians to use the technology effectively in neurorehabilitation.

In some studies on experimental devices, electromyography has been used in a promising way as a trigger for movements performed by an exoskeleton. In fact, each patient can use his own voluntary muscle activation to control impaired muscles through the use of electromyography in the rehabilitation setting [31].

The possibility of monitoring the activation of the muscles during different tasks has allowed the understanding of complex muscle synergies, in particular at the level of the upper limb, which enable to differentiate different patterns based on the intention of the movement. In this sense, surface electromyography can represent an interface between intention and movement execution, which can provide triggers aimed at functional use of a limb and monitor correct muscle activation.

## 11.5 Electroencephalogram

Electroencephalogram (EEG) is a non-invasive neuroimaging technique that records electrical activity in the brain through electrodes placed on the scalp [32]. EEG can be used to detect abnormalities in brain activity and to identify areas of the brain that are affected by injury or disease. In neurorehabilitation, EEG has been widely used to assess brain function, monitor neurological changes, and guide treatment interventions. Furthermore, EEG can provide valuable information on the effectiveness of rehabilitation interventions and can help guide treatment by mapping brain areas that need to be stimulated or suppressed [33].

EEG biofeedback, also known as neurofeedback, is a treatment approach that uses EEG to train individuals to control their brain activity [34]. This technique has been used to improve cognitive function, attention, and motor function in individuals with neurological disorders such as stroke and traumatic brain injury [35].

EEG is a safe and non-invasive tool that can be used to assess and guide neurorehabilitation interventions. Additional research is needed to fully understand the potential benefits of EEG in neurorehabilitation and to develop standardized protocols for its use.

## 11.6 Heart Rate Variability

A further evolution of the methods of use and management of sensors is represented by the possibility of being wearable and integrated into T-shirts, which are thus sensorized, with long-lasting batteries, and allow real-time communication of the parameters detected. In this way, useful conditions are created for monitoring and personalizing the treatment. With the progressive miniaturization of increasingly precise and specific sensors, it has been possible to give answers to the need to extract useful information from electrocardiographic data, respiratory parameters, and data related to the activation of the autonomic nervous system.

This leads to information of variable complexity, which contextualizes, in terms of biological multifactorial control of movement and interaction with the environment, the simple quantification of the number of movements in space. Consequently, it is possible to generate analyses that take into account the inference of cognitive components in the planning, implementation, and management of mobility.

In particular, a physiological phenomenon of great interest for rehabilitation is the Heart Rate Variability (HRV). HRV [36] is a measure of the variation in intervals between heartbeats, reflecting the degree of autonomic nervous system activity that is regulated by parasympathetic and sympathetic branches. HRV has been used as a marker of cardiovascular function and autonomic regulation in various clinical populations, including neurological disorders such as stroke, traumatic brain injury, and multiple sclerosis.

Different studies have highlighted the potential of HRV as a predictor for functional outcomes in stroke and traumatic brain injury patients [37, 38]. HRV can be used as an indicator of recovery and response to neurorehabilitation, as well as a risk factor for cardiovascular events and overall morbidity and mortality. Although the exact mechanisms by which HRV influences neurological recovery and functional outcomes are not fully understood yet, it is supposed that HRV plays a crucial role in maintaining physiological homeostasis and facilitating the adaptive response to stressors in various clinical settings.

Recent studies have investigated the use of HRV biofeedback training as a complementary treatment for neurological disorders. HRV biofeedback is a non-invasive and patient-centered approach, which involves training individuals to modulate their heart rate variability, for example, by breathing exercises, relaxation techniques, or slow-paced yoga. HRV biofeedback training has been shown to improve cognitive, physical, and psychological outcomes in patients with various neurological conditions such as traumatic brain injury, stroke, cerebral palsy, and Parkinson's disease.

Furthermore, HRV has been found to be associated with cognitive flexibility [39]. Cognitive flexibility refers to the ability to switch between different tasks, ideas, or concepts as the situation demands. A higher level of HRV has been linked with better cognitive flexibility. Several studies have found that increased HRV is associated with better performance on tasks that require cognitive flexibility. For example, one study found that individuals with higher HRV were better able to switch between different cognitive tasks and demonstrate greater cognitive flexibility. Another study found that HRV was positively correlated with the ability to shift attention and maintain attention over longer periods of time. Higher HRV was also related to better performance on working memory tasks, which require the ability to hold and manipulate information in the mind.

Overall, these findings suggest that HRV may be an important indicator of cognitive flexibility and that interventions aimed at improving HRV may have positive effects on cognitive functions such as attention, working memory, and overall cognitive flexibility.

## 11.7 Challenges and Future Directions

While soft robotics holds great promise for neurorehabilitation, several challenges need to be addressed for widespread adoption:

It is crucial to enhance robust control algorithms that can adapt to the complex and variable dynamics of soft robotic systems, and both must be adapted to the complexity of the motor control relearning of the human brain after a central nervous system damage. Advances in machine learning and artificial intelligence can help this huge data analysis to enable more sophisticated and adaptable control strategies.

Despite the technical possibilities, soft robotic devices need to be user-friendly, affordable, and accessible to a wide range of patients. Research should focus on reducing the cost, improving the usability, and addressing the specific needs of different patient populations.

Integrating soft robotics into clinical practice requires collaboration between engineers, clinicians, and researchers. Efforts should be made to bridge the gap between technological advancements and their translation into effective and evidence-based rehabilitation protocols.

To maximize neuroplasticity in UMNS patients while avoiding maladaptive plasticity, we still need to understand how much a soft robot is able to give more than the other devices available for rehabilitation in terms of embodiment and empowerment. It has not yet been understood how important the patient's feeling of "belonging to one's own body" is present for the devices used to recover functions and abilities.

It still remains to be understood how early it is useful to propose a treatment with precision devices, such as soft robots, to allow patients to "corticalize" a correct sequence of movements that could be effective in their daily life activities, promoting adaptive plasticity and avoiding the early acquisition of maladaptive compensatory strategies after a brain injury.

Long-term efficacy studies are essential to establish the effectiveness of soft robotics in neurorehabilitation. Such studies should focus on functional outcomes, patient satisfaction, and quality of life improvements, but also on the possibility to modify biological parameters detectable with wearable sensors, also integrated in soft robots.

The great adventure of the next decade will be the possibility of transforming, thanks to soft robotics, the patient's home into a real rehabilitation gym, with precise monitoring of the activities and a constant feedback on the time of use, the kinematics used by the limbs during active and passive movements, or biomarkers able to detect involvement of the patient in the treatment and functionality of the cardiovascular system.

The change in the methods of evaluation, monitoring, and definition of outcome measures will soon impose a revolution in the methods of structuring clinical trials, due to the possibility of creating "adaptive design" in "decentralized clinical trials," with a focus on the "home as first place of care," in particular for the chronic patient.

The use of complex measures, which derive from the integration of detectable data obtained from multiparametric temporal-spatial dynamics of observation, could be analyzed thanks to the development of artificial intelligence and will allow the identification of highly personalized rehabilitation treatments. Furthermore, with an innovative monitoring of the functional evolution and response to treatment, the issue of the possibility of predicting recovery after an acute injury in a more definite way must be taken into consideration, in order to direct rehabilitation resources in the best possible way.

## References

1. Hobbs B, Artemiadis P. A review of robot-assisted lower-limb stroke therapy: unexplored paths and future directions in gait rehabilitation. Front Neurorobot. 2020;14:19.
2. Holanda LJ, Silva PMM, Amorim TC, et al. Robotic assisted gait as a tool for rehabilitation of individuals with spinal cord injury: a systematic review. J Neuroeng Rehabil. 2017;14:126.
3. Kwakkel G, Kollen BJ, Krebs HI. Effects of robot-assisted therapy on upper limb recovery after stroke: a systematic review. Neurorehabil Neural Repair. 2008;22(2):111–21.
4. Mehrholz J, Thomas S, Elsner B, Kugler J. Electromechanical and robotic-assisted arm training for improving activities of daily living, arm function, and arm muscle strength after stroke. Cochrane Database Syst Rev. 2017;(11):CD006876.

5. Fareh R, Elsabe A, Baziyad M, Kawser T, Brahmi B, Rahman MH. Will your next therapist be a robot?—a review of the advancements in robotic upper extremity rehabilitation. Sensors (Basel). 2023;23(11):5054.
6. Molteni F, Gasperini G, Cannaviello G, Guanziroli E. Exoskeleton and end-effector robots for upper and lower limbs rehabilitation: narrative review. PM R. 2018;10(9 Suppl 2):S174–88.
7. Sale A, Berardi N, Maffei L. Enrich the environment to empower the brain. Trends Neurosci. 2009;32(4):233–9.
8. Feys H, De Weerdt W, Verbeke G, et al. Early and repetitive stimulation of the arm can substantially improve the long-term outcome after stroke: a 5-year follow-up study of a randomized trial. Stroke. 2004;35(4):924–9.
9. Klamroth-Marganska V, Blanco J, Campen K, et al. Three-dimensional, task-specific robot therapy of the arm after stroke: a multicentre, parallel-group randomised trial. Lancet Neurol. 2014;13(2):159–66.
10. Lo AC, Guarino PD, Richards LG, et al. Robot-assisted therapy for long-term upper-limb impairment after stroke. N Engl J Med. 2010;362(19):1772–83.
11. Freivogel S, Mehrholz J, Husak-Sotomayor T, Schmalohr D. Gait training with the newly developed 'LokoHelp'-system is feasible for non-ambulatory patients after stroke, spinal cord and brain injury. A feasibility study. Brain Inj. 2008;22(7–8):625–32.
12. Hesse S, Uhlenbrock D. A mechanized gait trainer for restoration of gait. J Rehabil Res Dev Clin Suppl. 2000;37(6):701–8.
13. Hesse S, Waldner A, Tomelleri C. Innovative gait robot for the repetitive practice of floor walking and stair climbing up and down in stroke patients. J Neuroeng Rehabil. 2010;7:30.
14. Ochieze C, Zare S, Sun Y. Wearable upper limb robotics for pervasive health: a review. Prog Biomed Eng. 2023;5:032003.
15. Chu CY, Patterson RM. Soft robotic devices for hand rehabilitation and assistance: a narrative review. J Neuroeng Rehabil. 2018;15:9.
16. Borboni A, Mor M, Faglia R. Gloreha—hand robotic rehabilitation: design, mechanical model, and experiments. J Dyn Syst Meas Control. 2016;138(11):111003.
17. Barak S, Duncan PW. Issues in selecting outcome measures to assess functional recovery after stroke. NeuroRx. 2006;3(4):505–24.
18. Geroin C, Mazzoleni S, Smania N, Gandolfi M, Bonaiuti D, Gasperini G, et al. Systematic review of outcome measures of walking training using electromechanical and robotic devices in patients with stroke. J Rehabil Med. 2013;45(10):987–96.
19. Mehrholz J, Wagner K, Rutte K, Meissner D, Pohl M. Predictive validity and responsiveness of the functional ambulation category in hemiparetic patients after stroke. Arch Phys Med Rehabil. 2007;88:1314–9.
20. Salter K, Jutai JW, Teasell R, Foley NC, Bitensky J. Issues for selection of outcome measures in stroke rehabilitation: ICF body functions. Disabil Rehabil. 2005;27:191–207.
21. Kavanagh JJ, Menz HB. Accelerometry: a technique for quantifying movement patterns during walking. Gait Posture. 2008;28:1–15.
22. Wüest S, Aminian F, Aminian K, Gonzenbach R, de Bruin E. Reliability and validity of the inertial sensor-based timed "up and go" test in individuals affected by stroke. J Rehabil Res Dev. 2016;53:599–610.
23. Campanini I, Disselhorst-Klug C, Rymer WZ, Merletti R. Surface EMG in clinical assessment and neurorehabilitation: barriers limiting its use. Front Neurol. 2020;11:934.
24. Pollind M, Soangra R. Development and validation of wearable inertial sensor system for postural sway analysis. Measurement. 2020;165:108101.
25. Alugubelli N, Abuissa H, Roka A. Wearable devices for remote monitoring of heart rate and heart rate variability—what we know and what is coming. Sensors (Basel). 2022;22(22):8903.
26. Bowman T, Gervasoni E, Arienti C, Lazzarini SG, Negrini S, Crea S, Cattaneo D, Carrozza MC. Wearable devices for biofeedback rehabilitation: a systematic review and meta-analysis to design application rules and estimate the effectiveness on balance and gait outcomes in neurological diseases. Sensors (Basel). 2021;21(10):3444.

27. Amjadi E, Park YJ, Jeong IS, Park I. Highly stretchable and sensitive strain sensor based on silver nanowire–elastomer nanocomposite. ACS Nano. 2014;8(5):5154–63.
28. Cram J, Kasman G, Holtz J. Introduction to surface electromyography. Gaithersburg: Aspen Publisher; 1998.
29. Kotov-Smolenskiy AM, Khizhnikova AE, Klochkov AS, et al. Surface EMG: applicability in the motion analysis and opportunities for practical rehabilitation. Hum Physiol. 2021;47:237–47.
30. Balbinot G, Li G, Wiest MJ, et al. Properties of the surface electromyogram following traumatic spinal cord injury: a scoping review. J Neuroeng Rehabil. 2021;18(1):105.
31. Ambrosini E, Gasperini G, Zajc J, Immick N, Augsten A, Rossini M, Ballarati R, Russold M, Ferrante S, Ferrigno G, Bulgheroni M, Baccinelli W, Schauer T, Wiesener C, Gfoehler M, Puchinger M, Weber M, Weber S, Pedrocchi A, Molteni F, Krakow K. A robotic system with EMG-triggered functional electrical stimulation for restoring arm functions in stroke survivors. Neurorehabil Neural Repair. 2021;35(4):334–45.
32. Biasiucci A, Franceschiello B, Murray MM. Electroencephalography. Curr Biol. 2019;29(3):R80–5.
33. Milani G, Antonioni A, Baroni A, Malerba P, Straudi S. Relation between EEG measures and upper limb motor recovery in stroke patients: a scoping review. Brain Topogr. 2022;35(5–6):651–66.
34. Grosse-Wentrup M, Mattia D, Oweiss K. Using brain-computer interfaces in neurorehabilitation. Springer; 2011.
35. Daly JJ, Wolpaw JR. Brain-computer interfaces in neurological rehabilitation. Lancet Neurol. 2008;7(4):1032–43.
36. McCraty R, Shaffer F. Heart rate variability: new perspectives on physiological mechanisms, assessment of self-regulatory capacity, and health risk. Glob Adv Health Med. 2015;4(1):46–61.
37. Grilletti JVF, Scapini KB, Bernardes N, Spadari J, Bigongiari A, de Andrade e Souza Mazuchi F, De Angelis K. Impaired baroreflex sensitivity and increased systolic blood pressure variability in chronic post-ischemic stroke. Clinics. 2018;73(11):1–4.
38. Palamarchuk I, Kimpinski K, Lippert C, Hachinski V. Nocturnal deterioration after ischemic stroke and autonomic dysfunction: hypothesis and implications. Cerebrovasc Dis. 2013;36(5–6):454–61.
39. Gevirtz R. The promise of heart rate variability biofeedback: evidence-based applications. Biofeedback. 2013;41(3):110–20.

# 12 Telerehabilitation in the Neurological Field

Paolo Tonin, Loris Pignolo, and Marianna Contrada

## 12.1 Introduction

In the last few years, the length of hospitalization for neurological patients has shortened, and a significant portion of the rehabilitation phase has increasingly focused on outpatient settings. However, a high percentage of discharged patients still do not receive adequate rehabilitation: resources are limited, and an individual's ability to receive proper rehabilitation services often depends on non-clinical factors such as resource availability, geographic location, age, and level of wealth. The benefits gained during hospital-based rehabilitation often are not sustained long term, as individuals reduce activity and cease engagement in exercises, leading to functional decline. To address these challenges, Telerehabilitation (TR) can provide care to individuals confined to their homes without moving therapists or patients. This concept has more relevance to Teleneurorehabilitation (TNR), as neurological diseases are the most severe cause of disability and the leading cause of Years Lived with Disability (YLD).

Many studies have shown that rehabilitative treatment produces equivalent results whether delivered in person or remotely through telerehabilitation systems, confirming that TNR can effectively meet the growing demand for improved home care. Despite these data, TNR has not yet been widely integrated into routine territorial rehabilitation treatment, as many aspects still need to be clarified. The first critical point is the lack of common and shared methodologies for TNR treatment delivery. Another aspect that needs to be better defined is the wide range of possible TNR services, from teleconsultation to telehealth, from telemonitoring to teletherapy, which accentuates the organizational complexity of TNR. Another critical

P. Tonin (✉) · L. Pignolo · M. Contrada
S'Anna Institute, Crotone, Italy
e-mail: l.pignolo@isakr.it



R. S. Calabrò (ed.), *Translational Neurorehabilitation*,
https://doi.org/10.1007/978-3-031-63604-2_12

point is the lack of substantial clinical data: most TNR studies are pilot studies or small projects that cannot be extended to a large population [1].

The need for studies to further investigate the effectiveness of many different clinical interventions such as motor, cognitive, and language in many different clinical pathologies is urgent.

TNR can be a valuable tool for improving the effectiveness, accessibility, and efficiency of neurorehabilitation treatments for patients with brain injuries, allowing:

- Access to highly qualified specialists in remote geographic areas.
- Receiving neurorehabilitation treatments at home, reducing the need for frequent hospital or outpatient visits.
- Neurorehabilitation professionals could monitor patients' progress regularly and accurately.

## 12.2 Methodology of Teleneurorehabilitation

The term TNR includes a wide range of rehabilitative interventions different not only for pathologies, programs, duration, and involved health disciplines, but also for technologies applied including the telephone, the Internet, virtual reality (VR), and sensors or wearable devices [2]. Reported there are many differences among the models of TNR (telephone calls, videoconferencing, chats and VR systems, etc.), as among the duration and the frequency of rehabilitation programs. Unfortunately, the not sufficient data in the literature do not allow us to determine which model or TNR tool is optimal [3]. However, the majority of the most recent devices of TNR combine technologies to support the communication between the patient and the operator and VR-based tasks [4].

The presence of a camera is the minimum requirement (videoconference) for a TNR system because allows one to see and directly monitor the treatment. However, the existing TNR systems cannot do without the sensors that record the movements of the patient. Especially motor TNR is closely dependent on sensoristics for the acquisition of movement data [5]: movement kinematics, in fact, is crucial to provide real-time feedback to the patient on the quality of the performed gesture. Feedback, moreover, has been recognized to be a key element in the effectiveness of relearning processes during neuromotor rehabilitation [6, 7].

## 12.3 Teleneurorehabilitation Services

TNR can be characterized by three different services:

1. **Teleconsultation and Teleassistance**. A counselling center should be organized as a service that is open 12 h/a day to receive patient calls and to verify them, but also to contact patients; the counselling center should not only be a call center but should be organized with regular hours. The service may limit activity to

phone calls, but it would be advisable to use current and free applications such as Skype or WhatsApp to offer visual contact from a computer or smartphone as well. These devices can also be used to telecockpit for performance (movements and posture), cognitive performance (attention, memory), language (conversation), swallowing (head posture), etc., of the patient. Teleconsultation and teleassistance are currently considered supportive measures. In this phase, where patients' loneliness at home can seriously affect the clinical outcome, these remote interventions can play an important role.

2. **Telemonitoring**. The service can be implemented using wearable devices. There are many different sensors currently available at low cost to remotely monitor patient movement and physiological data such as heart rate, respiratory rate, and oxygen saturation. Many different sensors can allow service personnel to monitor the status of a large number of patients.
3. **Teletherapy**. For this level of intervention, there are greater difficulties. Numerous teletherapy experiences have been published in the literature, especially based on virtual reality systems; unfortunately, the vast majority of these studies in the past, used closed systems, which were useful for research activities but very limited for a complete TNR service. Currently, the potential of teletherapy has greatly increased, as will be highlighted in the following paragraphs.

## 12.4 Teletherapy

Teletherapy (TLT) is certainly the most innovative form of TLT. The first scientific publications with teletherapy experiences date from 2000 for orthopedic rehabilitation [8] to 2002 for motor neurorehabilitation and aphasia treatment [9, 10] and 2003 for cognitive neurorehabilitation [11].

However, for many years, TLT remained an under-investigated topic: in 2013, about 10 years after the first articles, only 46 publications could be found in Pubmed. The picture has completely changed since 2019 with the emergence of the pandemic, and by 2021 there were about 450 articles in Pubmed. And at the same time as the increase in scientific interest in TLT, the interest in public and private health services has also exploded. This rapid evolution has been very helpful for the spread and evolution of TLT, but, as we will see below, it has also opened up some problems.

Although TLT treatments, as well as standard treatments, often require the co-presence of motor and cognitive rehabilitation treatments, it is important to illustrate the two TLT scenarios separately.

### 12.4.1 Motor Teletherapy

Since the first article in 2002, studies of TLT have been based on the results of neurophysiologist studies, which have highlighted the role of reinforced feedback in promoting motor learning/relearning. Particular mention should be made of studies from the neurophysiology laboratory at MIT Boston [12]. Subsequently, almost all motor TLT studies and experiments based on reinforced feedback have used VR systems. A VR system can facilitate motor learning of complex exercises, anyway, the rules indicated by neurophysiology studies should always be observed [6].

To increase motor learning of a complex movement, a VR system must include three different levels of information: (a) knowledge of the correct reference trajectory made by an expert; (b) knowledge of the actual results made by the subject invited to repeat the expert's movement; and (c) knowledge of the differences between the subject's movements and the reference movement. If any of these critical components of the virtual environment are removed, reinforcement of motor learning is no longer achieved. In other words, a generic activity in a virtual environment is not sufficient to increase motor learning: it should be emphasized that comprehensive and specific VR exercises must be delivered for each motor task.

This rigorousness has stimulated a debate about the type of exercises delivered for TLT. Some initial comparisons led to suspicions that VR is effective if delivered through specific exercises that adhere to rules to promote motor learning but is not if the exercises are delivered through standardized commercial games. However, these uncertainties stemmed from small studies with few patients, and the uncertainty lasted for a long time. Finally, in 2019 a major meta-analysis involving many articles and many patients resolved the uncertainty and solidly showed positive results for the group of articles with treatment delivered by specific exercises and indifferent results for the group with treatment delivered by nonspecific exercises [13].

### 12.4.2 Cognitive and Aphasia Teletherapy

The treatment of aphasia with TLT systems has followed a similar path to that of motor treatment.

Many applications of TLT have been used since 2002 in the management of child and adult speech disorders, stuttering, voice disorders, laryngectomy, and swallowing dysfunction [14]. These authors early on highlighted that TLT has the potential to provide services at home or in the local community through computer-based interactive therapeutic activities. This form of service delivery can optimize outcomes by facilitating the generalization of treatment effects within the person's everyday environment. In the context of TLT studies on aphasia, various modes of operation have been tested; among them, the possibility of delivering TLT treatment either in a synchronous mode, i.e., a two-way videoconference in the presence of the speech therapist, or in an asynchronous mode, i.e., without the presence of the speech therapist, but with self-administration of the exercises by the patient [15].

As far as cognitive teletherapy is concerned, the picture is less defined: there is still a lack of a sufficient amount of studies to characterize the specificities of enriched virtual environments, useful for different cognitive disorders; moreover, in this field, unlike the motor area, there is still a lack of information on the usefulness or otherwise of the use of commercially available or otherwise nonspecific systems. The treatment discouraged for motor recovery might acquire a different meaning in one or more of the different cognitive domains [16]. The first systematic review to analyze the effectiveness of cognitive TLT interventions, which was very recent, was published 20 years after the first TLT article [15]. This review aimed to analyze and synthesize the evidence in patients with neurological diseases, compared with traditional face-to-face rehabilitation. Improvements in the global cognitive domain, learning, memory, language functions, and executive functions were evaluated. Performance improvement comparable to that achieved through face-to-face treatment was observed in all domains analyzed; significant differences between the two groups in favor of TLT were also shown for working memory and total executive function, however, with little data. Thus, it seems that the trend of cognitive TLT is like that of motor TLT, although more supporting data are still needed.

### 12.4.3 The Efficacy of Teletherapy

Since the earliest experiences with TNR, uncertainty has fluctuated among scholars as to whether or not remote TNR treatment is effective, compared with standard direct-contact treatment. After numerous studies, a Cochrane Library review was published in 2020 that defined the validity of TNR services for stroke [17]. The data from this landmark study showed that the same treatment, administered in person or remotely, produces the same result. Therefore, TNR is not inferior to usual rehabilitation delivered in person and therefore should be considered a viable way of providing rehabilitation treatments for patients who need rehabilitation after the acute or subacute phase. This implies that at home, these patients can receive the same number and quality of specific treatments as during a hospital stay. This Cochrane review was very important and based on which TNR was accepted by health care management and administrative systems and entered the daily rehabilitation routine in the Covid era. Indeed, there are numerous benefits derived from TNR for the disadvantaged population and neurological patients with disabilities that adversely affect their mobility and ability to reach rehabilitation centers, and the reduced travel time and cost, and access to services otherwise unavailable should be emphasized [18].

Anyway, it should be emphasized that from a scientific point of view, this Cochrane finding was validated only for post-stroke patients. However, this should not be taken as a negative finding but is only a consequence of the limited number of articles on TNR delivered for other conditions, which therefore cannot be adequately evaluated for now. Therefore, it is reasonable to assume that in time the same conclusion can be reached for other conditions such as Parkinson's disease, multiple sclerosis, etc.

## 12.5 Teletherapy Experiences

### 12.5.1 Stroke

Many post-stroke patients still have a residual functional disability that impairs their ability to perform activities of daily living, even after standard rehabilitation treatment is completed. Although the need for rehabilitation services for stroke patients after discharge from acute hospitals is widely recognized, inpatient and outpatient rehabilitation is often compromised for reasons such as lack of accessibility, increased costs, and poor compliance [19]. Within this framework, TNR offers an essential opportunity to provide continuous rehabilitation services for stroke patients. TNR is increasingly gaining a key role in improving functional recovery, which explains why the majority of TNR experiences (65%), are for patients with stroke outcomes. Numerous results show that, in patients with stroke outcomes, the continuation of rehabilitation treatment through TNR is effective, for motor function, cognitive disorders, and speech deficits. And an improvement in depressive symptoms and the perception of a better general quality of life (QoL) has also been found because patients feel more cared for [20]. Customarily, TNR is used very frequently for patients with upper limb deficits. Remote rehabilitation treatment also for the lower limb and balance is more recent and less widespread. The most advanced TNR systems allow remote, motor, cognitive, and speech rehabilitation treatments to be delivered on a rotating basis to the same patient as needed [21].

### 12.5.2 Parkinson's Disease

TNR for patients with Parkinson's disease (PD) is the most widely used modality after the treatment of post-stroke patients. The rehabilitation demands of these patients differ from post-stroke patients, and obviously, the TNR interventions are also different [22]. The most common TNR treatments in patients with PD are concern balance, postural stability, and gait [23]. As for non-motor disorders, the most frequent TNR interventions concern speech speed and dysphagia. Remote treatment of motor speech and language disorders is one of the earliest TNR experiences, using a specific intensive speech therapy technique (Lee Silverman Voice Treatment -LSVT-) [24]. For patients with PD, many studies have shown that providing remote specialist care improves health care, QoL, and social participation. Taking into account the time course of the disease and the opportunities of LSVT, it is reasonable to assume that periodic LSVT treatment may become a guideline for patients with PD.

## 12.6 Multiple Sclerosis

Teletherapy experiences for patients with Multiple Sclerosis (MS) represent the third group, after post-stroke patients and those with PD, which is catalogued in this chapter. The number of patients is very small compared to patients with Parkinson's and even more so compared to those with stroke outcomes. The rehabilitation needs of patients with MS also differ from those of other patients and TNR interventions are different; the most frequent concern is the general domain of QoL and motor symptoms. However, it has already been possible to define some operational addresses; for these patients, TLT has been aimed primarily at reducing fatigue and improving functional activities and QoL. This is a different goal that paves the way for TLT as a tool for periodic support of the quality of activities of daily living. Some positive results have already emerged; however, due to the scarcity of patients studied, the results obtained are still weak. A Cochrane review has been launched on this subject, similar to the one mentioned above for patients with stroke outcomes. In this case, insufficient evidence has emerged to support the remote rehabilitation interventions [25]. However, it should be remembered that this reference Cochrane review for MS is many years ahead of the 2020 Cochrane review which recorded the effectiveness of TNR for patients with stroke outcomes.

## 12.7 Alzheimer's Disease

Studies on TNR applied to patients with Alzheimer's disease have also been published in recent years, limited to subjects with mild to moderate cognitive impairment. The rehabilitation goal consists of an early intervention of cognitive stimulation, to strengthen the remaining cognitive abilities for as long as possible. The results showed significant improvements, particularly in memory and attentive abilities. However, these experiences of TNR are mostly limited to pilot studies, with very small numbers of patients and centers involved [26]. Although the results of these studies are still uncertain, it is appropriate to mention them because TNR can become the periodic reinforcement of cognitive conditions and support the quality of daily life of patients and caregivers.

## 12.8 The Future of Teleneurorehabilitation

The TNR is in the process of great scientific, clinical, and organizational development, however, from now on we can glimpse two new paths that can represent the future of the TNR. The first involves deepening virtual reality details that can implement and enhance the variety and quality of enhanced feedback. Already today interesting indications are emerging on the use of different feedbacks, which could produce different effects; in particular, two indications have emerged with regard to motor learning, which suggests that verbal feedback seems more effective in recognizing and correcting errors, while visual feedback seems more useful in the complete realization of a task [27]. The second, which calls for a much broader horizon, starts from the observation that Internet availability is high in cities and poorer in rural areas. This imbalance is particularly severe in developing countries, but the trend is the same throughout the world. The strength of socio-economic interests and the weakness of sociocultural interests make development very difficult in rural areas of telemedicine, and in particular of TNR, which often requires faster connections [28]. Only a strong sociocultural change in the use and meaning of the internet could rebalance the distribution of the Internet between urban/suburban and rural areas. It should never be forgotten that without connectivity, it is impossible to activate TR services.

## References


1. Hill AJ, Paik NJ, Kiran S, Tonin P. Tele-NeuroRehabilitation. Front Neurol. 2021;1901.
2. Kim H, Kwon YT, Lim HR, Kim JH, Kim YS, Yeo WH. Recent advances in wearable sensors and integrated functional devices for virtual and augmented reality applications. Adv Funct Mater. 2021;31(39):2005692.
3. Tchero H, Tabue Teguo M, Lannuzel A, Rusch E. Telerehabilitation for stroke survivors: systematic review and meta-analysis. J Med Internet Res. 2018;20(10):e10867.
4. Li L, Yu F, Shi D, Shi J, Tian Z, Yang J, et al. Application of virtual reality technology in clinical medicine. Am J Transl Res. 2017;9(9):3867.
5. Zheng H, Black ND, Harris ND. Position-sensing technologies for movement analysis in stroke rehabilitation. Med Biol Eng Comput. 2005;43(4):413–20.
6. De Bartolo D, Spitoni GF, Iosa M, Morone G, Ciancarelli I, Paolucci S, Antonucci G. From movement to thought and back: a review on the role of cognitive factors influencing technological neurorehabilitation. Funct Neurol. 2019;34(3):131–41.
7. Morone G, Ghanbari Ghooshchy S, Palomba A, Baricich A, Santamato A, Ciritella C, et al. Differentiation among bio-and augmented-feedback in technologically assisted rehabilitation. Expert Rev Med Devices. 2021;18(6):513–22.
8. Burdea G, Popescu V, Hentz V, Colbert K. Virtual reality-based orthopedic telerehabilitation. IEEE Trans Rehabil Eng. 2000;8(3):430–2.
9. Brennan D, Georgeadis A, Baron C. Telerehabilitation tools for the provision of remote speech-language treatment. Top Stroke Rehabil. 2002;8(4):71–8.
10. Piron L, Tonin P, Atzori AM, Zanotti E, Massaros C, Trivello E, Dam M. Virtual environment system for motor tele-rehabilitation. In: Medicine meets virtual reality 02/10. Amsterdam: IOS Press; 2002. p. 355–61.
11. Diamond BJ, Shreve GM, Bonilla JM, Johnston MV, Morodan J, Branneck R. Telerehabilitation, cognition and user-accessibility. NeuroRehabilitation. 2003;18(2):171–7.
12. Todorov E, Shadmehr R, Bizzi E. Augmented feedback presented in a virtual environment accelerates learning of a difficult motor task. J Mot Behav. 1997;29(2):147–58.
13. Maier M, Rubio Ballester B, Duff A, Duarte Oller E, Verschure PF. Effect of specific over nonspecific VR-based rehabilitation on poststroke motor recovery: a systematic meta-analysis. Neurorehabil Neural Repair. 2019;33(2):112–29.
14. Hill A, Theodoros D. Research into telehealth applications in speech-language pathology. J Telemed Telecare. 2002;8(4):187–96.
15. Cacciante L, Pietà CD, Rutkowski S, Cieślik B, Szczepańska-Gieracha J, Agostini M, Kiper P. Cognitive telerehabilitation in neurological patients: systematic review and meta-analysis. Neurol Sci. 2022;43:847–62.
16. Larson EB, Feigon M, Gagliardo P, Dvorkin AY. Virtual reality and cognitive rehabilitation: a review of current outcome research. NeuroRehabilitation. 2014;34(4):759–72.
17. Laver KE, Adey-Wakeling Z, Crotty M, Lannin NA, George S, Sherrington C. Telerehabilitation services for stroke. Cochrane Database Syst Rev. 2020;(1).
18. Housley SN, Garlow AR, Ducote K, Howard A, Thomas T, Wu D, et al. Increasing access to cost effective home-based rehabilitation for rural veteran stroke survivors. Austin J Cerebrovasc Dis Stroke. 2016;3(2):1.
19. Knepley KD, Mao JZ, Wieczorek P, Okoye FO, Jain AP, Harel NY. Impact of telerehabilitation for stroke-related deficits. Telemed eHealth. 2021;27(3):239–46.
20. Stephenson A, Howes S, Murphy PJ, Deutsch JE, Stokes M, Pedlow K, McDonough SM. Factors influencing the delivery of telerehabilitation for stroke: a systematic review. PLoS One. 2022;17(5):e0265828.
21. Anil K, Freeman JA, Buckingham S, Demain S, Gunn H, Jones RB, et al. Scope, context and quality of telerehabilitation guidelines for physical disabilities: a scoping review. BMJ Open. 2021;11(8):e049603.

22. León-Salas B, González-Hernández Y, Infante-Ventura D, de Armas-Castellano A, García-García J, García-Hernández M, et al. Telemedicine for neurological diseases: a systematic review and meta-analysis. Eur J Neurol. 2023;30(1):241–54.
23. Vellata C, Belli S, Balsamo F, Giordano A, Colombo R, Maggioni G. Effectiveness of telerehabilitation on motor impairments, non-motor symptoms and compliance in patients with Parkinson's disease: a systematic review. Front Neurol. 2021;12:627999.
24. Hill AE, Davidson BJ, Theodoros DG. A review of standardized patients in clinical education: implications for speech-language pathology programs. Int J Speech Lang Pathol. 2010;12(3):259–70.
25. Khan F, Amatya B, Kesselring J, Galea MP. Telerehabilitation for persons with multiple sclerosis. A Cochrane review. Eur J Phys Rehabil Med. 2015;51(3):311–25.
26. Jelcic N, Agostini M, Meneghello F, Bussè C, Parise S, Galano A, et al. Feasibility and efficacy of cognitive telerehabilitation in early Alzheimer's disease: a pilot study. Clin Interv Aging. 2014;9:1605–11.
27. Jaszczur-Nowicki J, Romero-Ramos O, Rydzik Ł, Ambroży T, Biegajło M, Nogal M, et al. Motor learning of complex tasks with augmented feedback: modality-dependent effectiveness. Int J Environ Res Public Health. 2021;18(23):12495.
28. Kemp S. Digital 2021: global overview report. 2021.

GPSR Compliance

*The European Union's (EU) General Product Safety Regulation (GPSR) is a set of rules that requires consumer products to be safe and our obligations to ensure this.*

*If you have any concerns about our products, you can contact us on ProductSafety@springernature.com*

In case Publisher is established outside the EU, the EU authorized representative is:

Springer Nature Customer Service Center GmbH
Europaplatz 3
69115 Heidelberg, Germany

**Batch number: 10732044**

Printed by Printforce, the Netherlands